Electrical Installation Work

Level 2

Updated in line with the 18th Edition of the Wiring Regulations and written specifically for the EAL Diploma in Electrical Installation, this book has a chapter dedicated to each unit of the EAL syllabus, allowing you to master each topic before moving on to the next. This new edition also includes information on LED lighting. End-of-chapter revision questions help you to check your understanding and consolidate the key concepts learned in each chapter.

This is the number one textbook for all EAL Level 2 courses in electrical installation. It sets out the core facts and principles with solid explanation – not just to pass the exam but to confidently work as an electrician with a proper understanding of the regulations. Ideal for both independent and tutor-based study.

Trevor Linsley was formerly Senior Lecturer in Electrical Engineering at Blackpool and the Fylde College of Technology. There he taught subjects at all levels from first year trainee to first year undergraduate courses; and was also head of the multidiscipline NVQ Assessment Centre and responsible for establishing and running the AM2 Electrical Skills Assessment Centre. He has had 27 books published and has also written many bespoke training packages for local SME electrical engineering companies in the North West of England.

Electrical Installation Work Level 2

EAL Edition

Second Edition

Trevor Linsley

Routledge
Taylor & Francis Group

LONDON AND NEW YORK

Second edition published 2019
by Routledge
2 Park Square, Milton Park, Abingdon, Oxon, OX14 4RN

and by Routledge
52 Vanderbilt Avenue, New York, NY 10017

Routledge is an imprint of the Taylor & Francis Group, an informa business

© 2019 Trevor Linsley

First edition published by Routledge 2015

British Library Cataloguing-in-Publication Data
A catalogue record for this book is available from the British Library

Library of Congress Cataloging-in-Publication Data
Names: Linsley, Trevor, author.
Title: Electrical installation work. Level 2 / Trevor Linsley.
Description: EAL Edition. | Second edition. | Abingdon, Oxon : Routledge,
2019. | Includes bibliographical references and index. |
Identifiers: LCCN 2018059032 (print) | LCCN 2018060584 (ebook) |
ISBN 9780429511752 (Adobe PDF) | ISBN 9780429518614 (Mobipocket) |
ISBN 9780429515187 (ePub) | ISBN 9780367195625 (hardback) | ISBN 9780367195618 (pbk.) |
ISBN 9780429203176 (ebook)
Subjects: LCSH: Electric apparatus and appliances--Installation--Textbooks. |
Electric wiring, Interior--Textbooks.
Classification: LCC TK452 (ebook) | LCC TK452 .L564 2019 (print) |
DDC 621.319/24--dc23
LC record available at https://lccn.loc.gov/2018059032

ISBN: 978-0-367-19562-5 (hbk)
ISBN: 978-0-367-19561-8 (pbk)
ISBN: 978-0-429-20317-6 (ebk)

DOI: 10.1201/9780429203176

Visit the companion website: www.routledge.com/cw/linsley

Printed in Great Britain by Bell and Bain Ltd, Glasgow

Contents

CHAPTER

Health and safety in electrical installation

EAL Unit Elec2/01

Learning outcomes

When you have completed this chapter you should:

1. Understand how health and safety applies to electrotechnical operations.
2. Understand health and safety procedures in the work environment.
3. Understand the basic electrical safety requirements.
4. Know the safety requirements for using access equipment.
5. Understand the importance of establishing a safe working environment. (i)
6. Understand the importance of establishing a safe working environment. (ii)

EAL Electrical Installation Work – Level 2, 2nd Edition 978 0 367 19562 5
© 2019 Linsley. Published by Taylor & Francis. All rights reserved.
www.routledge.com/9780367195618

DOI: 10.1201/9780429203176-1

Introduction

Understanding how health and safety is legislated and put into practice is essential for anyone operating within the electrical installation industry. Defining how statutory regulation ties in with non-statutory codes of practice such as application of wiring regulations is essential, as is understanding the roles and responsibilities within health and safety. These range from employers to the Health and Safety Executive (HSE) and from employees to safety representatives, all with a common aim: to learn and inform so that such events are not repeated.

Understanding how reduced electrical systems bring about a reduction in electric shocks is paramount: in essence the likelihood of being exposed to live parts is lessened. Such measures are complemented through a regime that inspects and tests all electrical equipment and ensures that all work is made safe through incorporating safe isolation procedures. Safety is also applied when using access equipment, so that the user can choose the most appropriate equipment through current Prefabricated Access Suppliers' and Manufacturers' Association (PASMA) requirements. Understanding how employers implement a safe system of work is fundamental in instigating risk assessments in order to recognize hazards, control through permit-to-work systems and define operations through method statements which signpost exactly how a task is to be implemented. It is essential to understand that personal protective equipment (PPE) only lessens the effects of accidents and does not eliminate them, and is only necessary when hazards cannot be eliminated. The exposure to hazards could also be minimized if less hazardous substances are used, the hazard is enclosed or employees are fully trained with only authorized personnel permitted to operate certain hazardous environments and equipment, in particular when asbestos is involved. The use of signage, fire-fighting equipment and training, also ensure measures are in place to deal with situations during emergencies.

Figure 1.1 Wearing PPE lessens the impact of accidents but does not eliminate them.

Assessment criteria 1.1

State the general aims of health and safety legislation

Safety regulations and laws

At the beginning of the nineteenth century children formed a large part of the working population of Great Britain. They started work early in their lives and they worked long hours for unscrupulous employers or masters.

The Health and Morals of Apprentices Act of 1802 was introduced by Robert Peel in an attempt to reduce apprentice working hours to 12 hours per day and to improve the conditions of their employment. The Factories Act of 1833 restricted the working week for children aged 13–18 years to 69 hours in any working week.

With the introduction of the Factories Act of 1833, the first four full-time Factory Inspectors were appointed. They were allowed to employ a small number of assistants and were given the responsibility of inspecting factories throughout England, Scotland, Ireland and Wales. This small, overworked band of men were the forerunners of the modern HSE Inspectorate, enforcing the safety laws passed by Parliament. As the years progressed, new Acts of Parliament increased the powers of the Inspectorate and the growing strength of the trade unions meant that employers were increasingly being pressed to improve health, safety and welfare at work.

The most important recent piece of health and safety law, the Health and Safety at Work Act, was passed by Parliament in 1974. This Act gave added powers to the Inspectorate and is the basis of all modern statutory health and safety laws. This law not only increased the employer's liability for safety measures, but also put the responsibility for safety on employees too.

Health, safety and welfare legislation has increased the awareness of everyone to the risks involved in the workplace. All statutes within the Acts of Parliament must be obeyed and, therefore, we all need an understanding of the laws as they apply to the electrical industry. This is the fundamental aim of health and safety legislation.

Figure 1.2 Both workers and managers are responsible for health and safety on-site.

Assessment criteria 1.2

Recognize the legal status of health and safety documents

Statutory laws

Acts of Parliament are made up of statutes. Statutory regulations have been passed by Parliament and have, therefore, become laws. Non-compliance with the laws of this land may lead to prosecution by the courts and possible imprisonment for offenders.

We shall now look at some of the statutory regulations as they apply not only to employers, employees and contractors within the electrical industry, but equally to visitors on-site.

The Health and Safety at Work Act 1974

Many governments have passed laws aimed at improving safety at work, but the most important recent legislation has been the Health and Safety at Work Act 1974. This Act should be thought of as an umbrella Act that other statutory legislation sits under. The purpose of the Act is to provide the legal framework for stimulating and encouraging high standards of health and safety at work; the Act puts the responsibility for safety at work on both workers and managers.

Duty of Care

The employer has a duty to care for the health and safety of employees (Section 2 of the Act). To do this, he or she must ensure that:

- the working conditions and standard of hygiene are appropriate;
- the plant, tools and equipment are properly maintained;
- the necessary safety equipment – such as PPE, dust and fume extractors and machine guards – is available and properly used;
- the workers are trained to use equipment and plant safely.

Definition

Statutory regulations have been passed by Parliament and have, therefore, become laws.

Top tip

A statutory law is an Act of Parliament.

Top tip

HSE applies to both employers and employees to secure the welfare of everyone involved.

Failure to comply with the Health and Safety at Work Act is a criminal offence and any infringement of the law can result in heavy fines, a prison sentence or both. This would apply to an employer who could be prosecuted if knowingly allowing an employee to work and that employee then places other people at risk or possible injury.

Employees have a duty to care for their own health and safety and that of others who may be affected by their actions including fellow employees and members of the public (Section 7 of the Act). To do this they must:

- take reasonable care to avoid injury to themselves or others as a result of their work activity;
- co-operate with their employer, helping him or her to comply with the requirements of the Act;
- not interfere with or misuse anything provided to protect their health and safety.

The Electricity at Work Regulations 1989 (EWR)

This legislation came into force in 1990 and replaced earlier regulations such as the Electricity (Factories Act) Special Regulations 1944. The regulations are made under the Health and Safety at Work Act 1974, and enforced by the Health and Safety Executive. The purpose of the regulations is to 'require precautions to be taken against the risk of death or personal injury from electricity in work activities'.

Section 4 of the EWR tells us that 'all systems must be constructed so as to prevent danger …, and be properly maintained … Every work activity shall be carried out in a manner which does not give rise to danger … In the case of work of an electrical nature, it is preferable that the conductors be made dead before work commences.'

The EWR do not tell us specifically how to carry out our work activities but they can be used in a court of law as evidence to claim compliance with other statutory requirements. If proceedings were brought against an individual for breaking the EWR, the only acceptable defence would be 'to prove that all reasonable steps were taken and all diligence exercised to avoid the offence' (Regulation 29).

An electrical contractor could reasonably be expected to have 'exercised all diligence' if the installation was wired according to the IET Wiring Regulations (see below). However, electrical contractors must become more 'legally aware' following the conviction of an electrician for manslaughter at Maidstone Crown Court in 1989. The court accepted that an electrician had caused the death of another man as a result of his shoddy work in wiring up a central heating system. He received a nine-month suspended prison sentence. This case has set an important legal precedent, and in future any tradesman or professional who causes death through negligence or poor workmanship risks prosecution and possible imprisonment.

The EWR is split into 16 regulations of its own. These regulations apply to any person who is engaged with electrical work: employers, the self-employed and employees, including apprentices.

Figure 1.3 This kind of wiring is unacceptable in the UK as a result of the IET Wiring Regulations.

Regulation 1 Citation and commencement

The first regulation puts the EWR into context and cites that the EWR came into force on 1 April 1990.

Regulation 2 Interpretation

Introduces certain terms used in the EWR, such as how we define system and conductor and even what we mean by 'danger'.

A system, for instance, is defined as: an electrical system in which all the electrical equipment is, or may be, electrically connected to a common source of electrical energy.

This means that the term 'system' includes all the constituent parts of a system, including the conductors and all the electrical equipment that fits within it.

'Electrical equipment' as defined in the regulations includes every type of electrical equipment from, for example, a 400kV overhead line to a battery-powered hand lamp. The reason that the EWR applies to even low-powered equipment is that, although the risk of electric shock may be low, there could still be a risk of explosion, for example.

Figure 1.4 The socket is one of the few parts of the electrical system that our customers see every day.

Figure 1.5a and 1.5b The Wiring Regulations define everything from overhead lines to hand lamps as electrical equipment.

A very important distinction is made regarding the terms 'charged' and 'live'. This is because when electricians carry out safe isolation procedures, they must ensure that all forms of energy are removed from a circuit, including any batteries or other devices such as capacitors that can store charge.

Consequently, the term 'dead' means a conductor that is not 'live' or 'charged'.

Regulation 3 Persons on whom duties are imposed by these Regulations

This regulation gives a clear statement of who the EWR applies to and makes a statement:

> It shall be the duty of every employee while at work to comply with the provisions of these regulations in so far as they relate to matters which are within his/her control.

This means that trainee electricians, although not fully qualified, must adhere to the EWR as well as always cooperating with their employer. The EWR also applies to non-technical personnel, such as office workers for instance, who must realize what they can or cannot do regarding electrical equipment and the importance of not interfering with safety equipment.

The EWR also defines the distinction between the terms 'absolute' and 'reasonably practicable'. 'Absolute' means that something must be met irrespective of time or cost. While 'reasonably practicable' means that a duty holder must decide the extent of the risks involved with the job in question against the costs involved as well as the actual difficulty in implementing safety measures.

Regulation 4 Systems, work activities and protective equipment

'System' has already been defined above, but this regulation ensures that systems are designed so far as is reasonably practicable to not pose a danger to anybody. This includes scheduling maintenance activities so that the equipment selected for any system is fit for purpose.

Regulation 5 Strength and capability of electrical equipment

This regulation ensures that the system and all related equipment can withstand certain electromechanical/chemical stresses and temperature rise during normal operation, overload conditions and even fault current.

Regulation 6 Adverse or hazardous environments

The regulation draws attention to the kinds of adverse conditions where danger could arise if equipment is not designed properly or fit for purpose. These sorts of considerations include: impact damage, weather conditions, temperature and even explosive conditions such as dust-rich environments.

Regulation 7 Insulation, protection and placing of conductors

This regulation looks at the danger surrounding electric shock, and is concerned with insulating conductors or placing and/or shielding them so that people cannot touch them directly and receive an electric shock or burn.

Regulation 8 Earthing or other suitable precautions

This regulation looks at how systems are protected to ensure that danger and, specifically, electric shock is minimized if faults occur. Both basic and fault protection measures apply, including earthing, bonding, separation and insulation of live parts.

Figure 1.6 Insulators provide protection from high voltages, but in this instance the electricity will be turned off to ensure safety.

Regulation 9 Integrity of referenced conductors

The object of this regulation is to prevent certain conductors that are designed for electrical safety from being altered or to impose restrictions in their use to stop dangerous situations arising. This is because any interference or fault in combined conductors can lead to possibly high, dangerous voltages being developed across parts that are not normally live, therefore bringing about danger of electric shock.

Regulation 10 Connections

The object of this regulation is to define the requirement regarding joints and connections. It specifies that all electrical connections need to be both mechanically and electrically strong as well as being suitable for use. In essence, all electrical connections need to be low in resistance, but a problem occurs when joints are not formed properly because they create high resistance joints, which in turn creates areas where power is not normally dissipated and electrical fires are created.

Regulation 11 Means for protecting from excess of current

The object of this regulation is the requirement to include protective devices to interrupt the supply when excess current is drawn. This is normally provided through fuses and circuit breakers as well as additional protection through residual current devices (RCDs).

Figure 1.7 Residual current devices add additional protection against excess current.

Regulation 12 Means for cutting off the supply and for isolation

The object of this regulation is to install where necessary a suitable means of electrical isolation. For instance, if the control equipment of an electrical motor is in a different room to the motor, then the control equipment must be encased in a lockable enclosure.

It is worth noting that 'isolation' means that all forms of energy are removed from a circuit, including any batteries or other devices such as capacitors that can store charge.

Regulation 13 Precautions for work on equipment made dead

This regulation ensures that adequate precautions shall be taken to prevent electrical equipment, which has been made dead, from becoming live when work is carried out on or near that equipment. Essentially, what this regulation is proposing is that electricians always carry out a safe isolation procedure. The regulation also informs that where reasonable, a written procedure known as a permit-to-work should be used to authorize and control electrical maintenance.

Regulation 14 Work on or near live conductors

This regulation ensures that no person shall be engaged in any work activity on or so near any live conductor (other than one suitably covered with insulating material so as to prevent danger) that danger may arise unless:

a. it is unreasonable in all the circumstances for it to be dead; and
b. it is reasonable in all the circumstances for the person to be at work on or near it while it is live; and
c. suitable precautions (including where necessary the provision of suitable protective equipment) are taken to prevent injury.

In other words there is an expectation that electricians only work on dead supplies unless there is a reason against it and you can justify that reason.

The regulation also specifies that the test instrument used to establish that a circuit is dead must be fit for purpose or 'approved'. It also mentions procedures regarding working in and around overhead power lines as well as quoting the Health and Safety at Work Act 1974 with regard to the provision of suitably trained first-aiders in the workplace.

Regulation 15 Working space, access and lighting

This regulation ensures that for the purposes of preventing injury, adequate working space, adequate means of access and adequate lighting shall be provided when working with electrical equipment. For instance, when live conductors are in the immediate vicinity, adequate space allows electricians to pull back away from the conductors without hazard as well as allowing space for people to pass one another safely without hazard.

Regulation 16 Persons to be competent to prevent danger and injury

This regulation ensures that people practising with electrical work are: technically knowledgeable, experienced and competent to carry out that work activity. Electrical apprentices can engage in electrical work activity but must be supervised accordingly.

The main object of the regulation is to ensure that people are not placed at risk due to a lack of skills on their part or on that of others in dealing with electrical equipment.

Remember the EWR 1989 regulations, if defied, can be used in any proceedings for an offence under this regulation.

Six Pack Regulations

As previously highlighted, the Health and Safety at Work Act 1974 is the main umbrella Act that other statutory Acts are drawn from. Alongside it certain statutory Acts form what is known as the Six Pack Regulations, as shown below:

The Management of Health and Safety at Work Regulations 1999

The Health and Safety at Work Act 1974 places responsibilities on employers to have robust health and safety systems and procedures in the workplace. Directors and managers of any company who employ more than five employees can be held personally responsible for failures to control health and safety. The Management of Health and Safety at Work Regulations 1999 tell us that employers must systematically examine the workplace, the work activity and the management of safety in the establishment through a process of 'risk assessments'. A record of all significant risk assessment findings must be kept in a safe place and be available to an HSE inspector if required. Information based on these findings must be communicated to relevant staff and, if changes in work behaviour patterns are recommended in the interests of safety, they must be put in place. The process of risk assessment is considered in detail later in this chapter.

Risks, which may require a formal assessment in the electrical industry, might be:

* working at height;
* using electrical power tools;
* falling objects;
* working in confined places;
* electrocution and personal injury;
* working with 'live' equipment;
* using hire equipment;
* manual handling, pushing, pulling and lifting;
* site conditions, falling objects, dust, weather, water, accidents and injuries.

And any other risks which are particular to a specific type of workplace or work activity.

Personal Protective Equipment (PPE) at Work Regulations 1998

PPE is defined as all equipment designed to be worn, or held, to protect against a risk to health and safety. This includes most types of protective clothing, and equipment such as eye, foot and head protection, safety harnesses, life-jackets and high-visibility clothing. Under the Health and Safety at Work Act, employers must provide free of charge any PPE and employees must make full and proper use of it.

Safety first

PPE

Always wear or use the PPE (personal protective equipment) provided by your employer for your safety.

Figure 1.8 Personal protective equipment should be worn at all times on-site.

Provision and Use of Work Equipment Regulations 1998

These regulations tidy up a number of existing requirements already in place under other regulations such as the Health and Safety at Work Act 1974, the Factories Act 1961 and the Offices, Shops and Railway Premises Act 1963. The Provision and Use of Work Equipment Regulations 1998 place a general duty on employers to ensure minimum requirements of plant and equipment. If an employer has purchased good-quality plant and equipment which is well maintained, there is little else to do. Some older equipment may require modifications to bring it into line with modern standards of dust extraction, fume extraction or noise, but no assessments are required by the regulations other than those generally required by the Management Regulations 1999, discussed previously.

Manual Handling Operations Regulations 1992 (as amended)

In effect, any activity that requires an individual to lift, move or support a load will be classified as a manual handling task. More than a third of all reportable injuries are believed to involve incorrect lifting techniques or carrying out manual handling operations without using mechanical lifting devices. Companies will also train their personnel with specific persons appointed as manual handling advisors or safety representatives who will carry out both induction and refresher training in order to educate their workforce on correct lifting techniques and manual handling procedures.

Health and Safety (display screen equipment) Regulations 1992

These regulations are concerned with providing specific parameters on the expected safety and health requirements and implications for personnel who work with display screen equipment.

The equipment must be scrutinized so that it is not a source of risk for operators and should include the:

- display screen
- keyboard
- user space
- chair
- lighting (glare)
- software.

Workplace Health, Safety and Welfare Regulations 1992

This regulation specifies the general requirements and expectation of accommodation standards for nearly all workplaces. Breach of this regulation would be seen as a crime, punishable following any successful conviction.

Other statutory regulations include:

The Control of Substances Hazardous to Health Regulations 2002 (COSHH)

The original COSHH Regulations were published in 1988 and came into force in October 1989. They were re-enacted in 1994 with modifications and improvements, and the latest modifications and additions came into force in 2002.

The COSHH Regulations control people's exposure to hazardous substances in the workplace. Regulation 6 requires employers to assess the risks to health from working with hazardous substances, to train employees in techniques which will reduce the risk and provide personal protective equipment (PPE) so that employees will not endanger themselves or others through exposure to hazardous substances. Employees should also know what cleaning, storage and disposal procedures are required and what emergency procedures to follow. The necessary information must be available to anyone using hazardous substances as well as to visiting HSE inspectors. Hazardous substances include:

1 any substance which gives off fumes causing headaches or respiratory irritation;
2 man-made fibres which might cause skin or eye irritation (e.g. loft insulation);
3 acids causing skin burns and breathing irritation (e.g. car batteries, which contain dilute sulphuric acid);
4 solvents causing skin and respiratory irritation (strong solvents are used to cement together PVC conduit fittings and tubes);
5 fumes and gases causing asphyxiation (burning PVC gives off toxic fumes);
6 cement and wood dust causing breathing problems and eye irritation;
7 exposure to asbestos – although the supply and use of the most hazardous asbestos material is now prohibited, huge amounts were installed between 1950 and 1980 in the construction industry and much of it is still in place today.

In their latest amendments, the COSHH Regulations focus on giving advice and guidance to builders and contractors on the safe use and control of asbestos products. These can be found in Guidance Notes EH 71 or visit www.hse.uk/hiddenkiller.

Remember: where PPE is provided by an employer, employees have a duty to use it.

Working at Height Regulations

Working above ground level creates added dangers and slows down the work rate of the electrician. New Work at Height Regulations came into force on 6 April 2005. Every precaution should be taken to ensure that the working platform is appropriate for the purpose and in good condition. This is especially important since the main cause of industrial deaths comes from working at height.

Figure 1.9 Asbestos removal requires very strict procedures to be followed.

The Electricity Safety, Quality and Continuity Regulations 2002 (formerly Electricity Supply Regulations 1989)

The Electricity Safety, Quality and Continuity Regulations 2002 were issued by the Department of Trade and Industry. They are statutory regulations which are enforceable by the laws of the land. They are designed to ensure a proper and safe supply of electrical energy up to the consumer's terminals.

These regulations impose requirements upon the regional electricity companies regarding the installation and use of electric lines and equipment. The regulations are administered by the Engineering Inspectorate of the Electricity Division of the Department of Energy and will not normally concern the electrical contractor, except that it is these regulations which lay down the earthing requirement of the electrical supply at the meter position.

The regional electricity companies must declare the supply voltage and maintain its value between prescribed limits or tolerances.

The government agreed on 1 January 1995 that the electricity supplies in the United Kingdom would be harmonized with the rest of Europe. Thus the voltages used previously in low-voltage supply systems of 415 V and 240 V have become 400 V for three-phase supplies and 230 V for single-phase supplies. The permitted tolerances to the nominal voltage have also been changed from ±6% to +10% and −6%. This gives a voltage range of 216–253 V for a nominal voltage of 230 V and 376–440 V for a nominal supply voltage of 400 V.

Regulation 29 gives the area boards the power to refuse to connect a supply to an installation which in their opinion is not constructed, installed and protected to an appropriately high standard. This regulation would only be enforced if the installation did not meet the requirements of the IET Regulations for Electrical Installations.

Control of Asbestos at Work Regulations

In October 2010 the HSE launched a national campaign to raise awareness among electricians and other trades of the risk to their health of coming into contact with asbestos. It is called the 'Hidden Killer Campaign' because

Figure 1.10 Signs should be placed around the site whenever public safety may be put at risk.

approximately six electricians will die each week from asbestos-related diseases. For more information about asbestos hazards, visit www.hse.uk/hiddenkiller.

Non-statutory regulations

Statutory laws and regulations are written in a legal framework; some don't actually tell us how to comply with the laws at an everyday level.

Non-statutory regulations and codes of practice interpret the statutory regulations, telling us how we can comply with the law.

They have been written for every specific section of industry, commerce and situation, to enable everyone to comply with or obey the written laws.

When the Electricity at Work Regulations (EWR) tell us to 'ensure that all systems are constructed so as to prevent danger' they do not tell us how to actually do this in a specific situation. However, the IET Regulations tell us precisely how to carry out electrical work safely in order to meet the statutory requirements of the EWR. Part 1 of the IET Regulations, at 114.1, states: 'the Regulations are non-statutory. They may, however, be used in a court of law in evidence to claim compliance with a statutory requirement.' If your electrical installation work meets the requirements of the IET Regulations, you will also meet the requirements of EWR.

Over the years, non-statutory regulations and codes of practice have built upon previous good practice and responded to changes by bringing out new editions of the various regulations and codes of practice to meet the changing needs of industry and commerce.

Definition

Statutory laws and regulations are written in a legal framwork.

Non-statutory regulations and codes of practice interpret the statutory regulations, telling us how we can comply with the law.

The IET Wiring Regulations 18th Edition requirements for electrical installations to BS 7671: 2018

We will now look at one non-statutory regulation, which is sometimes called 'the electrician's bible' and is the most important set of regulations for anyone working in the electrical industry: the BS 7671: 2018 Requirements for Electrical Installations, IET Wiring Regulations 18th Edition.

The Institution of Engineering and Technology Requirements for Electrical Installations (the IET Regulations) are non-statutory regulations. They relate principally to the design, selection, erection, inspection and testing of electrical installations, whether permanent or temporary, in and about buildings generally and to agricultural and horticultural premises, construction sites and caravans and their sites.

Figure 1.11 The IET Wiring Regulations must be followed for any electrical installation regardless of where it is or how short term it may be.

Paragraph 7 of the introduction to the EWR says: 'the IET Wiring Regulations is a code of practice which is widely recognized and accepted in the United Kingdom and compliance with them is likely to achieve compliance with all relevant aspects of the Electricity at Work Regulations.' The IET Wiring Regulations are the national standard in the United Kingdom and apply to installations operating at a voltage up to 1000 V a.c. They do not apply to electrical installations in mines and quarries, where special regulations apply because of the adverse conditions experienced there.

The current edition of the IET Wiring Regulations is the 18th Edition 2018. The main reason for incorporating the IET Wiring Regulations into British Standard BS 7671: 2018 was to create harmonization with European Standards.

The IET Regulations take account of the technical intent of the CENELEC European Standards, which in turn are based on the IEC International Standards.

The purpose of harmonizing British and European Standards is to help develop a single European market economy so that there are no trade barriers to electrical goods and services across the European Economic Area.

A number of guidance notes have been published to assist electricians in their understanding of the regulations. The guidance notes which are frequently referred to in this book are contained in the *On-Site Guide.* Eight other guidance notes booklets are also currently available. These are:

- Selection and Erection;
- Isolation and Switching;
- Inspection and Testing;
- Protection against Fire;
- Protection against Electric Shock;
- Protection against Overcurrent;
- Special Locations;
- Earthing and Bonding.

These guidance notes are intended to be read in conjunction with the regulations.

The IET Wiring Regulations are the electrician's bible and provide an authoritative framework for anyone working in the electrical industry.

Assessment criteria 1.3

State their own roles and responsibilities and those of others with regard to current relevant legislation

Health and safety responsibilities

We have now looked at statutory and non-statutory regulations which influence working conditions in the electrical industry today. So, who has *responsibility* for these workplace health and safety regulations?

In 1970, a Royal Commission was set up to look at the health and safety of employees at work. The findings concluded that the main cause of accidents at work was apathy on the part of *both* employers and employees.

The Health and Safety at Work Act 1974 was passed as a result of Recommendations made by the Royal Commission and, therefore, the Act puts legal responsibility for safety at work on *both* the employer and employee.

In general terms, the employer must put adequate health and safety systems in place at work and the employee must use all safety systems and procedures responsibly.

In specific terms the employer must:

- provide a Health and Safety Policy Statement if there are five or more employees, such as that shown in Figure 1.13;
- display a current employers' liability insurance certificate as required by the Employers' Liability (Compulsory Insurance) Act 1969;
- report certain injuries, diseases and dangerous occurrences to the enforcing authority (HSE area office;
- provide adequate first-aid facilities (see Tables 1.1 and 1.2);

Key fact

IET Regulations
- They are the UK National Standard for all electrical work.
- They are the 'electrician's bible'.
- Comply with the IET Regulations and you also comply with Statutory Regulations (IET Regulation 114).

- provide PPE;
- provide information, training and supervision to ensure staff health and safety;
- provide adequate welfare facilities;
- put in place adequate precautions against fire, provide a means of escape and means of fighting fire;
- ensure plant and machinery are safe and that safe systems of operation are in place;
- ensure articles and substances are moved, stored and used safely;
  make the workplace safe and without risk to health by keeping dust, fumes and noise under control.

In specific terms the employee must:

- take reasonable care of his/her own health and safety and that of others who may be affected by what they do;
- co-operate with his/her employer on health and safety issues by not interfering with or misusing anything provided for health, safety and welfare in the working environment;
- report any health and safety problem in the workplace to, in the first place, a supervisor, manager or employer.

Duty of care

The Health and Safety at Work Act and the Electricity at Work Regulations make numerous references to employer and employees having a 'duty of care' for the health and safety of others in the work environment. For instance, specific statutory regulations are put in place to ensure that **employers safeguard that plant and systems so far as is reasonably practicable are safe and without risk to health**. Such expectation would also extend to the:

- handling, storage and transport of goods including substances;
- installation of training and supervision.

In essence, there is an expectation that employers conduct their business in such a way that, *so far as is reasonably practicable*, their employees and members of the public are not put at risk. For instance, an employer under the management of health and safety regulations could be prosecuted if they knowingly allowed an employee to work and they in turn placed others at risk.

In this context, the Electricity at Work Regulations refer to a person as a 'duty holder'. This phrase recognizes the level of responsibility which electricians are expected to take on as part of their job in order to control electrical safety in the work environment. Everyone has a duty of care, but not everyone is a duty holder. The regulations recognize the amount of control that an individual might exercise over the whole electrical installation. The person who exercises 'control over the whole systems, equipment and conductors' and is the electrical company's representative on-site, is the *duty holder*. They might be a supervisor or manager, but they will have a duty of care on behalf of their employer for the electrical, health, safety and environmental issues on that site. Duties referred to in the regulations may have the qualifying terms 'reasonably practicable' or 'absolute'.

If the requirement of the regulation is absolute, then that regulation must be met regardless of cost or any other consideration. If the regulation is to be met 'so far as is reasonably practicable', then risks, cost, time, trouble and difficulty can be considered.

Definition

'*Duty holder*' - this phrase recognizes the level of responsibility which electricians are expected to take on as a part of their job in order to control electrical safety in the work environment.

Everone has a duty of care, but not everyone is a *duty holder*.

The person who exercises 'control over the whole systems, equipment and conductors' and is the electrical company's representative on-site is a *duty holder*.

'*Reasonably practicable*' or '*absolute*' – if the requirement of the regulation is absolute, then that regulation must be met regardless of cost or any other consideration. If the regulation is to be met 'so far as is reasonably practicable', then risks, cost, time, trouble and difficulty can be considered.

Often there is a cost-effective way to reduce a particular risk and prevent an accident from occurring. For example, placing a fireguard in front of the fire at home when there are young children in the family is a reasonably practicable way of reducing the risk of a child being burned. If a regulation is not qualified with 'so far as is reasonably practicable', then it must be assumed that the regulation is absolute. In the context of the Electricity at Work Regulations, where the risk is very often death by electrocution, the level of duty to prevent danger more often approaches that of an absolute duty of care.

As already discussed, the term 'responsibility' also extends to employees since being ignorant of either the law or your responsibilities is not a viable excuse. This is why employees are expected to take reasonable care of both themselves and other people, including the general public, which is why embracing a positive attitude is so important. Moreover, interfering with safety systems or engaging in horseplay or using recreational drugs including alcohol at work would be deemed gross misconduct resulting in instant dismissal.

Safety first

Duty holder
This person has the responsibility to control electrical safety in the work environment.

Figure 1.12 The duty holder is overseeing this work, but both of these workers are responsible for safety.

Safety documentation

Under the Health and Safety at Work Act, the employer is responsible for ensuring that adequate instruction and information is given to employees to make them safety-conscious. Part 1, Section 3 of the Act instructs all employers to prepare a written health and safety policy statement and to bring this to the notice of all employees. Figure 1.13 shows a typical Health and Safety Policy Statement of the type which will be available within your company. Your employer must let you know who your safety representatives are, and the new Health and Safety poster shown in Figure 1.14 has a blank section onto which the names and contact information of your specific representatives can be added. This is a large, laminated poster, 595–415 mm, suitable for wall or noticeboard display. All workplaces employing five or more people have to display the poster.

FLASH-BANG ELECTRICAL

Statement of Health and Safety at Work Policy in accordance with the Health and Safety at Work Act 1974

Company objective

The promotion of health and safety measures is a mutual objective for the Company and for its employees at all levels. It is the intention that all the Company's affairs will be conducted in a manner which will not cause risk to the health and safety of its members, employees or the general public. For this purpose it is the Company policy that the responsibility for health and safety at work will be divided between all the employees and the Company in the manner outlined below.

Company's responsibilities

The Company will, as a responsible employer, make every endeavour to meet its legal obligations under the Health and Safety at Work Act to ensure the health and safety of its employees and the general public. Particular attention will be paid to the provision of the following:

1 Plant equipment and systems of work that are safe.
2 Safe arrangements for the use, handling, storage and transport of articles, materials and substances.
3 Sufficient information, instruction, training and supervision to enable all employees to contribute positively to their own safety and health at work and to avoid hazards.
4 A safe place of work, and safe access to it.
5 A healthy working environment.
6 Adequate welfare services.

Note: Reference should be made to the appropriate safety etc. manuals.

Employees' responsibilities

Each employee is responsible for ensuring that the work which he/she undertakes is conducted in a manner which is safe to himself or herself, other members of the general public, and for obeying the advice and instructions on safety and health matters issued by his/her superior. If any employee considers that a hazard to health and safety exists it is his/her responsibility to report the matter to his/her supervisor or through his/her Union Representative or such other person as may be subsequently defined.

Management and supervisors' responsibilities

Management and supervisors at all levels are expected to set an example in safe behaviour and maintain a constant and continuing interest in employee safety, in particular by:

1 acquiring the knowledge of health and safety regulations and codes of practice necessary to ensure the safety of employees in the workplace,
2 acquainting employees with these regulations on codes of practice and giving guidance on safety matters,
3 ensuring that employees act on instructions and advice given.

General Managers are ultimately responsible to the Company for the rectification or reporting of any safety hazard which is brought to their attention.

Joint consultations

Joint consultation on health and safety matters is important. The Company will agree with its staff, or their representatives, adequate arrangements for joint consultation on measures for promoting safety and health at work, and make and maintain satisfactory arrangements for the participation of their employees in the development and supervision of such measures. Trade Union representatives will initially be regarded as undertaking the role of Safety Representatives envisaged in the Health and Safety at Work Act. These representatives share a responsibility with management to ensure the health and safety of their members and are responsible for drawing the attention of management to any shortcomings in the Company's health and safety arrangements. The Company will in so far as is reasonably practicable provide representatives with facilities and training in order that they may carry out this task.

Review

A review, addition or modification of this statement may be made at any time and may be supplemented as appropriate by further statements relating to the work of particular departments and in accordance with any new regulations or codes of practice.

This policy statement will be brought to the attention of all employees.

Figure 1.13 Typical health and safety policy statement.

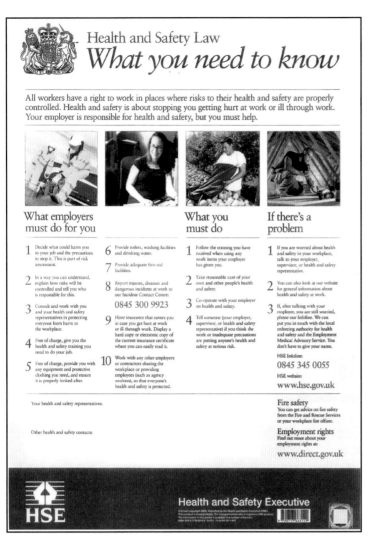

Figure 1.14 Health and Safety law poster. Source: HSE © Reproduced by permission of HSE.

Assessment criteria 1.4

State the role of enforcing authorities under health and safety legislation

Enforcement of health and safety regulations

Laws and rules must be enforced if they are to be effective. The system of control under the Health and Safety at Work Act comes from the Health and Safety Executive (HSE) which is charged with enforcing the law. The HSE is divided into a number of specialist inspectorates or sections which operate from local offices throughout the United Kingdom. From the local offices the inspectors visit individual places of work.

The HSE inspectors have been given wide-ranging powers to assist them in the enforcement of the law. They can:

1 Enter premises unannounced and carry out investigations, take measurements or photographs.

2 Take statements from individuals.

3 Check the records and documents required by legislation.

4 Give information and advice to an employee or employer about safety in the workplace.

5 Demand the dismantling or destruction of any equipment or material or issue an improvement notice which will require an employer to put right, within a specified period of time, a minor infringement of the legislation.

6 Issue an improvement notice which will require an employer to put right, within a specified period of time, a minor infringement of the legislation.

7 Issue a prohibition notice which will require an employer to stop immediately any activity likely to result in serious injury, and which will be enforced until the situation is corrected.

8 Prosecute all persons who fail to comply with their safety duties, including employers, employees, designers, manufacturers, suppliers and the self-employed.

Assessment criteria 2.1

State the procedures that should be followed in the case of accidents which involve injury

Accidents at work

Despite all the safety precautions taken on construction sites to prevent injury to the workforce, accidents do happen and *you* may be the only other person able to take action to assist a workmate. This section is not intended to replace a first-aid course but to give learners the knowledge to understand the types of injuries they may come across in the workplace. If you are not a qualified first-aider, limit your help to obvious common-sense assistance and call for help, *but* do remember that if a workmate's heart or breathing has stopped as a result of an accident they have only minutes to live unless you act quickly.

The Health and Safety (first-aid) Regulations 1981 and relevant approved codes of practice and guidance notes place a duty of care on all employers to provide *adequate* first-aid facilities appropriate to the type of work being undertaken. *Adequate* facilities will relate to a number of factors such as:

* How many employees are employed?
* What type of work is being carried out?
* Are there any special or unusual hazards?
* Are employees working in scattered and/or isolated locations?
* Is there shift work or 'out of hours' work being undertaken?
* Is the workplace remote from emergency medical services?
* Are there inexperienced workers on-site?
* What were the risks of injury and ill health identified by the company's hazard risk assessment?

The regulations state that:

> Employers are under a duty to provide such numbers of suitable persons as is adequate and appropriate in the circumstances for rendering first aid to his employees if they are injured or become ill at work. For this purpose a person shall not be suitable unless he or she has undergone such training and has such qualifications as the Health and Safety Executive may approve.

Definition

An *accident* may be defined as an uncontrolled event causing injury or damage to an individual or properly.

Figure 1.15 Every workplace should have one of these.

This is typical of the way in which the health and safety regulations are written. The regulations and codes of practice do not specify numbers, but set out guidelines in respect of the number of first-aiders needed, dependent upon the type of company, the hazards present and the number of people employed.

Let us now consider the questions 'what is first aid?' and 'who might become a first-aider?' The regulations give the following definitions of first aid. '*First aid* is the treatment of minor injuries which would otherwise receive no treatment or do not need treatment by a doctor or nurse'; *or* 'In cases where a person will require help from a doctor or nurse, first aid is treatment for the purpose of preserving life and minimizing the consequences of an injury or illness until such help is obtained'. A more generally accepted definition of first aid might be as follows: first aid is the initial assistance or treatment given to a casualty for any injury or sudden illness before the arrival of an ambulance, doctor or other medically qualified person.

Now, having defined first aid, who might become a first-aider? A first-aider is someone who has undergone a training course to administer first aid at work and holds a current first-aid certificate. The training course and certification must be approved by the HSE. The aims of a first-aider are to preserve life, to limit the worsening of the injury or illness and to promote recovery. A first-aider may also undertake the duties of an appointed person. An appointed person is someone who is nominated to take charge when someone is injured or becomes ill, including calling an ambulance if required. The appointed person will also look after the first-aid equipment, including restocking the first-aid box.

Appointed persons should not attempt to give first aid for which they have not been trained, but should limit their help to obvious common-sense assistance and summon professional assistance as required.

Suggested numbers of first-aid personnel are given in Table 1.1. The actual number of first-aid personnel must take into account any special circumstances such as remoteness from medical services, the use of several separate buildings and the company's hazard risk assessment. First-aid personnel must be available at all times when people are at work, taking into account shift-working patterns and providing cover for sickness absences. Every company must have

Definition

First aid is the initial assistance or treatment given to a casualty for any injury or sudden illness before the arrival of an ambulance, doctor or other medically qualified person.

Definition

A *first-aider* is someone who has undergone a training course to administer first aid at work and holds a current first aid certificate.

Definition

An *appointed person* is someone who is nominated to take charge when someone is injured or becomes ill, including calling an ambulance if required. The appointed person will also look after the first-aid equipment, including restocking the first-aid box.

Figure 1.16 Every workplace should have appointed first-aiders who are trained to deal with medical emergencies.

at least one first-aid kit under the regulations. The size and contents of the kit will depend upon the nature of the risks involved in the particular working environment and the number of employees. Tables 1.1 and 1.2 give information concerning suggested numbers of first-aid personnel and the contents of any first-aid box in order to comply with the HSE Regulations.

There now follows a description of some first-aid procedures which should be practised under expert guidance before they are required in an emergency.

Asphyxiation

Asphyxiation is a condition caused by lack of air in the lungs leading to suffocation. Suffocation may cause discomfort by making breathing difficult or it may kill by stopping the breathing. There is a risk of asphyxiation to workers when:

* working in confined spaces;
* working in poorly ventilated spaces;
* working in paint stores and spray booths;
* working in the petrochemical industry;
* working in any environment in which toxic fumes and gases are present.

Under the Management of Health and Safety at Work Regulations, a risk assessment must be made if the environment may be considered hazardous to health. Safety procedures, including respiratory protective equipment, must be in place before work commences. The treatment for fume inhalation or asphyxia is to get the patient into fresh air but only if you can do this without putting yourself

Definition

Asphyxiation is a condition caused by lack of air in the lungs leading to suffocation. Suffocation may cause discomfort by making breathing difficult or it may kill by stopping the breathing.

Table 1.1 Suggested numbers of first-aid personnel

Category of risk	Numbers employed at any location	Suggested number of first-aid personnel
Lower risk e.g. shops and offices, libraries	Fewer than 50 50–100 More than 100	At least one appointed person At least one first-aider One additional first-aider for every 100 employed
Medium risk e.g. light engineering and assembly work, food processing, warehousing	Fewer than 20 20–100 More than 100	At least one appointed person At least one first-aider for every 50 employed (or part thereof) One additional first-aider for every 100 employed
Higher risk e.g. most construction, slaughterhouses, chemical manufacture, extensive work with dangerous machinery or sharp instruments	Fewer than five 5–50 More than 50	At least one appointed person At least one first-aider One additional first-aider for every 50 employed

Table 1.2 Contents of first-aid boxes

Item	Number of employees				
	1–5	6–10	11–50	51–100	101–150
Guidance card on general first aid	1	1	1	1	1
Individually wrapped sterile adhesive dressings	10	20	40	40	40
Sterile eye pads, with attachment (Standard Dressing No. 16 BPC)	1	2	4	6	8
Triangular bandages	1	2	4	6	8
Sterile covering for serious wounds (where applicable)	1	2	4	6	8
Safety-pins	6	6	12	12	12
Medium-sized sterile unmedicated dressings (Standard Dressings No. 9 and No. 14 and the Ambulance Dressing No. 1)	3	6	8	10	12
Large sterile unmedicated dressings (Standard Dressings No. 9 and No. 14 and the Ambulance Dressing No. 1)	1	2	4	6	10
Extra-large sterile unmedicated dressings (Ambulance Dressing No. 3)	1	2	4	6	8
Where tap water is not available, sterile water or sterile normal saline in disposable containers (each holding a minimum of 300 ml) must be kept near the first-aid box. The following minimum quantities should be kept:					
Number of employees	1–10	11–50	51–100	101–150	
Quantity of sterile water	1 × 300 ml	3 × 300 ml	6 × 300 ml	6 × 300 ml	

at risk. If the patient is unconscious, proceed with resuscitation as described below.

Bleeding

If the wound is dirty, rinse it under clean running water. Clean the skin around the wound and apply a plaster, pulling the skin together. If the bleeding is severe, apply direct pressure to reduce the bleeding and raise the limb if possible. Apply a sterile dressing or pad and bandage firmly before obtaining professional advice.

To avoid possible contact with hepatitis or the AIDS virus when dealing with open wounds, first-aiders should avoid contact with fresh blood by wearing plastic or rubber protective gloves, or by allowing the casualty to apply pressure to the bleeding wound.

Burns

Remove heat from the burn to relieve the pain by placing the injured part under clean, running cold water. Do not remove burnt clothing sticking to the skin. Do not apply lotions or ointments. Do not break blisters or attempt to remove loose skin. Cover the injured area with a clean, dry dressing.

Broken bones

Make the casualty as comfortable as possible by supporting the broken limb either by hand or with padding. Do not move the casualty unless by remaining in that position they are likely to suffer further injury. Obtain professional help as soon as possible.

Contact with chemicals

Wash the affected area very thoroughly with clean, cold water. Remove any contaminated clothing. Cover the affected area with a clean, sterile dressing and seek expert advice. It is a wise precaution to treat all chemical substances as possibly harmful; even commonly used substances can be dangerous if contamination is from concentrated solutions. When handling dangerous substances, it is also good practice to have a neutralising agent to hand.

Disposal of dangerous substances must not be into the main drains since this can give rise to an environmental hazard, but should be undertaken in accordance with local authority regulations.

Exposure to toxic fumes

Get the casualty into fresh air quickly and encourage deep breathing if he/she is conscious. Resuscitate if breathing has stopped. Obtain expert medical advice as fumes may cause irritation of the lungs.

Sprains and bruising

A cold compress can help to relieve swelling and pain. Soak a towel or cloth in cold water, squeeze it out and place it on the injured part. Renew the compress every few minutes.

Breathing stopped – resuscitation

Remove any restrictions from the face and any vomit, loose or false teeth from the mouth. Loosen tight clothing around the neck, chest and waist. To ensure a good airway, lay the casualty on their back and support the shoulders on some padding. Tilt the head backwards and open the mouth. If the casualty is faintly breathing, lifting the tongue and clearing the airway may be all that is necessary to restore normal breathing.

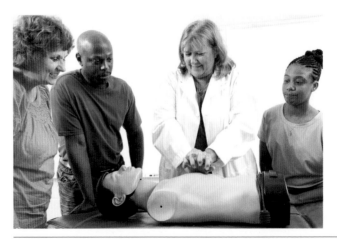

Figure 1.17 Following a few simple rules can save lives.

Heart stopped beating – CPR (Cardio pulmonary resuscitation)

This sometimes happens following a severe electric shock. If the casualty's lips are blue, the pupils of their eyes widely dilated and the pulse in their neck cannot be felt, then they may have gone into cardiac arrest. Act quickly and lay the casualty on their back. Kneel down beside them and place the heel of one hand in the centre of their chest. A good tip for finding the heart is to follow the ribs up to the top of the rib cage and then place two fingers, above which is the target for compression.

Then cover this hand with your other hand and interlace the fingers. Straighten your arms and press down on their chest sharply with the heel of your hands and then release the pressure. Continue to do this 15 times at the rate of one push per second. 'Push hard and fast, on the chest, to the rhythm of Ha, Ha, Ha, Ha, stayin' alive, stayin' alive'. That is, the music from *Saturday Night Fever*, as demonstrated by Vinnie Jones in a TV advertisement. An alternative would be the chorus of 'Nelly the elephant'. Next, check the casualty's pulse. If none is felt, give a further 15 chest compressions. Continue this procedure until the heartbeat is restored. Pay close attention to the condition of the casualty while giving heart massage. When a pulse is restored the blueness around the mouth will quickly go away and you should stop the heart massage. Treat the casualty for shock, place them in the recovery position and obtain professional help. Be aware for any first responder signs, which indicate that the location contains not only a heart start machine but also that a qualified first responder is nearby.

Shock

Everyone suffers from shock following an accident. The severity of the shock depends upon the nature and extent of the injury. In cases of severe shock the casualty will become pale and their skin will become clammy from sweating. They may feel faint, have blurred vision, feel sick and complain of thirst.

It is vital that you reassure the casualty that everything that needs to be done is being done. Loosen tight clothing and keep them warm and dry until help arrives. *Do not* move them unnecessarily or give them anything to drink.

Safety first

DURING AN EMERGENCY
- Raise the alarm.
- Walk to the assembly point.
- Ensure everyone is present.

Assessment criteria 2.2

Recognize appropriate procedures which should be followed when emergency situations occur in the workplace

Evacuation procedures

When the fire alarm sounds you must leave the building immediately by any one of the escape routes indicated. **Exit routes** are usually indicated by a green and white 'running man' symbol. Evacuation should be orderly; do not run but walk purposefully to your designated assembly point. The purpose of an **assembly point** is to get you away from danger to a place of safety where you will not be in the way of the emergency services. It also allows people to be accounted for and to make sure that no one is left in the building. You must not re-enter the building until a person in authority gives permission to do so.

An evacuation in a real emergency can be a frightening experience, especially if you do not really familiarize yourself with your company's fire safety procedures;

Definition

Exit routes are usually indicated by a green and white 'running man' symbol. Evacuation should be orderly; do not run but walk purposefully to your designated assembly point.

Definition

The purpose of an *assembly point* is to get you away from danger to a place of safety where you will not be in the way of the emergency services.

many companies carry out an induction programme in order to highlight key points and procedures.

Company induction

When an emergency situation happens it is all very well to say 'do not panic', but in truth, people do. A better course of action is to think fast but act slow, so that you give yourself time to think of a plan of action. One of the reasons why new personnel are given an induction process into a company is that you are then equipped with the knowledge to use in an emergency. For instance, knowing where the emergency stop buttons are so that the electrical supply can be isolated. This might be linked to an induction lecture to remind people that isolating a casualty from the supply can be done using a non-conducting item such as a wooden brush. Personnel also need to be aware of the relevant first-aid points and who the first-aiders or first responders are so that first aid can be administered quickly, efficiently and correctly

New personnel also need to know any designated escape routes thus avoiding any hazardous areas as well as where the assembly points are. Some companies employ a hat system, whereby someone in authority will don a green hat thus indicating that they are responsible for establishing manpower control, whilst a red hat would signal the person responsible for commanding a fire-fighting team. It is also very important that when summoning help you give as much information as possible. For instance: where the fire is; which building; or area outside of a building; type of fire and any suspected casualties or persons unaccounted for. Such information can better inform the fire service, especially if their response team is expected to enter a building complete with BA (breathing apparatus).

That said, the phone call must not be made in the building that is at risk just in case you are endangering yourself. Sounding the alarm is not always straightforward either, since some alarm systems are local and are not necessarily linked to the emergency services. Certain establishments such as military camps also coordinate their telephone exchanges, therefore despite dialling locally you might be speaking to someone in Scotland. In all situations, common sense should be used:

- raise the alarm;
- put out the fire if you can tackle a fire safely and are trained to do so, but never endanger yourself;
- always walk to your assembly point;
- always carry out a roll call to see if anyone is missing and report your findings to the incident commander or senior person present.

Site visitors

A construction site will have many people working on a building project throughout the construction period. The groundwork people lay the foundations, the steel work is erected, the bricklayers build the walls, the carpenters put on the roof and, only when the building is waterproof do the electricians and mechanical service trades begin to install the electrical and mechanical systems.

If there was an emergency and the site had to be evacuated, how would you know who, or how many people were on-site? You can see that there has to be a

Figure 1.18 Exit routes and assembly points should be clearly signposted.

procedure for logging people in and out, so that the main contractor can identify who is on-site at any one time.

If the architect pops in to make a quick inspection or resolve a problem, they must first 'sign in'. If the managing director of your electrical company drops by to see how work is progressing they must first 'sign in'. When they leave, they must 'sign out'. How else would you know who is on-site at any one time?

A formal site visitor procedure must be in place to meet the requirements of the health and safety regulations and to maintain site security. The site visitor procedure must meet the needs of the building project. For example, a high-profile building project, one which pushes the architectural boundaries such as the new Wembley Football Stadium in London, would have attracted many more visitors during its construction period than, say, a food distribution warehouse for a supermarket.

The site visitor procedure is only required to be 'fit for purpose', but whatever system is put in place, here are some suggestions for consideration:

- you will want your visitors to be safe when they are on-site and so you might insist that they wear a hard hat and a high-visibility jacket;
- in some cases, you might want 'ordinary people' visitors to be escorted for their own safety while on-site. Of course this would not apply to professional visitors such as the architect or quantity surveyor;
- the procedure should identify who the visitor is and who they wish to visit;
- you will want to know the time of their arrival and the time of their departure.

Many of these requirements can be met with a simple logbook divided into columns with headings such as, date and time in, visitor's name, company name, name of the person to be visited and time out or time of leaving.

Assessment criteria 2.3

State the procedures for recording near misses and accidents at work

Accident reports

Causes of accidents

Most accidents are caused by either human error or environmental conditions. Human errors include behaving badly or foolishly, being careless and not paying attention to what you should be doing at work, doing things that you are not competent to do or have not been trained to do. You should not work when tired or fatigued and should never work when you have been drinking alcohol or taking drugs.

Environmental conditions include unguarded or faulty machinery, damaged or faulty tools and equipment, poorly illuminated or ventilated workplaces and untidy, dirty or overcrowded workplaces.

How high are the risks? Think about what might be the worst result; is it a broken finger or someone suffering permanent lung damage or being killed? How likely is it to happen? How often is that type of work carried out and how close do people get to the hazard? How likely is it that something will go wrong?

Figure 1.19 First-aid logbook/accident book with data protection-compliant removable sheets.

The most common categories of risk and causes of accidents at work are:

* slips, trips and falls;
* manual handling, that is, moving objects by hand;
* using equipment, machinery or tools;
* storage of goods and materials which then become unstable;
* fire;
* electricity;
* mechanical handling.

More importantly perhaps is to state that the main cause of deaths in the construction industry is working at height.

Figure 1.20 Working at height is the main cause of deaths on building sites. http://www.hse.gov.uk/statistics/industry/construction/index.htm

Despite new legislation, improved information, education and training, accidents at work do still happen. An **accident** may be defined as an uncontrolled event causing injury or damage to an individual or property. An accident can nearly always be avoided if correct procedures and methods of working are followed.

Any accident which results in an absence from work for more than three days or causes a major injury or death is notifiable to the HSE. There are more than 40,000 accidents reported to the HSE each year which occur as a result of some building-related activity. To avoid having an accident you should:

1 follow all safety procedures (e.g. fit safety signs when isolating supplies and screen off work areas from the general public);
2 not misuse or interfere with equipment provided for health and safety;
3 dress appropriately and use PPE when appropriate;
4 behave appropriately and with care;
5 avoid over-enthusiasm and foolishness;
6 stay alert and avoid fatigue;
7 not use alcohol or drugs at work;
8 work within your level of competence;
9 attend safety courses and read safety literature;
10 take a positive decision to act and work safely.

Every accident, however, must be reported to an employer and minor accidents reported to a supervisor, safety officer or first-aider and the details of the accident and treatment given suitably documented. A first-aid logbook or accident book such as that shown in Figure 1.19 containing first-aid treatment record sheets could be used to effectively document accidents which occur in the workplace and the treatment given. Failure to do so may influence the payment of compensation at a later date if an injury leads to permanent disability.

To comply with the Data Protection Act, from 31 December 2003 all first-aid treatment logbooks or accident report books must contain perforated sheets which can be removed after completion and filed away for personal security. If the accident results in death, serious injury or an injury that leads to an absence from work of more than three days, your employer must report the accident to the local office of the HSE. The quickest way to do this is to call the Incident Control Centre on 0345 300 9923. They will require the following information:

- the name of the person injured;
- a summary of what happened;
- a summary of events prior to the accident;
- information about the injury or loss sustained;
- details of witnesses;
- date and time of accident;
- name of the person reporting the incident.

The Incident Control Centre will forward a copy of every report they complete to the employer for them to check and hold on record. However, good practice would recommend an employer or their representative to make an extensive report of any serious accident that occurs in the workplace. In addition to recording the above information, the employer or their representative should:

- sketch diagrams of how the accident occurred, where objects were before and after the accident, where the victim fell, etc;
- take photographs or video that show how things were after the accident, for example, broken stepladders, damaged equipment, etc.;
- collect statements from witnesses. Ask them to write down what they saw;
- record the circumstances surrounding the accident. Was the injured person working alone – in the dark – in some other adverse situation or condition – was PPE being worn – was PPE recommended in that area?

The above steps should be taken immediately after the accident has occurred and after the victim has been sent for medical attention. The area should be made safe and senior management informed so that any actions to prevent a similar occurrence can be put in place. Taking photographs and obtaining witness statements immediately after an accident happens means that evidence may still be around and memories still sharp.

RIDDOR

RIDDOR stands for Reporting of Injuries, Diseases and Dangerous Occurrences Regulations 2013, which is sometimes referred to as RIDDOR 95, or just RIDDOR for short. The HSE requires employers to report some work-related accidents or diseases so that they can identify where and how risks arise, investigate serious accidents and publish statistics and data to help reduce accidents at work. What needs reporting? Every work-related death, major injury, dangerous occurrence and disease.

Safety first

RIDDOR

Any major injury must be reported which will be followed by an independent investigation by the HSE.

For example, instances of:

- fractures, other than to fingers, thumbs and toes;
- amputations;
- any injury likely to lead to permanent loss of sight or reduction in sight;
- any crush injury to the head or torso causing damage to the brain or internal organs;
- serious burns (including scalding) which:
  - covers more than 10% of the body
  - causes significant damage to the eyes, respiratory system or other vital organs;
- any scalping requiring hospital treatment;
- any loss of consciousness caused by head injury or asphyxia;
- any other injury arising from working in an enclosed space which:
  - leads to hypothermia or heat-induced illness
  - requires resuscitation or admittance to hospital for more than 24 hours.

Where an employee or member of the public is killed as a result of an accident at work the employer or his representative must report the accident to the environmental health department of the local authority by telephone that day and give brief details. Within 10 days this must be followed up by a completed accident report form (Form No. F2508). Major injuries sustained as a result of an accident at work include amputations, loss of sight (temporary or permanent), fractures to the body other than to fingers, thumbs or toes and any other serious injury. Once again, the environmental health department of the local authority must be notified by telephone on the day that the serious injury occurs and the telephone call followed up by a completed Form F2508 within 10 days.

Dangerous occurrences are listed in the regulations and include the collapse of a lift, an explosion or injury caused by an explosion, the collapse of a scaffold over 5 m high, the collision of a train with any vehicle, the unintended collapse of a building and the failure of fairground equipment. Depending on the seriousness of the event, it may be necessary to immediately report the incident to the local authority. However, the incident must be reported within 10 days by completing Form F2508. If a doctor notifies an employer that an employee is suffering from a

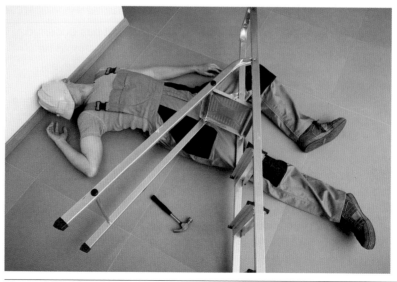

Figure 1.21 All accidents should be reported.

work-related disease then form F2508A must be completed and sent to the local authority. Reportable diseases include certain poisonings, skin diseases, lung disease, infections and occupational cancer. The full list is given within the pad of report forms.

An accident at work resulting in an over three day injury (that is, an employee being absent from work for over three days as a result of an accident at work) must be recorded but not reported.

An accident, however, that results in the injured person being away from work for more than seven days must be both recorded and reported; the number of days does not include the day the injury occurred, but does include weekends and rest days. The report must be made within 15 days of the accident.

Who are the reports sent to? They are sent to the environmental health department of the local authority or the area HSE offices. Accident report forms F2508 can also be obtained from them or by ringing the HSE Info-line, or by ringing the incident contact centre on 0345 300 9923. On-line forms can now be obtained through: hse.gov.uk/riddor.

For most businesses, a reportable accident, dangerous occurrence or disease is a very rare event. However, if a report is made, the company must keep a record of the occurrence for three years after the date on which the incident happened. The easiest way to do this would probably be to file a photocopy of the completed accident report form F2508, but a record may be kept in any form which is convenient.

Dangerous occurrences and hazardous malfunctions

A dangerous occurrence is a 'near miss' that could easily have led to serious injury or loss of life. Dangerous occurrences are defined in the Reporting of Injuries, Diseases and Dangerous Occurrences Regulations (RIDDOR) 1995.

Near-miss accidents occur much more frequently than injury accidents and are, therefore, a good indicator of hazard, which is why the HSE collects this data. In January 2008 a BA passenger aeroplane lost power to both engines as it prepared to land at Heathrow airport. The pilots glided the plane into a crash landing on the grass just short of the runway. This is one example of a dangerous occurrence which could so easily have been a disaster.

Consider another example: on a wet and windy night a large section of scaffold around a town centre building collapses. Fortunately this happens about midnight when no one is around because of the time and the bad weather. However, if it had occurred at midday, workers would have been using the scaffold and the streets would have been crowded with shoppers. This would be classified as a dangerous occurrence and must be reported to the HSE, who will investigate the cause and, using their wide range of powers, would either:

- stop all work;
- demand the dismantling of the structure;
- issue an Improvement Notice;
- issue a Prohibition Notice; or
- prosecute those who have failed in their health and safety duties.

Other reportable dangerous occurrences are:

- the collapse of a lift;
- plant coming into contact with overhead power lines;

Definition

Dangerous occurrence – a 'near miss' that could easily have led to serious injury or loss of life. Near-miss accidents occur much more frequently than injury accidents and are, therefore, a good indicator of hazard, which is why the HSE collects this data.

- any unexpected collapse which projects material beyond the site boundary;
- the overturning of a road tanker;
- a collision between a car and a train.

Assessment criteria 2.4

Define the limitations of their responsibilities in terms of health and safety in the workplace

Assessment criteria 2.5

State the actions to be taken in situations which exceed their level of responsibility for health and safety in the workplace

Assessment criteria 2.6

State the procedures to be followed in accordance with the relevant health and safety regulations for reporting health, safety and/or welfare issues at work

Roles and responsibilities in health and safety

There are various roles and responsibilities involved in both the implementation, regulations and policing of health and safety. It has already been stated that employees have a duty of care towards others and must at all times respect and implement all procedures and policies that impact them.

That said, if any employee finds themselves in a position of being unsure or unauthorized, or perhaps realize that they have not been sufficiently trained for a given situation or circumstance, then they must not put themselves or others at risk. Instead they should voice their concern by deferring the matter to their immediate supervisor. Some companies operate a Health and Safety Matters Suggestion Scheme, which aims to encourage their employees to contribute constructive ideas in order to improve the health and safety in their area.

Such schemes can be in written format or electronic and can extend to any welfare issue, therefore it is vital that employees are aware of how responsibility is assigned and shared and how they fit within the health and safety chain of command. This means that any standing procedures or order can be challenged, amended or even introduced as new to counter any perceived: weaknesses, omissions, faults or deficiencies within the system. It is also important to note that if you believe any issue that you have raised has not been addressed and the hazards or risk still exist, then the matter should be passed on to the next level or person in authority.

It is important to remember that if you observe a hazardous situation at work, your first action should be to report the issue to your safety representative or supervisor in order to raise the alarm and alert others of the hazards involved. If nobody in authority is around, then if possible, make the hazard safe using an appropriate method such as screening it off, but only if you can do so without putting yourself or others at risk.

Assessment criteria 2.7

Recognize the appropriate responsible persons to whom health and safety- and welfare-related matters should be reported

Health and safety management

Within the structure of a large company a nominated senior manager is ultimately responsible for coordinating day-to-day issues surrounding health and safety. He or she will be heavily involved in implementing and monitoring their safety policy but will delegate responsibility to other persons. For instance, very large organisations will employ a **health and safety officer**, who will coordinate health and safety training, influence policy and monitor day-to-day operations, especially liaising with department managers. It is normally a requirement that health and safety officers are qualified to NEBOSH (National Examination Board in Occupational Safety and Health) standard, largely because of the legality of the issues involved.

The health and safety officer will also appoint a department or section **health and safety representative** who will in turn coordinate all relevant business within their department. They may be required to carry out an IOSH course (Institution of Occupational Safety and Health), IOSH being a British organisation for health and safety professionals. Therefore, it will be the responsibility of the health and safety officer to ensure that new workers are inducted into a company and the responsibility of the health and safety representative to carry out induction of their processes within their department.

Safety first

If you are unsure: ask your supervisor.

Safety first

Notice a safety related issue on-site: report the matter to the Site H&S Officer directly or through your boss.

Figure 1.22 Discussing all accidents, near misses and dangerous occurences in a frank and honest way helps bring about an open culture and not a blame culture.

Health and safety champions may also be appointed; they are given responsibility for championing issues such as safe lifting techniques through manual handling training or coordinating the use of chemicals and substances through COSHH (control of substances hazardous to health).

It is important that the whole team involved hold both section- and company-wide health and safety meetings in order to discuss any developments, issues, accidents including near misses, dangerous occurrences or instances of RIDDOR being invoked. Some companies set up health and safety committees in order to challenge, educate and correct issues that are brought up at every level. Furthermore, it is equally important that personnel recognize that everyone is responsible for health and safety but reporting of any issues should be directly through their representative who in turn will progress it upwards. This should be done through an open and honest culture and not through a 'blame culture'. Lapses regarding safety need to be challenged and corrected before injury or serious injuries occur. The actual process involved needs to be transparent so that all correspondence involving the reporting of all incidents or accidents has to be visible as well as any relevant feedback being passed on to the individuals concerned.

Environmental Health Officers (EHOs)

This role is concerned with policing and investigating issues surrounding and affecting health such as pollution, accidents at work, noisy neighbours, toxic contamination, pest infestations, food poisoning and illegal dumping of waste. Any issues in relation to the above should be directed to the EHO.

HSE

The Health and Safety Executive (HSE) coordinates all issues with regard to health and safety. They appoint inspectors to actively police how health and safety is interpreted and put into practice. Inspectors, if warranted, can enter any workplace without warning and can engage any of the employees in order to question their understanding of the company policies and procedures within their day-to-day operations.

Following a serious incident, the HSE should be contacted immediately and the area in question will be regarded as a crime scene. This means the HSE will interview those involved, any actual witnesses, take photographs and piece together all the evidence in order to work out when, what and where the accident happened. Their sole purpose is to learn how this can be avoided in the future. They can also issue two kinds of notices, namely:

* Improvement notice

  Where an inspector is of the opinion that a person is violating a legal requirement in circumstances that make it likely that the violating will continue or be repeated.

* Prohibition notice

  Where an inspector is of the opinion that an activity carried on (or likely to be carried on) involves the risk of serious personal injury. Note that companies and individuals can appeal against any decision.

Trade unions

Trade unions are groups of employees who join together to maintain and improve their conditions of employment. They will represent their workers and defend them if they are being disciplined. Equally, if they believe that a worker has been dismissed unfairly, then they will fund their case through various appeal court actions. It is important, therefore, that workers know their rights and can refer to the Union if they feel that they being blamed by management for certain failings regarding health and safety. The trade union that currently represents electrical workers is called UNITE (http://www.unitetheunion.org/).

Assessment criteria 2.8

State the procedure for manual handling and lifting

Safe manual handling

Manual handling is lifting, transporting or supporting loads by hand or by bodily force. The load might be any heavy object, a printer, a VDU, a box of tools or a stepladder. Whatever the heavy object is, it must be moved thoughtfully and carefully, using appropriate lifting techniques if personal pain and injury are to be avoided. *Many people hurt their back, arms and feet, and over one third of all three-day reported injuries submitted to the HSE each year are the result of manual handling.*

When lifting heavy loads, correct lifting procedures must be adopted to avoid back injuries. Figure 1.23 demonstrates the technique.

Do not lift objects from the floor with the back bent and the legs straight as this causes excessive stress on the spine. Always lift with the back straight and the legs bent so that the powerful leg muscles do the lifting work. Bend at the hips and knees to get down to the level of the object being lifted, positioning the body as close to the object as possible. Grasp the object firmly and, keeping the back straight and the head erect, use the leg muscles to raise in a smooth movement. Carry the load close to the body. When putting the object down, keep the back straight and bend at the hips and knees, reversing the lifting procedure. A bad lifting technique will result in sprains, strains and pains.

There have been too many injuries over the years resulting from bad manual handling techniques. The problem has become so serious that the HSE has introduced new legislation under the Health and Safety at Work Act 1974, the Manual Handling Operations Regulations 1992.

Publications such as *Getting to Grips with Manual Handling* can be obtained from HSE Books.

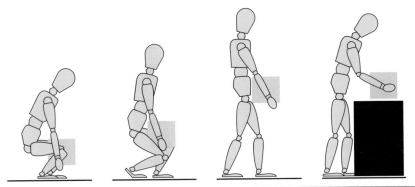

Figue 1.23 Correct manual lifting and carrying procedure.

Where a job involves considerable manual handling, employers must now train employees in the correct lifting procedures and provide the appropriate equipment necessary to promote the safe manual handling of loads:

Definition

Manual handling is lifting, transporting or supporting loads by hand or by bodily force.

Safety first

Lifting
- bend your legs;
- keep your back straight;
- use the leg muscles to raise the weight in a smooth movement.

Figure 1.24 Always use a mechanical aid to transport a load when available.

Consider some 'good practice' when lifting loads:

- Do not lift the load manually if it is more appropriate to use a mechanical aid.
- Only lift or carry what you can easily manage.
- Always use a trolley, wheelbarrow or truck such as that shown in Figure 1.24 when these are available.
- Plan ahead to avoid unnecessary or repeated movement of loads.
- Take account of the centre of gravity of the load when lifting – the weight acts through the centre of gravity.
- Never leave a suspended load unsupervised.
- Always lift and lower loads gently.
- Clear obstacles out of the lifting area.
- Use the manual lifting techniques described above and avoid sudden or jerky movements.
- Use gloves when manual handling to avoid injury from rough or sharp edges.
- Take special care when moving loads wrapped in grease or bubble wrap.
- Never move a load over other people or walk under a suspended load.
- Never approach a load to couple or uncouple it without a crane operator's permission.

Safety first

If someone was receiving an electric shock a wooden implement could be used to remove the person away from the supply.

Safety first

Isolation
- never work 'live';
- isolate;
- secure the isolation;
- prove the supply 'dead' before starting work.

Assessment criteria 3.1

Recognize how electric shock occurs

Secure electrical isolation

Electric shock above the 50 mA level is fatal unless the person is quickly separated from the supply. Below 50 mA only an unpleasant tingling sensation may be experienced or you may be thrown across a room or shocked enough to fall from a roof or ladder, but the resulting fall may lead to serious injury.

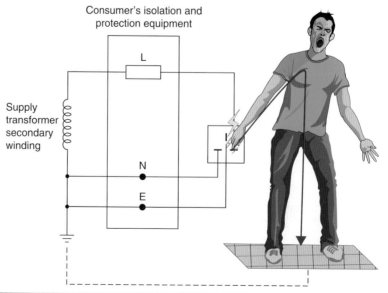

Figure 1.25 Touching live and earth or live and neutral makes a person part of the electrical circuit and can lead to an electric shock.

Various elements prevent people from receiving an electric shock. For instance, basic protection is provided through enclosing conductors within insulation, positioning live parts out of reach or separating live parts through barriers or enclosures. That said, all circuits contain protective devices. All exposed metal is earthed; protective devices are designed to trip under fault conditions and residual current devices (RCDs) are designed to trip below the fatal level as described in Chapter 4. Although there are special methods used in industrial applications, in combination with earthing and bonding these measures bring about fault protection.

> **Definition**
>
> Insulating a cable is one method of bringing about basic protection.

Assessment criteria 3.2

Recognize potential electrical hazards to be aware of on-site

Assessment criteria 3.3

Identify the different supply sources for electrical equipment and power tools

Assessment criteria 3.4

State why a reduced low voltage electrical supply is used for portable and handheld power tools on-site

Electrical hazards

Construction workers and particularly electricians do receive electric shocks, usually as a result of carelessness or unforeseen circumstances. This might come about due to a person actually touching live parts, or perhaps the insulation surrounding a live part is damaged or degraded which allows a person to become part of the circuit. This is particularly relevant when

Figure 1.26 Why is this dangerous?

temporary supplies are used since, by their very nature, they are not as robust and consequently hidden or buried cables in walls or in the ground respectively are especially dangerous. Considering that damp conditions are also common then any exposure to electrical supplies is that much more hazardous.

Furthermore, supplies are not necessarily identified to the same extent which is particularly relevant to neutral conductors within lighting circuits. Another potential hazard lies directly above through possible exposure to overhead cables, with which workers could come into contact by moving scaffolding, scaffold towers or operating elevated work platforms.

On-site electrical supplies and tools

Temporary electrical supplies on construction sites can save many person hours of labour by providing energy for fixed and portable tools and lighting. However, as stated previously in this chapter, construction sites are dangerous places and the temporary electrical supplies must be safe.

Given the personal contact that comes with operating mobile or transportable units, a reduced voltage system is utilized. This is with particular regard to Regulation 110.1 which tells us that *all* the regulations apply to temporary electrical installations such as construction sites. The frequency of inspection of construction sites is increased to every three months because of the inherent dangers. Regulation 704.410 Note 1 recommends the following voltages for distributing to plant and equipment on construction sites:

* 400 V – fixed plant such as cranes
* 230 V – site offices and fixed flood lighting robustly installed
* 110 V – portable tools and hand lamps
* SELV – portable lamps used in damp or confined places.

Portable tools must be fed from a 110 V socket outlet unit (see Figure 1.27(a)) incorporating splash-proof sockets and plugs with a keyway which prevents a tool from one voltage from being connected to the socket outlet of a different voltage.

The socket outlet and plugs are also colour coded for voltage identification:

25 V violet, 50 V white, 110 V yellow, 230 V blue and 400 V red, as shown in Figure 1.27(b).

While 110 V is strongly preferred for portable and handheld tools, a SELV system is preferred when operating in confined or damp conditions. SELV stands for

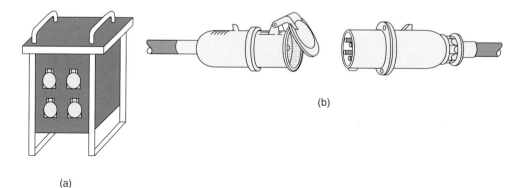

(b)

(a)

Figure 1.27 110 volts distribution unit and cable connector suitable for construction site electrical supplied: (a) reduced-voltage distribution unit incorporating industrial sockets to BS 4343; (b) industrial plug and connector.

'separated extra low voltage' and operating voltages are less than 50V a.c., typically being set at 12V. The SELV system also ensures that all live parts are covered by two layers of insulation which means that this type of equipment does not use a circuit protective conductor, since it should not be possible for a user to touch any live parts.

Assessment criteria 3.5

Outline the user checks on portable electrical equipment before use

Portable electric appliance testing

A quarter of all serious electrical accidents involve portable electrical appliances; that is, equipment which has a cable lead and plug and which is normally moved around or can easily be moved from place to place. This includes, for example, floor cleaners, kettles, heaters, portable power tools, fans, televisions, desk lamps, photocopiers, fax machines and desktop computers. There is a requirement under the Health and Safety at Work Act for employers to take adequate steps to protect users of portable appliances from the hazards of electric shock and fire. The responsibility for safety applies equally to small as well as large companies. The Electricity at Work Regulations 1989 also place a duty of care upon employers to ensure that the risks associated with the use of electrical equipment are controlled.

Against this background, the HSE has produced guidance notes HS(G) 107 *Maintaining Portable and Transportable Electrical Equipment* and leaflets *Maintaining Portable Electrical Equipment in Offices* and *Maintaining Portable Electrical Equipment in Hotels and Tourist Accommodation.* In these publications the HSE recommends that a three-level system of inspection can provide cost-effective maintenance of portable appliances. These are:

1 User checking.
2 Visual inspection by an appointed person.
3 Combined inspection and testing by a skilled or instructed person or contractor.

A user visually checking the equipment is probably the most important maintenance procedure. About 95% of faults or damage can be identified by just looking. The user should check for obvious damage using common sense. The use of potentially dangerous equipment can then be avoided.

Possible dangers to look for are as follows:

* damage to the power cable or lead which exposes the colours of the internal conductors, which are brown, blue and green with a yellow stripe;
* damage to the plug itself. The plug pushes into the wall socket, usually a square pin 13A socket in the United Kingdom, to make an electrical connection. With the plug removed from the socket the equipment is usually electrically 'dead'. If the Bakelite plastic casing of the plug is cracked, broken or burned, or the contact pins are bent, do not use it;
* non-standard joints in the power cable, such as taped joints;
* poor cable retention. The outer sheath of the power cable must be secured and enter the plug at one end and the equipment at the other. The coloured internal conductors must not be visible at either end;

Safety first

Remember, aside from formal portable appliance testing, a user check needs to be carried out before use.

Safety first

A user check ensures that the:
* cables
* labels
* case
* face
* extras

are all ok.

Figure 1.28 Portable appliance testing should be carried out regularly to avoid things like this happening in the workplace.

- damage to the casing of the equipment such as cracks, pieces missing, loose or missing screws or signs of melted plastic, burning, scorching or discolouration;
- equipment which has previously been used in unsuitable conditions such as a wet or dusty environment.

If any of the above dangers are present, the equipment should not be used until the person appointed by the company to make a 'visual inspection' has had an opportunity to do so.

A visual inspection will be carried out by an appointed person within a company, such person having been trained to carry out this task. In addition to the user checks described above, an inspection could include the removal of the plug top cover to check that:

- a fuse of the correct rating is being used and also that a proper cartridge fuse is being used and not a piece of wire, a nail or silver paper;
- the cord grip is holding the sheath of the cable and not the coloured conductors;
- the wires (conductors) are connected to the correct terminals of the plug top as shown in Figure 1.29;
- the coloured insulation of each conductor wire goes right up to the terminal so that no bare wire is visible;
- the terminal fixing screws hold the conductor wires securely and the screws are tight;
- all the conductor wires are secured within the terminal;
- there are no internal signs of damage such as overheating, excessive 'blowing' of the cartridge fuse or the intrusion of foreign bodies such as dust, dirt or liquids.

The above inspection cannot apply to 'moulded plugs', which are moulded onto the flexible cable by the manufacturer in order to prevent some of the bad

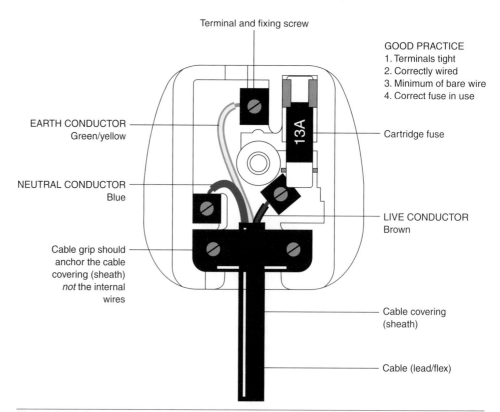

Terminal and fixing screw

GOOD PRACTICE
1. Terminals tight
2. Correctly wired
3. Minimum of bare wire
4. Correct fuse in use

EARTH CONDUCTOR
Green/yellow

Cartridge fuse

13A

NEUTRAL CONDUCTOR
Blue

LIVE CONDUCTOR
Brown

Cable grip should anchor the cable covering (sheath) *not* the internal wires

Cable covering (sheath)

Cable (lead/flex)

Figure 1.29 Correct connection of plug.

practice described above. In the case of a moulded plug top, only the fuse can be checked. The visual inspection checks described above should also be applied to extension leads and their plugs. The HSE recommends the writing of a simple procedure to give guidance to the 'appointed person' carrying out the visual inspection.

Combined inspection and testing is also necessary on some equipment because some faults cannot be seen by just looking – for example, the continuity and effectiveness of earth paths on class I equipment. For some portable appliances the earth is essential to the safe use of the equipment and, therefore, all earthed equipment and most extension leads should be periodically tested and inspected for these faults. All portable appliance test instruments (PAT testers) will carry out two important tests: earth bonding and insulation resistance.

Earth bonding tests apply a substantial test current, typically about 25 A, down the earth pin of the plug top to an earth probe, which should be connected to any exposed metalwork on the portable appliance being tested. The PAT tester will then calculate the resistance of the earth bond and either give an actual reading or indicate pass or fail. A satisfactory result for this test would typically be a reading of less than $0.1\,\Omega$. The earth bond test is, of course, not required for double-insulated portable appliances because there will be no earthed metalwork.

Insulation resistance tests apply a substantial test voltage, typically 500 V, between the live and neutral bonded together and the earth. The PAT tester then calculates the insulation resistance and either gives an actual reading or indicates pass or fail. A satisfactory result for this test would typically be a reading greater than $2\,M\Omega$.

Some PAT testers offer other tests in addition to the two described above. These are described below.

A flash test tests the insulation resistance at a higher voltage than the 500 V test described above. The flash test uses 1.5 kV for Class 1 portable appliances, that is, earthed appliances, and 3 kV for Class 2 appliances, which are double insulated. The test establishes that the insulation will remain satisfactory under more stringent conditions but must be used with caution, since it may overstress the insulation and will damage electronic equipment. A satisfactory result for this test would typically be less than 3 mA.

A fuse test tests that a fuse is in place and that the portable appliance is switched on prior to carrying out other tests. A visual inspection will be required to establish that the *size* of the fuse is appropriate for that particular portable appliance.

An earth leakage test measures the leakage current to earth through the insulation. It is a useful test to ensure that the portable appliance is not deteriorating and liable to become unsafe. It also ensures that the tested appliances are not responsible for nuisance 'tripping' of RCDs (see Chapter 3). A satisfactory reading is typically less than 3 mA. *An operation test* proves that the preceding tests were valid (i.e. that the unit was switched on for the tests), that the appliances will work when connected to the appropriate voltage supply and that they will not draw a dangerously high current from that supply.

A satisfactory result for this test would typically be less than 3.2 kW for 230 V equipment and less than 1.8 kW for 110V equipment. All PAT testers are supplied with an operating manual, giving step-by-step instructions for their use and pass and fail scale readings. The HSE suggested intervals for the three levels

Table 1.3 HSE suggested intervals for checking, inspecting and testing of portable appliances in offices and other low-risk environments

Equipment/ environment	User checks	Formal visual inspection	Combined visual inspection and electrical testing
Battery-operated (less than 20V)	No	No	No
Extra low-voltage (less than 50V a.c.): e.g. telephone equipment, low-voltage desk lights	No	No	No
Information technology: e.g. desktop computers, VDU screens	No	Yes, 2–4 years	No if double insulated – otherwise up to 5 years
Photocopiers, fax machines: *not* handheld, rarely moved	No	Yes, 2–4 years	No if double insulated – otherwise up to 5 years
Double-insulated equipment: *not* handheld, moved occasionally, e.g. fans, table lamps, slide projectors	No	Yes, 2–4 years	No
Double-insulated equipment: *handheld*, e.g. power tools	Yes	Yes, 6 months to 1 year	No
Earthed equipment (Class 1): e.g. electric kettles, some floor cleaners, power tools	Yes	Yes, 6 months to 1 year	Yes, 1–2 years
Cables (leads) and plugs connected to the above	Yes	Yes, 6 months to 4 years depending on the type of equipment it is connected to	Yes, 1–5 years depending on the type of equipment it is connected to
Extension leads (mains voltage)	Yes	As above	As above

of checking and inspection of portable appliances in offices and other low-risk environments are given in Table 1.3.

Who does what?

When actual checking, inspecting and testing of portable appliances takes place will depend upon the company's safety policy and risk assessments. In low-risk environments such as offices and schools, the three-level system of checking, inspection and testing recommended by the HSE should be carried out. Everyone can use common sense and carry out the user checks described earlier. Visual inspections must be carried out by a 'skilled or instructed person' but that person does not need to be an electrician or electronics service engineer.

Safety first

Power tools

- Look at the power tools that you use at work.
- Do they have a PAT Test label?
- Is it 'in date'?

Any sensible member of staff who has received training can carry out this duty. They will need to know what to look for and what to do, but more importantly, they will need to be able to avoid danger to themselves and to others. The HSE recommends that the appointed person follows a simple written procedure for each visual inspection. A simple tick sheet would meet this requirement. For example:

1 Is the correct fuse fitted? Yes/No.
2 Is the cord grip holding the cable sheath? Yes/No.

The tick sheet should incorporate all the appropriate visual checks and inspections described earlier.

Testing and inspection require a much greater knowledge than is required for simple checks and visual inspections. This more complex task need not necessarily be carried out by a qualified electrician or electronics service engineer.

However, the person carrying out the test must be trained to use the equipment and to interpret the results. In addition, greater knowledge will be required for the inspection of the range of portable appliances which might be tested.

Keeping records

Records of the inspecting and testing of portable appliances are not required by law but within the Electricity at Work Regulations 1989 it is generally accepted that some form of recording of results is required to implement a quality control system. The control system should:

- ensure that someone is nominated to have responsibility for portable appliance inspection and testing;
- maintain a log or register of all portable appliance test results to ensure that equipment is inspected and tested when it is due;
- label tested equipment with the due date for its next inspection and test as shown in Figure 1.30. If it is out of date, don't use it.

Title colour: White on green background

Title colour: White on green background

Title colour: White on red background

Figure 1.30 Typical PAT Test labels.

Assessment criteria 3.6

State the actions to be taken when portable electrical equipment is found to be damaged or faulty

Any piece of equipment which fails a PAT Test should be disabled and taken out of service (usually by cutting off the plug), labelled as faulty and sent for repair.

If the equipment is awaiting repair then it should be locked away from use. The register of PAT Test results will help managers to review their maintenance procedures and the frequency of future visual inspections and testing. Combined inspection and testing should be carried out where there is a reason to suspect that the equipment may be faulty, damaged or contaminated but this cannot be verified by visual inspection alone. Inspection and testing should also be carried out after any repair or modification to establish the integrity of the equipment or at the start of a maintenance system, to establish the initial condition of the portable equipment being used by the company.

Assessment criteria 3.7

State the need for safe isolation

Assessment criteria 3.8

Recognize the safe isolation procedure

Safe isolation

As an electrician working on electrical equipment you must always make sure that the equipment is switched off or electrically isolated before commencing work. Every circuit must be provided with a means of isolation (IET Regulation 422.3.13). When working on portable equipment or desk top units it is often simply a matter of unplugging the equipment from the adjacent supply. Larger pieces of equipment and electrical machines may require isolating at the local isolator switch before work commences.

Figure 1.31 Locking devices.

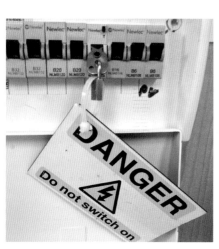

Figure 1.32 Isolated and locked off with a warning sign.

To deter anyone from reconnecting the supply while work is being carried out on equipment, a sign 'Danger – Electrician at Work' should be displayed on the isolator and the isolation 'secured' with a small padlock or the fuses removed so that no one can reconnect while work is being carried out on that piece of equipment.

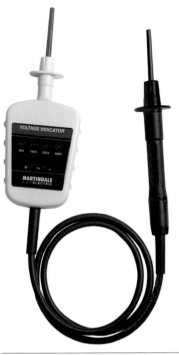

Figure 1.33 Typical voltage indicator.

Figure 1.34 Voltage proving unit.

Regulation 12(1) of the Electricity at Work Regulations 1989 is very specific that we must ensure the disconnection and separation of electrical equipment from every source of supply and that this disconnection and separation is secure. Where a test instrument or voltage indicator is used to prove the supply dead, Regulation 4(3) of the Electricity at Work Regulations 1989 recommends that the following procedure is adopted:

1 First, connect the test device such as that shown in Figure 1.33 to the supply which is to be isolated.

 The test device should indicate mains voltage.

2 Next, isolate the supply and observe that the test device now reads zero volts.

3 Then connect the same test device to a known live supply or proving unit such as that shown in Figure 1.34 to 'prove' that the tester is still working correctly.

4 Finally, secure the isolation and place warning signs; only then should work commence.

The test device being used by the electrician must incorporate safe test leads which comply with the Health and Safety Executive Guidance Note 38 on electrical test equipment. These leads should incorporate barriers to prevent the user from touching live terminals when testing and incorporating a protective fuse and be well insulated and robust, such as those shown in Figure 1.35.

To isolate a piece of equipment or individual circuit successfully, competently, safely and in accordance with all the relevant regulations, we must follow a procedure such as that given by the flow diagram in Figure 1.36. Start at the top and work down the flow diagram.

When the heavy outlined amber boxes are reached, pause and ask yourself whether everything is satisfactory up until this point. If the answer is 'yes', move on. If the answer is 'no', go back as indicated by the diagram.

Figure 1.35 Voltage indicator being proved.

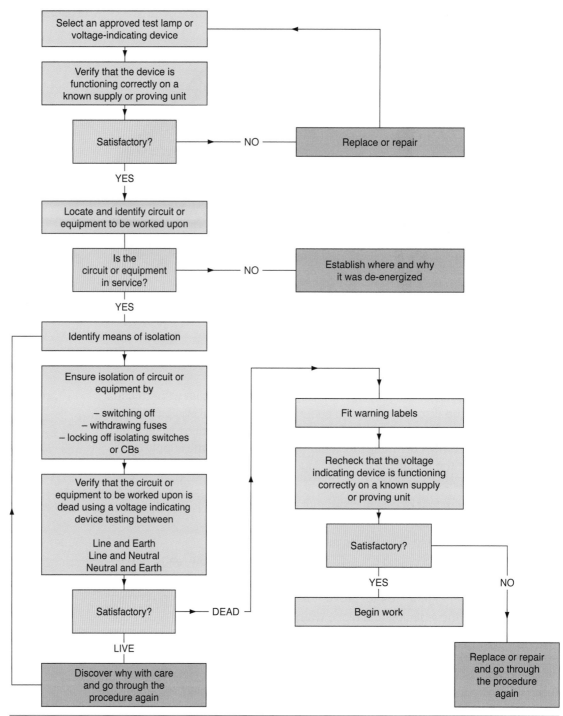

Figure 1.36 Flowchart for a secure isolation procedure.

Assessment criteria 4.1

Identify the various types of access equipment available

1

Assessment criteria 4.2

Determine the most appropriate access equipment to gain access

Assessment criteria 4.3

Identify safety checks to be carried out on access equipment

Assessment criteria 4.4

Describe safe practices and procedures for using access equipment

Access equipment

Working at height restrictions

Working above ground level creates added dangers and slows down the work rate of the electrician. New Work at Height Regulations came into force on 6 April 2005 with a direct purpose of ensuring that every precaution should be taken to ensure that any working platform is appropriate for the purpose and in good condition.

Ladders

The term 'ladder' is generally taken to include stepladders and trestles. The use of ladders for working above ground level is only acceptable for access and work of short duration (Work at Height Regulations 2005).

It is advisable to inspect the ladder before climbing it. It should be straight and firm. All rungs and tie rods should be in position and there should be no cracks in the stiles. The ladder should not be painted since the paint may hide defects.

Extension ladders should be erected in the closed position and extended one section at a time. Each section should overlap by at least the number of rungs indicated below:

* ladder up to 4.8 m length – two rungs overlap;
* ladder up to 6.0 m length – three rungs overlap;
* ladder over 6.0 m length – four rungs overlap.

The angle of the ladder to the building should be in the proportion four up to one out or 75° as shown in Figure 1.37. The ladder should be lashed at the top and bottom when possible to prevent unwanted movement and placed on firm and level ground. Footing is only considered effective for ladders smaller than 6 m and manufactured securing devices should always be considered. When ladders provide access to a roof or working platform the ladder must extend at least 1.05 m or five rungs above the landing place.

Short ladders may be carried by one person resting the ladder on the shoulder, but longer ladders should be carried by two people, one at each end, to avoid accidents when turning corners.

Safety first

Ladders

New Working at Height Regulations tell us:

* Ladders are only to be used for access.
* Must only be used for work of short duration.

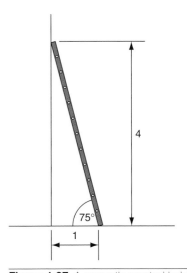

Figure 1.37 A correctly erected ladder.

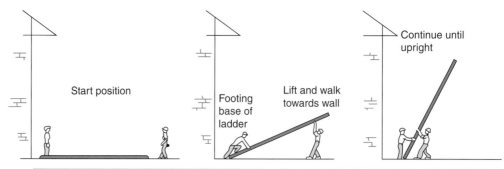

Figure 1.38 Correct procedure for erecting long or extension ladders.

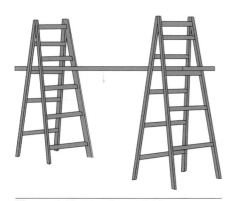

Figure 1.39 An example of safe ladder use.

Figure 1.40 A trestle scaffold.

Long ladders or extension ladders should be erected by two people as shown in Figure 1.38. One person stands on or 'foots' the ladder, while the other person lifts and walks under the ladder towards the wall. When the ladder is upright it can be positioned in the correct place, at the correct angle and secured before being climbed.

A very important consideration regarding the use of ladders is that when climbing or working from a ladder or stepladder, three points of contact should be maintained.

Stepladders

The rules that apply to ladders largely apply to stepladders, in that the same pre-use checks need to be used in order to ascertain whether the stepladder is serviceable. The stepladder, however, does include extra elements to ensure that both the cord is strong enough and that the actual hinge mechanism is secure and operates correctly.

It is important to stop and carry out a risk assessment regarding the use of a stepladder, since it is not always possible to maintain a handhold. When using, always refrain from overreaching and leaning out; it is far safer and correct to move the stepladder several times in order to carry out an activity safely.

Trestle scaffold

Figure 1.40 shows a trestle scaffold. Two pairs of trestles spanned by scaffolding boards provide a simple working platform. The platform must be at least two boards or 450 mm wide. At least one third of the trestle must be above the working platform. If the platform is more than 2 m above the ground, toe boards and guardrails must be fitted, and a separate ladder provided for access. The boards which form the working platform should be of equal length and not overhang the trestles by more than four times their own thickness. The maximum span of boards between trestles is:

- 1.3 m for boards 40 mm thick;
- 2.5 m for boards 50 mm thick.

Trestles which are higher than 3.6 m must be tied to the building to give them stability. Where anyone can fall more than 4.5 m from the working platform, trestles may not be used.

Mobile scaffold towers

Mobile scaffold towers may be constructed of basic scaffold components or made from light alloy tube. The tower is built up by slotting the sections together until the required height is reached. A scaffold tower is shown in Figure 1.41.

If the working platform is above 2 m from the ground it must be close boarded and fitted with guardrails and toe boards. When the platform is being used, all four wheels must be locked. The platform must not be moved unless it is clear of tools, equipment and workers, and should be pushed at the base of the tower and not at the top.

The stability of the tower depends on the ratio of the base width to tower height. A ratio of base to height of 1:3 gives good stability. Outriggers can be used to increase stability by effectively increasing the base width. If outriggers are used then they must be fitted diagonally across all four corners of the tower and not on one side only. The tower must not be built more than 12 m high unless it has

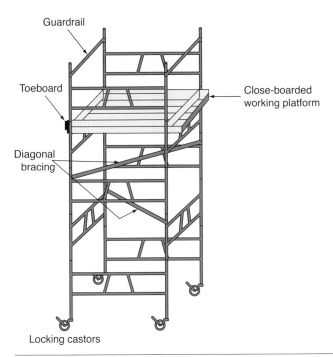

Figure 1.41 A mobile scaffold tower.

been specially designed for that purpose. Any tower higher than 9 m should be secured to the structure of the building to increase stability.

Access to the working platform of a scaffold tower should be by a ladder securely fastened vertically to the inside of the tower. Ladders must never be leaned against a tower since this might push the tower over.

Mobile elevated work platforms

A scissor lift, also known as an aerial work platform (AWP), is a type of mechanical device used to provide temporary access to trade persons to access areas at height. For instance, a scissor lift would be ideal to replace a light fitting that was located centrally in the middle of a warehouse ceiling. Only authorized personnel would be allowed to operate the scissor lift having being given specific training and their authority would be recorded and stored with their personal files.

Approved operators would have to be re-authorized usually every six months. Extreme caution must be taken when using any form of mechanical lifting equipment, since given their great extension they could come into contact with overhead power lines with potentially lethal results.

Figure 1.42 A scissor lift.

Figure 1.43 Mobile platform.

Another form of lifting device known as an elevated work platform is commonly called a cherry picker. The precautions involving overhead cable lines are particularly relevant to a cherry picker given the height to which it can be extended. Another consideration is the strength of the wind since it could be un-stabilized. The actual person in the basket should also wear and use a harness which should be attached to the safety cage when working at height. Again, only authorized and trained personnel are allowed to operate it.

Roof ladders

Roof ladders are used when a safe way is required for carrying out installation work on sloping roofs. Many have wheels fitted which means securing is only a matter of hooking them over the roof ridge.

Crawling boards

Working in attic spaces is potentially hazardous for many reasons. During hot summer days the heat can be stifling, made so much worse if the house has been insulated with thermal insulation. Access to the attic space is not guaranteed unless a proper roof ladder or stairs are in place. Also, adequate lighting might not be in place, therefore causing a further hazard. However, all these elements make working in attics dangerous especially if the floor has not been boarded over. It is not good practice to try and negotiate stepping over the rafter's floor boards, since one slip might not cause a serious injury but cause major damage to the trade person, severe inconvenience to the householder and a lot of embarrassment to the employer. Even though the employer might have liability insurance, the subscription will increase following any claim.

Scaffolding

The use of scaffolding should be thought of as a system that can provide all-round support and protection especially for extensive periods. Although it might be tied to the building, scaffolding must be structurally independent.

Applications include gaining access for work on:

Figure 1.44 Scaffolding isn't always used on such a grand scale as this.

- stonework;
- roof;
- external plastering, cladding and rendering;
- installing photovoltaic (solar) panels.

Only approved contractors are allowed to erect or dismantle scaffolding and according to PASMA, http://www.pasma.co.uk/ scaffolding must be checked every seven days.

Edge protection

The main cause of deaths in the construction industry is known to occur through working at height and according to the HSE, falls occur:

- from roof edges;
- through gaps or holes in roofs;
- through fragile roof materials.

In addition to this, people have been subjected to serious injuries when material is either deliberately thrown or accidentally kicked off of roofs. To combat this, the HSE recommends that some form of edge protection must be installed as follows:

- a main guardrail at least 950 mm above the edge;
- a toe board and brick guard where there is risk of objects being kicked off the edge of the platform;
- a suitable number of intermediate guardrails or suitable alternatives positioned so that there is no gap more than 470 mm.

Sometimes a roof parapet may provide equivalent protection, but if it does not, extra protection will be required. It is also essential to include cross bracing at every lift to ensure rigidity.

It is usual to access a scaffold from a ladder and it must be positioned so that this can be done easily and safely. It is important therefore to leave a suitable gap in the handrail and toe board arrangement to allow operatives access to the scaffold. Apart from edge protection, safety nets are sometimes used as well as rubbish chutes, which will reduce the possibility of debris falling and injuring passers-by.

Assessment criteria 5.2

Outline the procedures for working in accordance with provided safety instructions

Safe system of work

The Management of Health and Safety at Work Regulations 1999 tell us that employers must systematically examine the workplace, the work activity and the management of safety, ensuring that a safe system of work is in place.

Therefore safe systems of work (SSOW) is simply a means of ensuring employees operate safely. In the electrical installation industry, a safe system of work will include different elements which will outline how employees deal with hazardous work processes.

Risk assessments, method statement and a permit-to-work are ways of bringing about a safe system of work. They are designed to:

- identify hazards and exposure to the hazards (risk): **risk assessment**;
- identify proper procedures: **method statement**;
- authorize work in hazardous areas: **permit-to-work**;
- ensure workers are properly trained and that training is reviewed.

The whole process revolves around:

- identification of hazards;
- providing PPE when those hazards cannot be removed;
- training the workers with all necessary skills and techniques;
- reviewing all safety policies, procedures and protocols.

Let us now examine some of these elements in greater detail, starting with risk assessments.

In the establishment, a record of all significant risk assessment findings must be kept in a safe place and made available to an HSE inspector if required. Information based on the risk assessment findings must be communicated to relevant staff and, if changes in work behaviour patterns are recommended in the interests of safety, they must be put in place.

So, risk assessment must form part of any employer's robust policy of health and safety. However, an employer only needs to 'formally' assess the significant risks. He or she is not expected to assess the trivial and minor types of household risks.

Staff are expected to read and to act upon these formal risk assessments, and they are unlikely to do so enthusiastically if the file is full of trivia. An assessment of risk is nothing more than a careful examination of what in your work could cause harm to people. It is a record that shows whether sufficient precautions have been taken to prevent harm.

> **Definition**
>
> Employers of more than five people must document the risks at work and the process is known as *hazard risk assessment*.

> **Key fact**
>
> **Definition**
> - A hazard is something that might cause harm.
> - A risk is the chance that harm will be done.

Assessment criteria 5.1

State how to produce a risk assessment and method statement

Hazard risk assessment: the procedure

The HSE recommends five steps to any risk assessment.

Step 1

Look at what might reasonably be expected to cause harm. Ignore the trivial and concentrate only on significant hazards that could result in serious harm or injury. Manufacturers' data sheets or instructions can also help you spot hazards and put risks in their true perspective.

Step 2

Decide who might be harmed and how. Think about people who might not be in the workplace all the time: cleaners, visitors and contractors or maintenance personnel. Include members of the public or people who share the workplace. Is there a chance that they could be injured by activities taking place in the workplace?

Step 3

Evaluate the risk arising from an identified hazard. Is it adequately controlled or should more be done? Even after precautions have been put in place, some risk may remain. What you have to decide, for each significant hazard, is whether this remaining risk is low, medium or high. First of all, ask yourself if you have done all the things that the law says you have got to do.

For example, there are legal requirements on the prevention of access to dangerous machinery. Then ask yourself whether generally accepted industry standards are in place, but do not stop there – think for yourself, because the law also says that you must do what is reasonably practicable to keep the workplace safe. Your real aim is to make all risks small by adding precautions, if necessary. If you find that something needs to be done, ask yourself:

- Can I get rid of this hazard altogether?
- If not, how can I control the risk so that harm is unlikely?

Only use PPE when there is nothing else that you can reasonably do.

If the work that you do varies a lot, or if there is movement between one site and another, select those hazards which you can reasonably foresee, the ones that apply to most jobs and assess the risks for them. After that, if you spot any unusual hazards when you get on-site, take what action seems necessary.

Step 4

Record your findings and say what you are going to do about risks that are not adequately controlled. If there are fewer than five employees, you do not need to write anything down but if there are five or more employees, the significant findings of the risk assessment must be recorded. This means writing down the more significant hazards and assessing if they are adequately controlled and recording your most important conclusions. Most employers have a standard risk assessment form which they use, such as that shown in Figure 1.45, but any format is suitable. The important thing is to make a record.

Definition

Risk assessments need to be *suitable* and *sufficient*, not perfect.

There is no need to show how the assessment was made, provided you can show that:

1 a proper check was made;
2 you asked those who might be affected;
3 you dealt with all obvious and significant hazards;
4 the precautions are reasonable and the remaining risk is low;
5 you informed your employees about your findings.

Risk assessments need to be *suitable* and *sufficient*, not perfect. The two main points are:

1 Are the precautions reasonable?

2 Is there a record to show that a proper check was made?

File away the written assessment in a dedicated file for future reference or use.

It can help if an HSE inspector questions the company's precautions or if the company becomes involved in any legal action. It shows that the company has done what the law requires.

Step 5

Review the assessments from time to time and revise them if necessary. Ask yourself:

- Do the same hazards exist?
- Do the same risks apply to personnel?
- Are there new hazards?
- Are personnel exposed to these new hazards?
- Are the safety measures currently in place adequate?

Completing a risk assessment

When completing a risk assessment such as that shown in Figure 1.45, do not be over-complicated. In most firms in the commercial, service and light industrial sector, the hazards are few and simple. Checking them is common sense but necessary.

Step 1

List only hazards which you could reasonably expect to result in significant harm under the conditions prevailing in your workplace. Use the following examples as a guide:

- slipping or tripping hazards (e.g. from poorly maintained or partly installed floors and stairs);
- fire (e.g. from flammable materials you might be using, such as solvents);
- chemicals (e.g. from battery acid);
- moving parts of machinery (e.g. blades);
- rotating parts of hand tools (e.g. drills);
- accidental discharge of cartridge-operated tools;
- high-pressure air from airlines (e.g. air-powered tools);
- pressure systems (e.g. steam boilers);
- vehicles (e.g. forklift trucks);
- electricity (e.g. faulty tools and equipment); dust (e.g. from grinding operations or thermal insulation);
- fumes (e.g. from welding);
- manual handling (e.g. lifting, moving or supporting loads);
- noise levels too high (e.g. machinery);
- poor lighting levels (e.g. working in temporary or enclosed spaces);
- low temperatures (e.g. working outdoors or in refrigeration plant);
- high temperatures (e.g. working in boiler rooms or furnaces).

HAZARD RISK ASSESSMENT	FLASH-BANG ELECTRICAL CO.
For Company name or site: _____ Address: _____ _____	Assessment undertaken by: _____ Signed: _____ Date: _____
STEP 5 Assessment review date: _____	
STEP 1 List the hazards here _____ _____ _____ _____ _____ _____	STEP 2 Decide who might be harmed _____ _____ _____ _____ _____ _____
STEP 3 Evaluate (what is) the risk – is it adequately controlled? State risk level as low, medium or high _____ _____ _____ _____ _____	STEP 4 Further action – what else is required to control any risk identified as medium or high? _____ _____ _____ _____ _____

Figure 1.45 Hazard risk assessment standard form.

Step 2

Decide who might be harmed; do not list individuals by name. Just think about groups of people doing similar work or who might be affected by your work:

- office staff;
- electricians;
- maintenance personnel;
- other contractors on-site;
- operators of equipment;
- cleaners;
- members of the public.

Pay particular attention to those who may be more vulnerable, such as:

- staff with disabilities;
- visitors;

Figure 1.46 Even a poorly placed toy can lead to a significant injury if people aren't paying attention.

- young or inexperienced staff;
- people working in isolation or enclosed spaces.

Step 3

Calculate what the risk is – is it adequately controlled? Have you already taken precautions to protect against the hazards which you have listed in Step 1? For example:

- have you provided adequate information to staff?
- have you provided training or instruction?

Do the precautions already taken:

- meet the legal standards required?
- comply with recognized industrial practice?
- represent good practice?
- reduce the risk as far as is reasonably practicable?

If you can answer 'yes' to the above points then the risks are adequately controlled, but you need to state the precautions you have put in place. You can refer to company procedures, company rules, company practices, etc. in giving this information. For example, if we consider there might be a risk of electric shock from using electrical power tools, then the risk of a shock will be *less* if the company policy is to portable appliance test (PAT) all power tools each year and to fit a label to the tool showing that it has been tested for electrical safety.

If the stated company procedure is to use battery drills whenever possible, or 110 V drills when this is not possible, and to *never* use 230 V drills, then this again will reduce the risk. If a policy such as this is written down in the company safety policy statement, then you can simply refer to the appropriate section of the safety policy statement and the level of risk will be low. (Note: PAT testing is described in Advanced Electrical Installation Work.)

Step 4

Further action – what more could be done to reduce those risks which were found to be inadequately controlled?

Safety first

Safety procedures
- Hazard risk assessment is *an essential part* of any health and safety management system.
- The aim of the planning process is to minimize risk.
- HSE Publication HSG (65).

Figure 1.47 Warning signs must be read to be effective.

You will need to give priority to those risks that affect large numbers of people or which could result in serious harm. Senior managers should apply the principles below when taking action, if possible in the following order:

1 Remove the risk completely.
2 Try a less risky option.
3 Prevent access to the hazard (e.g. by guarding).
4 Organise work differently in order to reduce exposure to the hazard.
5 Issue PPE.
6 Provide welfare facilities (e.g. washing facilities for removal of contamination and first aid).

Any hazard identified by a risk assessment as *high risk* must be brought to the attention of the person responsible for health and safety within the company.

Ideally, in Step 4 of the risk assessment you should be writing 'No further action is required. The risks are under control and identified as low risk'. The assessor may use as many standard hazard risk assessment forms, such as that shown in Figure 1.45, as the assessment requires. Upon completion they should be stapled together or placed in a plastic wallet and stored in the dedicated file.

You might like to carry out a risk assessment on a situation you are familiar with at work, using the standard form in Figure 1.45, or your employer's standard forms.

Step 5

When reviewing the assessments ask yourself:

- Do the same hazards exist?
- Do the same risks apply to personnel?
- Are there new hazards?
- Are employees exposed to these new hazards?
- Are there new employees exposed to these new hazards?
- Are the safety measures currently in place adequate?

Method statement

The Construction, Design and Management Regulations and Approved Codes of Practice define a method statement as a written document laying out the work procedure and sequence of operations to ensure health and safety.

If the method statement is written as a result of a risk assessment carried out for a task or operation, then following the prescribed method will reduce the risk.

The safe isolation procedure described in Figure 1.48 is a method statement. Following this method meets the requirements of the Electricity at Work Regulations and the IET Regulations, and reduces the risk of electric shock to the operative and other people who might be affected by the operative's actions.

Shown in Figure 1.48 is an example of a method statement which indicates the sequence of events required so that an attic luminaire can be replaced safely. Notice that the method statement draws reference to:

- only approved equipment to be used;
- pre-use checks of equipment to be carried out;
- proper access equipment listed;
- safe isolation procedure to be used;
- crawling boards to be used in the attic;
- communication to keep the client informed;
- correct disposal of the damaged lamp.

Permit-to-work system

Definition

The *permit-to-work procedure* is a type of 'safe system to work' procedure used in specialised and potentially dangerous plant process situations.

The permit-to-work procedure is a type of 'safe system to work' procedure used in specialised and potentially dangerous plant process situations. The procedure was developed for the chemical industry, but the principle is equally applicable to the management of complex risk in other industries or situations.

For example:

- working on part of an assembly line process where goods move through a complex, continuous process from one machine to another (e.g. the food industry);
- repairs to railway tracks, tippers and conveyors;
- working in confined spaces (e.g. vats and storage containers);
- working on or near overhead crane tracks;
- working underground or in deep trenches;
- working on pipelines;
- working near live equipment or unguarded machinery;
- roof work;
- working in hazardous atmospheres (e.g. the petroleum industry);
- working near or with corrosive or toxic substances.

All the above situations are high-risk working situations that should be avoided unless you have received special training and will probably require the completion of a permit-to-work. Permits-to-work must adhere to the following eight principles:

1 Wherever possible the hazard should be eliminated so that the work can be done safely without a permit-to-work.

2 The site manager has overall responsibility for the permit-to-work even though he or she may delegate the responsibility for its issue.

Background information	
Company details	Flash Bang Electrical Company Contact street Tel Fax email
Site address	Contact name Address Contact No.
Activity – risk	To access and work safely in the attic space during the replacement of a fluorescent light fitting.

Implementation and control of risk	
Hazardous task – risk	**Method of control**
Access roof space	Access to the roof space will be via a suitably secured stepladder of the correct height for the task. All relevant PPE will be worn.
Access working area	Walk boards will be used to walk across ceiling joists. Electric lead lights shall illuminate access and work areas.
Replacement of existing light fitting	Safe isolation procedure to be carried out.
Removal of fluorescent light fitting	The fluorescent light fitting will be carefully lowered through the loft hatch. Use a secured strap and rope system if necessary.
Safe deposit of fluorescent light fitting	Fluorescent light fitting to be sealed in one piece and by prior arrangement, deposited at the council tip by the customer.

Site control	
Inspection and testing of equipment	All equipment such as stepladder, tooling and electrical equipment shall be regularly inspected before commencement of work. All test equipment to be calibrated annually.
Customer awareness	The customer will be made aware of all potential dangers throughout the contract. They will also be made aware if any electrical isolation is required.

Figure 1.48 Typical method statement.

3　The permit must be recognized as the master instruction, which, until it is cancelled, overrides all other instructions.

4　The permit applies to everyone on-site, other trades and subcontractors.

5   The permit must give detailed information; for example: (i) which piece of plant has been isolated and the steps by which this has been achieved; (ii) what work is to be carried out; (iii) the time at which the permit comes into effect.

6   The permit remains in force until the work is completed and is cancelled by the person who issued it.

7   No other work is authorized. If the planned work must be changed, the existing permit must be cancelled and a new one issued.

8   Responsibility for the plant must be clearly defined at all stages because the equipment that is taken out of service is released to those who are to carry out the work.

The people doing the work, the people to whom the permit is given, take on the responsibility of following and maintaining the safeguards set out in the permit, which will define what is to be done (no other work is permitted) and the timescale in which it is to be carried out.

The permit-to-work system must help communication between everyone involved in the process or type of work. Employers must train staff in the use of such permits and, ideally, training should be designed by the company issuing the permit, so that sufficient emphasis can be given to particular hazards present and the precautions which will be required to be taken. For further details see Permit-to-Work at www.hse.gov.uk.

Assessment criteria 5.3

Describe the procedures to be taken to eliminate or minimize risk before deciding personal protective equipment is needed

Assessment criteria 5.4

State the purpose of personal protective equipment

Personal protective equipment (PPE)

PPE is defined as all equipment designed to be worn, or held, to protect against a risk to health and safety. However a very important point must be made first in that PPE applies only when a specific hazard cannot within reason be removed. In other words, employers still have a duty of care to remove a hazard rather than work with it and dish out PPE. This is why the purpose of PPE is defined as equipment that reduces the risk of injury to the wearer.

This also means that employees need to be responsible for making sure they have the correct PPE for the specific job involved, paying particular attention to the environment, especially if it's one that is subject to change. There is no one size fits all situation regarding PPE; the equipment must be appropriate to the task, and appropriate regarding the hazard and the risks involved to the user. Employees need to report any concerns to their immediate supervisor or safety representative.

Assessment criteria 5.5

Specify the appropriate protective clothing and equipment that is required for specific work tasks

PPE includes most types of protective clothing, and equipment such as eye, foot and head protection, safety harnesses, life-jackets and high-visibility clothing.

Under the Health and Safety at Work Act, employers must provide free of charge any PPE and employees must make full and proper use of it. Safety signs such as those shown in Figure 1.50 are useful reminders of the type of PPE to be used in a particular area. The vulnerable parts of the body which may need protection are the head, eyes, ears, lungs, torso and hands and feet; in addition, protection from falls may need to be considered. Objects falling from a height present the major hazard against which head protection is provided. Other hazards include striking the head against projections and hair becoming entangled in machinery.

Typical methods of protection include helmets, light duty scalp protectors called 'bump caps' and hairnets. The eyes are very vulnerable to liquid splashes, flying particles and light emissions such as ultraviolet light, electric arcs and lasers. Types of eye protectors include safety spectacles, safety goggles and face shields. Screen-based workstations are being used increasingly in industrial and commercial locations by all types of personnel. Working with VDUs (visual display units) can cause eye strain and fatigue.

Noise is accepted as a problem in most industries and surprisingly there has been very little control legislation. The Health and Safety Executive have published a 'Code of Practice' and 'Guidance Notes' HSG 56 for reducing the exposure of employed persons to noise. A continuous exposure limit of below 85 dB for an eight-hour working day is recommended by the Code.

Noise may be defined as any disagreeable or undesirable sound or sounds, generally of a random nature, which do not have clearly defined frequencies.

 Safety first

PPE
Always wear or use the PPE (personal protective equipment) provided by your employer for your safety.

 Safety first

PPE
PPE only reduces the effect of an accident; it does not remove the danger.

Figure 1.49 All workers on-site must wear head protection. It will not stop accidents but will lessen the impact.

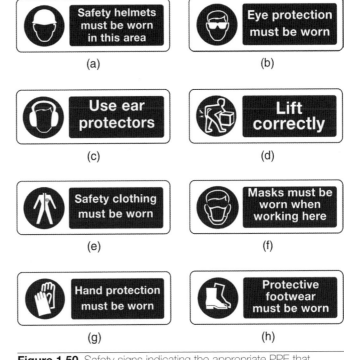

Figure 1.50 Safety signs indicating the appropriate PPE that should be worn.

The usual basis for measuring noise or sound level is the decibel scale. Whether noise of a particular level is harmful or not also depends on the length of exposure to it. This is the basis of the widely accepted limit of 85 dB of continuous exposure to noise for eight hours per day.

A peak sound pressure of above 200 Pascals or about 120 dB is considered unacceptable and 130 dB is the threshold of pain for humans. If a person has to shout to be understood at 2 metres, the background noise is about 85 dB. If the distance is only 1 metre, the noise level is about 90 dB. Continuous noise at work causes deafness, makes people irritable, affects concentration, causes fatigue and accident proneness and may mask sounds which need to be heard in order to work efficiently and safely.

It may be possible to engineer out some of the noise, for example, by placing a generator in a separate sound-proofed building. Alternatively, it may be possible to provide job rotation, to rearrange work locations or provide acoustic refuges.

Where individuals must be subjected to some noise at work, it may be reduced by ear protectors. These may be disposable ear plugs, reusable ear plugs or ear muffs. The chosen ear protector must be suited to the user and suitable for the type of noise, and individual personnel should be trained in its correct use. Breathing reasonably clean air is the right of every individual, particularly at work. Some industrial processes produce dust which may present a potentially serious hazard. The lung disease, asbestosis, is caused by the inhalation of asbestos dust or particles and the coal dust disease, pneumoconiosis, suffered by many coal miners, has made people aware of the dangers of breathing in contaminated air.

Safety first

Safety signs

Always follow the instructions given in the safety signs where you are working – it will help to keep you safe.

Some people may prove to be allergic to quite innocent products such as flour dust in the food industry or wood dust in the construction industry. The main effect of inhaling dust is a measurable impairment of lung function. This can be avoided by wearing an appropriate mask, respirator or breathing apparatus as recommended by the company's health and safety policy and indicated by local safety signs.

A worker's body may need protection against heat or cold, bad weather, chemical or metal splash, impact or penetration and contaminated dust. Alternatively, there may be a risk of the worker's own clothes causing contamination of the product, as in the food industry. Appropriate clothing will be recommended in the company's health and safety policy. Ordinary working clothes and clothing provided for food hygiene purposes are not included in the Personal Protective Equipment at Work Regulations.

Hands and feet may need protection from abrasion, temperature extremes, cuts and punctures, impact or skin infection. Gloves or gauntlets provide protection from most industrial processes, but should not be worn when operating machinery because they may become entangled in it. Care in selecting the appropriate protective device is required; for example, barrier creams provide only a limited protection against infection.

Try this

PPE

- Make a list of any PPE which you have used at work.
- What was this PPE protecting you from?

Boots or shoes with in-built toe-caps can give protection against impact or falling objects and, when fitted with a mild steel sole plate, can also provide protection from sharp objects penetrating through the sole. Special slip-resistant soles can also be provided for employees working in wet areas. Anyone working in a fuel-based environment would need shoes that are impervious to oils and petrol-based solutions.

Whatever the hazard to health and safety at work, the employer must be able to demonstrate that he or she has carried out a risk analysis, made recommendations which will reduce that risk and communicated these recommendations to the workforce. Where there is a need for PPE to protect against personal injury and to create a safe working environment, the employer must provide that equipment and any necessary training which might be required and the employee must make full and proper use of such equipment and training.

Assessment criteria 5.6

Define the first-aid facilities that must be available in accordance with health and safety regulations

Assessment criteria 5.7

Clarify why first-aid supplies and equipment must not be misused, and must be replaced when used

Providing a safe system of work also applies to actual first-aid facilities as well as ensuring that there are sufficiently qualified personnel authorized to the latest first-aid edition. These are covered in Tables 1.1 and 1.2.

Due to workplace hazards, it is imperative that companies ensure certain personnel are qualified to apply first aid treatment which possibly could keep a casualty alive until the emergency services arrive. Equally important is the replenishment of first aid supplies, since misuse of supplies or failure to replenish supplies after use could endanger someone in dire need of immediate treatment. Consequently if you use any first aid supplies or are aware that they have been used or exhausted, then you have a responsibility to inform the appointed person responsible for first aid provision. Even better, replenish them yourself and then inform them.

Assessment criteria 5.8

Describe safe practices and procedures using equipment, tools and substances

The principles which were laid down in the many Acts of Parliament and the regulations that we have already looked at in this chapter control our working environment. They make our workplace safer, but despite all this legislation, workers continue to be injured and killed at work or die as a result of a work-related injury. The number of deaths has consistently averaged about 200 each

Figure 1.51 First-aid kits must be checked regularly to ensure the contents are up to date and that everything on the checklist is there.

year for the past eight years. These figures only relate to employees. If you include the self-employed and members of the public killed in work-related accidents, the numbers almost double.

To combat this situation we need certain safe practices and procedures when using equipment, tooling and substances which should make the workplace a safer place to work.

Specific considerations would include:

- possible electrocution;
- pneumatic tools could cause injury through injection of an air bubble;
- over-pressurisation of equipment;
- cartridge guns firing through the body;
- injury through rotating machinery.

Risk involves exposure to equipment, therefore it goes without saying that operators need to be mindful of safety at all times.

As already explained, all employers are responsible for creating a safe workplace through the implementation and monitoring of safe systems of work. This includes processes that ensure only trained and authorized personnel can operate hazardous equipment, especially for certain types of rotating machinery. Thus, such machines would have accompanying written syllabi for operators to ensure the machine is never operated without the correct PPE or guarding in place. Such consideration would also apply to portable power tools such as cartridge guns, drills and grinders, since either noise, sparks emanating from grinders or even drills could impact on other employees or passers-by. Which is why, within reason, the operation of such equipment, given its inherent hazards, must also be controlled or limited in operation within a safe nominated area.

This applies to pneumatic (air) supplies, since inadvertent exposure could cause an air bubble to be injected into the skin, causing a possible embolism. These hazardous areas need to be signposted to alert employees or visitors as well as signalling if any mandatory PPE must be worn.

Typical signs include blue and white mandatory signs to highlight that PPE has to be worn, prohibition to stop unauthorized access and warning notices informing personnel of any specific hazards that may be present.

Another form of regulation and control is that of COSHH, whereby control of substances hazardous to health is regulated. We have already described the statutory regulation that obliges employers to put specific measures in place which will include appointing a duty holder to manage such a system, who in turn should notify all affected personnel about how their process works. For instance, all substances being used must be in line with manufacturers' recommendations, known as a data sheet. The information from the data sheet is then used to carry out a COSHH risk assessment, detailing exposure times, PPE required and whether operations or work activities have to be carried out in ventilated rooms. The details will then be recorded and filed in order that employees can be reminded of the hazards involved but also how to use the substances safely.

Assessment criteria 6.1

Recognize warning symbols of hazardous substances

Assessment criteria 6.2

Clarify what is meant by the term 'hazard'

Safe working environment

Multiple hazards exist within all working environments, therefore it is prudent to actually define both hazard and risk.

A hazard is something with the 'potential' to cause harm, for example, chemicals, electricity or working above ground.

A risk is the 'likelihood' of harm actually being done, or in other words, how likely you are to be harmed from being exposed to the hazards.

Working where toxic or corrosive substances exist

Hazardous substances exist in a variety of forms, both on a construction site and within the manufacturing sector. Acid for batteries, hydrogen given off by batteries when charging, fumes in tanks and many others are all substances that create a hazard. In all circumstances, however, the control process remains the same:

- identify the hazard;
- assess the risk;
- control the risk – better to eliminate the substance or substitute it for a less hazardous alternative. Only when this has been considered should other forms of control be used;
- train the staff;
- monitor the effectiveness of the measures;
- keep records.

The problem with toxic or corrosive substances is that there will never be a circumstance where it is safe to ignore safety measures and get away with it. Fumes exist, and no amount of thinking that things will never happen to me will stop you being overcome by those fumes.

When we consider the need for safe working within hazardous environments, we need to wear the clothes and the breathing apparatus as well as following the safe system of work in line with the permit-to-work scheme.

You should never work on your own in such an area and the precautions you take would be similar to those for working in confined spaces where toxic substances are present. This means that for hazardous situations it is essential that you have a safety person in a position where you have someone covering your back.

Hazardous substances

How substances and material are packed is governed by the acronym CHIP, which stands for Chemicals (Hazard Information and Packaging for Supply) Regulations. The HSE defines CHIP as helping protect people and the environment from the ill effects of chemicals by requiring suppliers to:

Definition

A *hazard* is something with the 'potential' to cause harm, for example, chemicals, electricity or working above ground.

Definition

A *risk* is the 'likelihood' of harm actually being done.

- identify the hazards (dangers) of the chemicals they supply;
- give information about the chemicals' hazards to their customers;
- package the chemicals safely.

The CHIP regulations were replaced by the European CLP Regulation in 2015 and apply to all Eu member states. Click on the HSE site for details: www.hse. gov.uk/chemical-classification/legal/chip-regulations.htmonto

Assessment criteria 6.3

Recognize hazards associated with the installation and maintenance of electrotechnical systems and equipment

Within electrical installation and, in general, the electrotechnical industry, there are specific hazards that apply to all those engaged in installation and maintenance activities.

For instance, electric shock would be high on that list given that all operators have to engage with electrical supplies. For example, certain industrial outlets do not use an earthing system, with fault protection coming about through specific bonding techniques, but have to be constantly maintained so that electric shock does not occur. On many sites until an advanced stage, the site is powered through temporary electrical supplies; therefore, such systems are not as robust as permanent supplies. Damp conditions increase the possibility that the effect of electric shock is greater, since a person's exposure to moisture lowers their body resistance.

Electric shock can also cause severe burns to the skin either through a deep seated burn by arcing or a thermal burn through an item overheating. All electrical connections and joints should be both mechanically and electrically sound and very low in resistance. Thus, no real power is developed across the connection. But badly formed connections cause the joint resistance to be high, causing power to be dissipated, which in turn manifests as heat. Bad resistance connections are a major factor in electrical fire; especially since the circuit resistance has increased through the badly formed joint and will therefore reduce the circuit current, which can delay the operation of circuit protection devices.

Figure 1.52 Garage explosions are relatively rare in spite of the highly flammable chemicals stored there.

Certain environments such as petrol stations, industrial/chemical plants and even flour mills can be potentially explosive, since vapour- or dust-rich conditions can explode if subjected to an ignition source.

This is why the EWR 1989 specified that systems are designed so far as is reasonably practicable in such a way that do not pose a danger to the user or beyond. This includes safeguarding that any equipment used is appropriate to the working environment and can withstand all related:

- electromechanical/chemical stresses;
- impact damage;
- weather conditions;
- temperature rise;
- potentially explosive environments.

Assessment criteria 6.4

Identify the main types of safety signs that may be used on-site

Categories of safety signs

The rules and regulations of the working environment are communicated to employees by written instructions, signs and symbols. All signs in the working environment are intended to inform. They should give warning of possible dangers and must be obeyed.

At first there were many different safety signs, but British Standard BS 5499 Part 1 and the Health and Safety (Signs and Signals) Regulations 1996 have

Figure 1.53 Text-only safety signs do not comply.

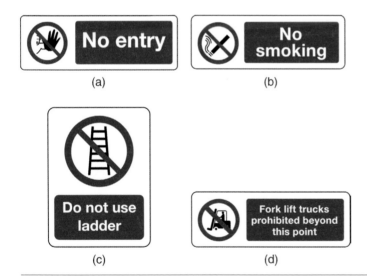

Figure 1.54 Prohibition signs. These are *must not do* signs.

introduced a standard system which gives health and safety information with the minimum use of words.

The purpose of the regulations is to establish an internationally understood system of safety signs and colours which draw attention to equipment and situations that do, or could, affect health and safety. Text-only safety signs became illegal from 24 December 1998. From that date, all safety signs have had to contain a pictogram or symbol such as those shown in Figure 1.53. Signs fall into four categories: prohibited activities, warnings, mandatory instructions and safe conditions.

Prohibition signs

These are *must not do* signs. They are circular white signs with a red border and red cross-bar, and are shown in Figure 1.54. They indicate an activity *which must not* be done.

Warning signs

These give safety information. These are triangular yellow signs with a black border and symbol, and are given in Figure 1.55. They *give warning* of a hazard or danger.

Mandatory signs

These are *must do* signs. These are circular blue signs with a white symbol, and are given in Figure 1.56. They *give instructions* which must be obeyed.

Figure 1.55 Warning signs. These give safety information.

Figure 1.56 Mandatory signs. These are *must do* signs.

Advisory or safe condition signs

These give safety information. They are square or rectangular green signs witha white symbol, and are shown in Figure 1.57. They *give information* about safety provision.

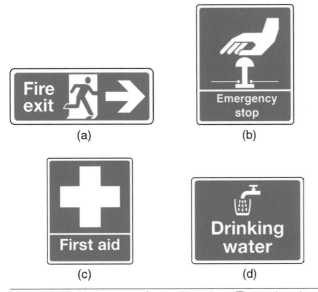

Figure 1.57 Advisory or safe condition signs. These also give safety information.

Assessment criteria 6.5

Identify hazardous situations in the workplace

Assessment criteria 6.6

Describe the practices and procedures for addressing hazards in the workplace

Getting injured at work is not a pleasant subject to think about but each year about 300 people in Great Britain lose their lives at work. In addition, there are about 158,000 non-fatal injuries reported to the Health and Safety Executive (HSE) each year and an estimated 2.2 million people suffer ill health caused by, or made worse by, work. It is a mistake to believe that these things only happen in dangerous occupations such as deep-sea diving, mining and quarrying, the fishing industry, tunnelling and fire-fighting, or that it only happens in exceptional circumstances such as would never happen in your workplace. This is not the case. Some basic thinking and acting beforehand could have prevented most of these accident statistics from happening.

Most accidents are caused by either human error or environmental conditions. Human errors include behaving badly or foolishly, being careless and not paying attention to what you should be doing at work, doing things that you are not competent to do or have not been trained to do, all of which are heightened when you are tired or fatigued. The effects of alcohol or recreational drugs are obvious but compulsory testing on-site is still a necessary evil in order to filter out any worker under the influence and therefore posing a danger to either themselves or others.

Statistically, the most common categories of risk and causes of accidents at work are linked to slips, trips and falls; whilst the main cause of deaths in the construction industry is working at height.

Think about your workplace and each stage of what you do, and then think about what might go wrong. Some simple activities may be hazardous. Here are some typical activities where accidents might happen.

Given the very nature of electricity, then electric shock and fire are very real hazards in the electrotechnical industry. Add to this certain environmental conditions including unguarded or faulty machinery, damaged or faulty tools and equipment, poorly illuminated or ventilated workplaces and untidy, dirty or overcrowded workplaces.

A rigorous system of assessment, prevention and control must therefore be instigated in order to highlight hazards and take precautions through site-specific procedures in order to:

Definition

A *skilled person* is anyone who has the necessary technical skills, training and expertise to safely carry out the particular activity.

- eliminate the cause;
- substitute a procedure or product with less risk;
- enclose the dangerous situation;
- put guards around the hazard;
- supervise, train and give information to staff;
- if the hazard cannot be removed or minimized then provide PPE.
- pose a question: can the job be done in another way that does not involve such risk.

This is why skilled or instructed persons are often referred to in the Health and Safety at Work Regulations 1974. In defining who is 'competent' and for the purposes of the Act, a competent person is anyone who has the necessary technical skills, training and expertise to safely carry out the particular activity. Therefore, a competent person dealing with a hazardous situation reduces the risk.

These sorts of measures are especially important on a construction site, since it is an extremely hazardous environment, which is why it is listed in the wiring regulations as a special location. Consequently entry on to site is controlled and regulated through certain provision such as the Construction Skills Certification Scheme (CSCS). The aim of the CSCS scheme is to assess and examine an individual's knowledge of health and safety policies and procedures in order to ensure that personnel on site recognize and understand their role and can fulfil their responsibilities safely.

These schemes are also designed to ensure that personnel are aware of not only the hazards that pervade workplaces but also how to address them. Examples are given below.

Temporary electrical supplies

The dangers of electricity when working on or near overhead power lines should be obvious; however, when electricity is being supplied to an area fraught with danger and inclusive of multiple hazards such as a construction site, then there is an enhanced risk of electric shock. Working in wet and damp conditions increases the danger even further, since the extra moisture content will reduce a person's body resistance, which will greatly increase the effect of an electric shock.

Figure 1.58 A temporary outdoor power supply.

Addressing the hazard

All hand held equipment on construction sites is supplied through a reduced voltage system such as 110V, and the input supplies are protected further by the use of residual current devices. Working alongside and additionally to other primary circuit protection devices, the RCD will ensure that automatic disconnection of the supply occurs if contact is made with any live part. Working in damp conditions requires that the electrical system reverts to being supplied by a SELV (separated extra low voltage), which makes the system even safer since its operating voltage will be <50V a.c., typically ranging between 25V and 12V. That said, the ultimate safety measure is to operate cordless or battery operated equipment. All electrical systems are further controlled through permit-to-work systems as well as specific method statements which will dictate when and how any electrical work is to be undertaken respectively.

Trailing leads

Trailing leads can cause an immediate tripping hazard, as well as possibly incurring a risk of electric shock.

Addressing the hazard

Highlighting any hazards through trailing leads should be addressed through a rigorous risk assessment, which means that control measures would include routing them away from any pathways or thoroughfares, preferably installing them along existing fence lines. Although cable ramps are used, they are not ideal since they can still form a trip hazard away from access routes. All electrical supplies should also be RCD protected and reduced voltage systems will also reduce the effect of an electric shock.

Slippery or uneven surfaces

Slippery or uneven surfaces can lessen the surface grip both when walking or when operating mechanical lifting devices. While lubricants such as petrol, diesel and any oil-based substances are a major cause of slip accidents, the same applies to uneven surfaces since they offer less friction and can also cause serious injuries. Both instances are a major consideration when bearing in mind that the main causes of accidents are attributed to slips, trips and falls.

Addressing the hazard

All floors should be regularly cleaned and any warning signs should be erected in order to alert personnel to any uneven surfaces. Moreover, any risk assessments carried out by scrutinising processes likely to cause spillage should be reviewed to ensure that control measures are effective. In addition, anti-slip paint and strips can be used in order to target specific hazardous locations.

Presence of dust and fumes

Dust and fumes not only present a hazard regarding inhalation but can also present a danger regarding explosion, since dust-rich environments, if exposed to an ignition source, could cause an explosion.

Addressing the hazard

Appropriate PPE should be used at all times; respirators for instance have filters and are therefore more appropriate and effective than face masks. Further protection can be given with full face-piece respirators that actually regulate and control air flow. The appropriate PPE should always be identified after a rigorous risk assessment has been carried out, whilst any technical operations would be documented on a method statement and controlled through a permit-to-work system. If work is to be done in areas where there is a risk of flammable vapours (such as in a petrochemical works), it will be necessary to select specially designed electrical equipment to prevent it acting as a source of ignition due to sparks and overheating. This would also include the positioning of appropriate fire-fighting equipment. Similar consideration must be given to the selection of any wiring system to ensure that the system being put in place is consistent with the environment and all its processes.

Handling and transporting equipment or materials

The delivery of materials and equipment will make full use of cranes, dumper trucks, forklift trucks, articulated lorries and even helicopters at some sites.

Addressing the hazard

The operation of machinery and equipment when required is only operated by trained and authorized personnel. Risk assessments would need to be instigated and remain both an active and proactive process inclusive of identifying where the use of mechanical lifting devices are required, as well as promoting good lifting techniques when lifting small loads. Certain areas must also be designated as points where goods and materials are delivered, dispatched and stored, so that they do not interfere with emergency routes. Furthermore, designated storage areas need to be allocated regarding materials, waste and even flammable substances, which need to be stored away from other materials in order to protect against accidental ignition. If materials are stored at height then control measures such as guardrails need to be in place.

All storage areas need to be kept tidy so that they do not become trip hazards in themselves. Deliveries also need to be planned so that only the amount of materials required is delivered on-site.

Contaminants and irritants

Dependent on the substance, but ill effects can range from mild irritation of the skin to severe pain and loss of limbs and even loss of life.

Addressing the hazard

COSHH risk assessments would need to be maintained in order to highlight any specific hazards and necessary control measures. Specific PPE would need to be allocated as well as implementing other control measures such as ventilation when required. The use of barrier cream can also reduce the effects of dermatitis.

Fire

Put simply, fire and smoke kill. Smoke tends to overcome the individual first and can travel extremely quickly. Sources of ignition include naked flames, external sparks, internal sparks, hot surfaces and static electricity.

Addressing the hazard

Any operation such as hot working processes would need to be subject to a risk assessment process and therefore certain control measures; fire-fighting equipment should be positioned in close proximity to the hazardous operation and certain procedures put in place to ensure that static electricity is not generated through such means as equipotential bonding measures. Specific and appropriate PPE should be identified and used. Because of the hazards, permit-to-work systems might also be in force in order to control and authorize the procedure. Lastly, a method statement would indicate how the procedure would be carried out safely, drawing on all the other points raised above. Evacuation procedures must also take such issues into account.

Working at height

The main cause of deaths occurs through working at height, either using unsupported or incorrect access equipment.

Addressing the hazard

Working at height is a hazardous activity and would need to be scrutinized through a full risk assessment procedure. The risk assessment might pinpoint control measures such as the use of harnesses and even control measures through a permit-to-work system. Operation of scissor lifts or cherry pickers would be controlled so that only authorized and trained personnel can operate them. The working at height activity would be subjected to a hierarchy of control measures, which means that for any activities carried out at height, precautions are required to prevent or minimize the risk of injury from a fall.

Hazardous malfunctions of equipment

If a piece of equipment was to fail in its function (that is, fail to do what it is supposed to do) and, as a result of this failure have the potential to cause harm, then this would be defined as a hazardous malfunction.

Consider an example: if a 'materials lift' on a construction site was to collapse when the supply to its motor failed, this would be a hazardous malfunction.

Figure 1.59 Workers should go through a full risk assessment procedure before working at height.

Addressing the hazard

All the regulations concerning work equipment state that it must be:

- suitable for its intended use;
- safe in use;
- maintained in a safe condition;
- used only by instructed persons;
- provided with suitable safety measures, protective devices and warning signs;
- properly used and stored.

Improper use of equipment could put the user at risk of serious injury. For instance, having isolated an electrical circuit, it is vital that the circuit is checked to ensure that it is 'dead'. Any electrician using non-approved equipment to check for dead, could be in serious trouble since the equipment might not be sensitive or accurate enough to sense low voltages. This means that a circuit could still be live and potentially present a lethal electric shock. Tools and equipment that are not stored correctly could also be damaged which in turn could render them ineffective or inaccurate.

Addressing the hazard

The only type of equipment that is authorized for carrying out safe isolation of an electrical circuit is categorised as approved equipment and listed within GS38. This document has been specifically designed by the Health and Safety Executive and is effectively a method statement. All tools and equipment should be regularly maintained, and it is essential that they are checked before and after use as well as being locked away in secure storage when not in use. Certain equipment, however, should be subjected to maintenance including calibration, where items of test equipment for example, are sent away to ensure their accuracy is within acceptable tolerances. Lastly the use of shadow boards can bring about control when issuing tools and equipment. This is because when returning the equipment they are laid against their shadow which is made from dayglow, a highly visible fluorescent material. When viewing the shadow board if dayglow is visible, it highlights that a certain tool or equipment has not been returned.

Assessment criteria 6.7

Define how combustion occurs

Assessment criteria 6.8

Identify the classes of fire

Fire control

Fire is a chemical reaction which will continue if fuel, oxygen and heat are present. To eliminate a fire *one* of these components must be removed. This is often expressed by means of the fire triangle shown in Figure 1.60; all three corners of the triangle must be present for a fire to burn.

Fuel

Fuel is found in the construction industry in many forms: petrol and paraffin for portable generators and heaters; bottled gas for heating and soldering. Most solvents are flammable. Rubbish also represents a source of fuel: offcuts of wood, roofing felt, rags, empty solvent cans and discarded packaging will all provide fuel for a fire.

To eliminate fuel as a source of fire, all flammable liquids and gases should be stored correctly, usually in an outside locked store. The working environment should be kept clean by placing rags in a metal bin with a lid. Combustible waste material should be removed from the work site or burned outside under controlled conditions by a skilled or instructed person.

Oxygen

Oxygen is all around us in the air we breathe, but can be eliminated from a small fire by smothering with a fire blanket, sand or foam. Closing doors and windows, but not locking them, will limit the amount of oxygen available to a fire in a building and help to prevent it from spreading.

Most substances will burn if they are at a high enough temperature and have a supply of oxygen. The minimum temperature at which a substance will burn is called the 'minimum ignition temperature' and for most materials this is considerably higher than the surrounding temperature. However, a danger does exist from portable heaters, blowtorches and hot airguns which provide heat and can cause a fire by raising the temperature of materials placed in their path above the minimum ignition temperature. A safe distance must be maintained between heat sources and all flammable materials.

Heat

Heat can be removed from a fire by dousing with water, but water must not be used on burning liquids since the water will spread the liquid and the fire. Some fire extinguishers have a cooling action which removes heat from the fire.

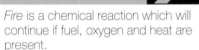

Definition

Fire is a chemical reaction which will continue if fuel, oxygen and heat are present.

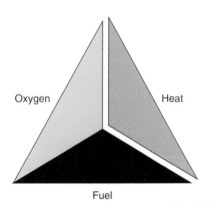

Figure 1.60 The fire triangle.

Oxygen   Heat

Fuel

Figure 1.61 Once started, fires can spread rapidly if the conditions are favourable.

Fires in industry damage property and materials, injure people and sometimes cause loss of life. Everyone should make an effort to prevent fires, but those which do break out should be extinguished as quickly as possible.

In the event of fire you should:

- raise the alarm;
- turn off machinery, gas and electricity supplies in the area of the fire;
- close doors and windows but without locking or bolting them;
- remove combustible materials and fuels away from the path of the fire, if the fire is small, and if this can be done safely;
- attack small fires with the correct extinguisher.

Only attack the fire if you can do so without endangering your own safety in any way. Always leave your own exit from the danger zone clear. Those not involved in fighting the fire should walk to a safe area or assembly point.

Fires are divided into four classes or categories:

- Class A Paper, wood and fabric;
- Class B Flammable liquids;
- Class C Flammable gases;
- Class D Metals such as sodium, aluminium, magnesium and titanium;
- Class E Fires involving electrical appliances;
- Class F Fires involving fat and cooking oil.

Definition

Fire extinguishers remove heat from a fire and are a first response for small fires.

Assessment criteria 6.9

Identify the correct type of fire extinguisher for a particular fire

Fire extinguishers are for dealing with small fires, and different types of fire must be attacked with a different type of extinguisher. Using the wrong type of extinguisher could make matters worse. For example, water must not be used on a liquid or electrical fire. The normal procedure when dealing with electrical fires is to cut off the electrical supply and use an extinguisher which is appropriate to whatever is burning. Figure 1.62 shows the correct type of extinguisher to be used on the various categories of fire. The colour coding shown is in accordance with BS EN3:1996.

Assessment criteria 6.10

State the danger of being exposed to asbestos

Control of asbestos at work regulations

In October 2010, the HSE launched a national campaign to raise awareness among electricians and other trades of the risk to their health of coming into contact with asbestos. It is called the 'Hidden Killer Campaign' because approximately six electricians will die each week from asbestos-related diseases.

Safety first

All forms of asbestos are dangerous.

Type of fire extinguisher / Type of fire	Water 	Foam 	Carbon dioxide gas 	Dry powder 
Class A Paper, wood and fabric	✓	✓	✗	✓
Class B Flammable liquids	✗	✓	✓	✓
Class C Flammable gases	✗	✗	✓	✓
Class D Metals such as sodium, aluminium, magnesium and titanium	✗	✗	✗	✗
Class E Fires involving electrical appliances	✗	✗	✓	✓
Class F Fires involving fat & cooking oil	✗	✗	✓	✓

Figure 1.62 Fire extinguishers and their applications (colour codes to BS EN3:1996). The base colour of all fire extinguishers is red, with a different-coloured flash to indicate the type.

Assessment criteria 6.11

Recognize situations where asbestos may be encountered

Assessment criteria 6.12

Outline the procedures for dealing with suspected presence of asbestos in the workplace

For more information about asbestos hazards, visit www.hse.uk/hiddenkiller. Asbestos is a mineral found in many rock formations. When separated it becomes a fluffy, fibrous material with many uses. It was used extensively in the construction industry during the 1960s and 1970s for roofing material, ceiling and floor tiles, fire-resistant board for doors and partitions, for thermal insulation and commercial and industrial pipe lagging.

There are three main types of asbestos:

* chysotile which is white and accounts for about 90% of the asbestos in use today;
* ammonite which is brown;
* crocidolite which is blue.

Asbestos cannot be identified by colour alone and a laboratory analysis is required to establish its type. Blue and brown are the two most dangerous forms of asbestos and have been banned from use since 1985. White asbestos was banned from use in 1999.

In the buildings where it was installed some 40 years ago, when left alone, it did not represent a health hazard, but those buildings are increasingly becoming in need of renovation and modernisation. It is in the dismantling, drilling and breaking up of these asbestos materials that the health hazard increases.

Asbestos is a serious health hazard if the dust is inhaled. The tiny asbestos particles find their way into delicate lung tissue and remain embedded for life, causing constant irritation and eventually, serious lung disease.

Asbestos materials may be encountered by electricians in decorative finishes such as Artex ceiling finishes, plaster and floor tiles. It is also found in control gear such as flash guards and matting in fuse carriers and distribution fuse boards, and in insulation materials in vessels, containers, pipework, ceiling ducts and wall and floor partitions.

Working with asbestos materials is not a job for anyone in the electrical industry. If asbestos is present in situations or buildings where you are expected to work, it should be removed by a specialist contractor before your work commences.

Specialist contractors, who will wear fully protective suits and use breathing apparatus, are the only people who can safely and responsibly carry out the removal of asbestos. They will wrap the asbestos in thick plastic bags and store the bags temporarily in a covered and locked skip. This material is then disposed of in a special landfill site with other toxic industrial waste materials and the site monitored by the local authority for the foreseeable future.

There is a lot of work for electrical contractors updating and improving the lighting in government buildings and schools.

This work often involves removing the old fluorescent fittings hanging on chains or fixed to beams, and installing a suspended ceiling and an appropriate number of recessed modular fluorescent fittings. So what do we do with the old fittings? Well, the fittings are made of sheet steel, a couple of plastic lamp holders, a little cable, a starter and ballast. All of these materials can go into the ordinary skip. However, the fluorescent tubes contain a little mercury and fluorescent powder with toxic elements, which cannot be disposed of in normal landfill sites.

Figure 1.63 Specialist contractors must be brought in to remove asbestos if it is found on-site.

Human factors

When an accident occurs, a major consideration is the 'human factor'. This means looking deeply into why the accident happened but more importantly, how it could have been avoided. Historically there had been a tendency to blame one specific group or individual, but a human factor investigation will look at:

* the training given to workers;
* any pressure exerted;
* any distractions;
* was the accident caused by bad or uncorrected habits;
* was there a lack of supervision;
* were the policies or procedures at fault;
* were the correct or approved tools used;
* were certain checks not carried out;
* were the correct materials used;

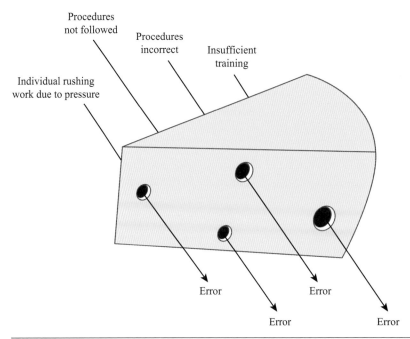

Figure 1.64 Unless systems are robust and continually monitored, holes can and do appear.

- was the test equipment calibrated;
- were the installers working shifts.

A very important consideration for electrical installation apprentices is the following statement:

> Some of the best people make the worst mistakes

Let's examine this statement, since why would 'some of the best people make the worst mistakes'?

The reason behind this is that some people will find a way of completing a job, even if there are glaring reasons why it cannot be completed. For instance, some supervisors will exert pressure on their workers stating that time is money. But on other occasions the pressure is self-imposed since the individual will have a 'can-do attitude'. However, we should always ask ourselves the question 'can I do it safely'? Another consideration for electrical installers is that faults do not always show themselves straight away. This type of fault is called latent, and will perhaps take days, weeks or even longer before it manifests itself. For example, an electrician damages the insulation of a cable when clipping it in place. When testing the cable the results will not necessarily show up instantly but the cable may deteriorate further over time. The same could be said of an electrician who installs a cable too close to floorboards which means that any give or movement could damage the cable when weight is applied to it. Another example would be an electrician who isn't careful when drilling holes through wooden joists, and whilst the cable might be electrically sound, the strength of the wooden joists in supporting the roof has been severely weakened; this error might take some time to show up.

Looking at the cheese wedge in Figure 1.64; we can see that error can get through certain holes and cause an accident.

The holes involved could be:

- insufficient training;
- training or good practice not always used;
- individual being stressed;
- procedures incorrect;
- procedures not always followed (takes longer if you do it correctly);
- individuals who dismiss their training or do not have a positive attitude.

The permit-to-work (PTW) system has already been described as it is one mechanism that brings about a safe system of work; essentially, it is a controlled document that authorizes and details the requirements and lists any special precautions and procedures to be followed.

The PTW will be raised by the 'authorized person', who will detail requirements and sign/date the document. The contractor will read the document, sign/date it and follow it when carrying out the work. When the work is complete, the 'permit to work' document is signed off (work completed and all systems are ready for use).

There are two famous examples of how a safe system of work broke down, which led to the loss of life. Both incidents include some element involving electrical workers.

The first is Piper Alpha, which was an oil platform that was destroyed on 6 July 1988, when an initial oil fire caused high pressure gas lines to explode, killing 167 people. It is still the worst offshore oilfield disaster to date in terms of human life lost. The main cause of the disaster surrounded reinstating a pump that was offline for maintenance. The engineers believed that the pump had only been electrically disconnected, but in actual fact a blanking plate had been fitted instead of a pressure valve. When they tried to operate it, the blanking plate failed and an explosion occurred. The work had been authorized on two permits, one for the maintenance and another regarding the fitment of the blanking plate. The latter was lost: not a safe working environment.

Figure 1.65 Accidents are always made up of a chain of events and not normally one single person failing in their duty.

The second incident is the Clapham Junction rail crash.

The Clapham Junction rail crash was a multiple train collision that occurred at roughly 08:10 on 12 December 1988. The enquiry found that a wiring fault

Figure 1.66 Any single person can break the chain of events; allowing an opportunity for people to learn from the experience.

stopped the alarm system which alerted an approaching train that another train lay ahead from operating. It was found that the technician had never been corrected regarding the bad practice of leaving redundant cables in a loom, despite the fact that the cable had been replaced. What made things worse was that the tradesman responsible did not insulate the ends of the redundant cables which then shorted out the alarm involved. The work of the technician was never checked, therefore a clear lack of supervision was also responsible for this disaster. Bad working practices, lack of supervision: not a safe working environment.

This is why it is important that electricians use certain 'Codes of Practice' such as referring to the Wiring Regulations which are written so that other statutory regulations such as the Electricity at Work Regulations are met. This is also why carrying out risk assessments, applying method statements such as safe isolation procedures, are so important.

When accidents happen it is usually a chain of events.

A single individual can break the chain: the incident or accident will not happen.

Put simply, health and safety regulations, examples of good practice and codes of practice are designed for a reason. Always follow them, speak up when they are not in place and speak up if you do not have the correct equipment or materials. Be that person and break the chain.

Test your knowledge

When you have completed the questions, check out the answers at the back of the book.

Note: more than one multiple-choice answer may be correct.

Learning outcome 1

1   Acts of Parliament are passed through:
     a.   legislation
     b.   common law
     c.   secondary regulations
     d.   technical guidance

2   The Electricity at Work Regulations require that:
     a.   only persons who have the appropriate level of knowledge and experience can work with electricity
     b.   only senior electricians can work with electricity
     c.   only electricians can work with electricity
     d.   only supervised tradespeople can work with electricity

3   The Health and Safety at Work Act:
     a.   affects employers only
     b.   affects employees only
     c.   highlights the duties of employers
     d.   highlights the duties of employers and employees

4   Which of the following is the most accurate regarding the Health and Safety at Work Act 1974?
     a.   It maintains safety policy
     b.   It maintains health and safety of persons at work
     c.   It maintains health, safety and welfare of persons in all places of work
     d.   It maintains health, safety and welfare of persons at their main place of work

5   Statutory regulation:
     a.   is the law of the land
     b.   must be obeyed
     c.   tells us how to comply with the law
     d.   is a code of practice

6   A non-statutory regulation:
     a.   is the law of the land
     b.   must be obeyed
     c.   tells us how to comply with the law
     d.   is a code of practice

7 The IET Wiring Regulations:
   a. are statutory regulations
   b. are non-statutory regulations
   c. are codes of good practice
   d. must always be complied with

8 Under the Health and Safety at Work Act 1974 an employee is responsible for:
   a. maintaining plant and equipment
   b. providing PPE
   c. wearing PPE
   d. taking reasonable care to avoid injury

9 Under the Health and Safety at Work Act 1974 an employer is responsible for:
   a. providing a safe working environment
   b. providing PPE
   c. providing counselling
   d. providing training

10 Employers of companies employing more than five people must:
    a. become a member of the NICEIC
    b. provide PPE if they want
    c. provide safe systems of work
    d. display a health and safety poster

11 Which of the following statements is correct?
    a. Health and safety falls mainly on the employer
    b. All employees must take reasonable care to protect themselves and their colleagues
    c. Employees have no responsibility regarding health and safety
    d. Health and safety falls mainly on the client

12 Who should ensure within reason that a workplace is risk free?
    a. Employer
    b. Employee
    c. Supervisor
    d. Client

Learning outcome 2

13 Following an accident on-site, what kind of investigation will the Health and Safety Executive carry out?
    a. Independent enquiry
    b. Immediate enquiry
    c. Partial enquiry
    d. Provisional enquiry

14 A near miss is:
    a. an accident that could have had minor consequences
    b. an accident that could have had serious consequences
    c. an incident that could have had minor consequences
    d. an incident that could have had serious consequences

15 An electrician has been hospitalized after an accident. As a witness to the event you need to:
  a. go through his valuables
  b. clock him off work
  c. fill out an accident report
  d. fill out his time sheet

16 When a health and safety problem is identified on-site, you should:
  a. report it to your supervisor, manager or employer
  b. report it in the site diary
  c. report it to the Health and Safety Executive
  d. report it to the council

17 You become aware that a safety notice has been damaged on-site. The most appropriate department to report this to would be:
  a. Health and Safety Executive
  b. site health and safety officer
  c. main contractor
  d. Harry Hill TV bloopers

18 Your first action on discovering a fire should be to:
  a. open the windows and make your escape
  b. close the windows and make your escape
  c. raise the alarm
  d. operate the nearest fire extinguisher

19 The abbreviation SWL means:
  a. stable weight load
  b. safe working load
  c. stable working load
  d. safe wide load

20 Which one of the following would be the best examples of good practice regarding manual handling?
  a. Use a mechanical lifting device if available
  b. Keep your head up
  c. Never bend your knees
  d. Always keep your heels off the ground

Learning outcome 3

21 Which of the following voltage systems would be the safest to use?
  a. 110 V
  b. 55 V
  c. 230 V
  d. Cordless equipment

22 Handheld equipment on-site should be powered through what kind of voltage system?
  a. 110 V
  b. 55 V
  c. 230 V
  d. 400 V

23 Handheld equipment on-site that is used in damp conditions should be powered through:
   a. 110 V
   b. SELV system
   c. RCD system
   d. three-phase system

24 110 V, 55 V and SELV are all examples of:
   a. low voltage systems
   b. high voltage systems
   c. SELV systems
   d. reduced voltage systems

Figure 1.67 Safety equipment label.

25 A piece of equipment which has the label shown means:
   a. it proves user safety for the whole of its lifetime
   b. it proves user safety for the date shown
   c. it is safe but must be re-tested before use
   d. it proves user safety for the date shown, but user must conduct pre-use check

26 Electrocution occurs when:
   a. someone touches a cable
   b. someone touches a conductor
   c. someone touches insulation
   d. someone becomes part of a circuit

27 Ensuring electrical safety is best achieved by:
   a. isolating and securing the supply
   b. switching off the supply
   c. working during a night shift
   d. fitting all live supplies with enclosures

28 An electric shock will depend on:
   a. the length of the conductor
   b. the length of the insulation
   c. amount of current and duration of contact
   d. amount of power and duration of contact

29 Unless absolutely necessary, an electrician should always work on:
   a. dead circuits
   b. live circuits
   c. RCD-protected circuits
   d. SELV circuits

30 The legislation which prohibits live working except in exceptional circumstances is:
   a. provision and use of Work Equipment Regulations (1992)
   b. Personal Protective Equipment Regulations (1992)
   c. BS 7671 Requirements for Electrical Installations
   d. Electricity at Work Regulations (1989)

31   Which of the following is correct regarding safe isolation?

a.   Begin wiring, verify circuit is dead, isolate and lock off, re-check voltage device

b.   Obtain permission, check voltage device, verify circuit is dead, isolate and lock off, re-check voltage device

c.   Begin work, lock off, verify circuit is dead, re-check voltage device, begin wiring

d.   Obtain permission, isolate and lock off, check voltage device, verify circuit is dead, re-check voltage device

32   Which of the following should be used when more than one person will be working on a circuit?

a.   Mortise clasp

b.   Multi-lock hasp

c.   Tywrap

d.   Interlocking lock and key

33   When dealing with faulty equipment which of the following is correct?

a.   Stop using it, remove and store securely, arrange repair, label it unserviceable

b.   Stop using it, arrange repair, remove from service, label it unserviceable

c.   Stop using it, arrange repair, label it serviceable, remove from service

d.   Stop using it, label it unserviceable, remove from service, arrange repair

Learning outcome 4

34   When working above ground for long periods of time, the most appropriate piece of equipment to use is:

a.   a ladder

b.   scaffolding

c.   a mobile scaffold tower

d.   a stepladder

35   The most appropriate piece of access equipment to walk around an attic space is:

a.   a ladder

b.   a crawling board

c.   a mobile scaffold tower

d.   a stepladder

36   If the base of a ladder is placed at a distance of 2 m from a wall, the top of the ladder should be positioned at a height of:

a.   2 m

b.   4 m

c.   6 m

d.   8 m

37   Never paint a ladder, because:

a.   it will make the ladder heavier

b.   defects can be hidden

c.   it will corrode the ladder

d.   it will weaken the ladder

38   Before use, a ladder inspection should be carried out by:
a.   a supervisor
b.   the user
c.   a health and safety representative
d.   a senior electrician

Learning outcome 5

39   Which of the following is correct?
a.   Employers wear PPE, employees provide it
b.   Employers provide PPE, employees wear it
c.   Employers provide first issue of PPE, employees provide it afterwards
d.   Employers provide first and second issue of PPE, employees provide it afterwards

40   Which of the following hazards applies to an electrician?
a.   Working at height
b.   Working with gas
c.   Working with electricity
d.   Working underground

41   Which of the following brings about a safe system of work?
a.   Using a permit-to-work
b.   Using a risk assessment
c.   Using a method statement
d.   Using common sense

42   Which of the following is correct?
a.   Method statement: technical procedure, RIDDOR: serious injury, PTW: gaining permission before working
b.   Method statement: serious injury, RIDDOR: technical procedure, PTW: gaining permission before working
c.   Method statement: non-technical procedure, RIDDOR: serious injury, PTW: technical procedure
d.   Method statement: non- technical procedure, RIDDOR: gaining permission before working, PTW: serious Injury

43   A respirator is better than a face mask because:
a.   it is more expensive
b.   it can stop asbestos
c.   it contains a filter
d.   it comes in more sizes

44   PPE can:
a.   limit the effects of accidents
b.   stop injuries
c.   stop accidents
d.   limit accidents

45   First-aid supplies must be replaced after use because:
a.   immediate first aid depends on supplies being available
b.   health and safety policies state so

c. health and safety insurance will become invalid

d. health and safety auditors will be pleased

46 Three components are necessary for a fire, they are:

a. fuel, wood and cardboard

b. petrol, oxygen and bottled gas

c. flames, fuel and heat

d. fuel, oxygen and heat

47 To highlight that people should not enter a secure area, you should use:

a. an information sign

b. a danger sign

c. a prohibition sign

d. a mandatory sign

48 What action should you take if you suspect that a building contains asbestos?

a. Carry on regardless

b. Stop work and wear the correct PPE

c. Stop work and get help

d. Finish what you are doing then get help

49 Leaving a work area untidy could cause:

a. a trip hazard

b. a dangerous occurrence

c. an incident

d. an accident

50 Which of the following shows the correct procedure regarding carrying out a risk assessment?

a. Identify risk, recognize who is at risk, evaluation of risk, further action

b. Identify risk, evaluation of risk, recognize who is at risk, further action

c. Recognize who is at risk, identify risk, evaluation of risk, further action

d. Identify risk, evaluation of risk, further action, recognize who is at risk

51 When carrying out a risk assessment, the main aim is to highlight:

a. dangers

b. risk

c. hazards

d. shocking behaviour

52 When working on an electrical circuit, it should be isolated, locked off and tagged in order to:

a. stop work before tea break

b. ensure all danger of electric shock is removed

c. trick the boss into thinking you are doing some work

d. all the above

53 Use bullet points to describe a safe isolation procedure of a 'live' electrical circuit.

54 How does the law enforce the regulations of the Health and Safety at Work Act?

55 List the responsibilities under the Health and Safety at Work Act of:
  a. an employer to his employees
  b. an employee to his employer

56 Safety signs are used in the working environment to give information and warning. Sketch and colour one sign from each of the four categories of signs and state the message given by that sign.

57 State the name of two important statutory regulations and one non-statutory regulation relevant to the electrical industry.

58 Define what is meant by PPE.

59 State five pieces of PPE which a trainee could be expected to wear at work and the protection given by each piece.

60 Describe the action to be taken upon finding a workmate apparently dead on the floor and connected to an electrical supply.

61 State how the Data Protection Act has changed the way in which we record accident and first-aid information at work.

62 List five common categories of risk.

63 List five common precautions which might be taken to control risk.

64 Use bullet points to list the main stages involved in lifting a heavy box from the floor, carrying it across a room and placing it on a worktop, using a safe manual handling technique.

65 Describe a safe manual handling technique for moving a heavy electric motor out of the stores, across a yard and into the back of a van for delivery to site.

66 Use bullet points to list a step-by-step safe electrical isolation procedure for isolating a circuit in a three-phase distribution fuse board.

67 Use bullet points to list each stage in the erection and securing of a long extension ladder. Identify all actions which would make the ladder safe to use.

68 Describe how you would use a mobile scaffold tower to re-lamp all the light fittings in a supermarket. Use bullet points to give a step-by-step account of re-lamping the first two fittings.

69 What is a proving unit used for?

70 The HSE Guidance Note GS 38 tells us about suitable test probe leads. Use a sketch to identify the main recommendations.

71 State how you would deal with the following materials when you are cleaning up at the end of the job:
  • pieces of conduit and tray;
  • cardboard packaging material;
  • empty cable rolls;
  • half-full cable rolls;
  • bending machines for conduit and tray;
  • your own box of tools;
  • your employer's power tools;
  • 100 old fluorescent light fittings;
  • 200 used fluorescent tubes.

Unit Elec2/01

Chapter 1 checklist

Learning outcome	Assessment criteria – the learner can	Page number
1. Understand how health and safety applies to electrotechnical operations.	1.1 State the general aims of health and safety legislation.	2
	1.2 Recognize the legal status of health and safety documents.	3
	1.3 State their own roles and responsibilities and those of others with regard to current relevant legislation.	15
	1.4 State the role of enforcing authorities under health and safety legislation.	19
2. Understand health and safety procedures in the work environment.	2.1 State the procedures that should be followed in the case of accidents which involve injury.	20
	2.2 Recognize appropriate procedures which should be followed when emergency situations occur in the workplace.	25
	2.3 State the procedures for recording near misses and accidents at work.	27
	2.4 Define the limitations of their responsibilities in terms of health and safety in the workplace.	32
	2.5 State the actions to be taken in situations which exceed their level of responsibility for health and safety in the workplace.	32
	2.6 State the procedures to be followed in accordance with the relevant health and safety regulations for reporting health, safety and/or welfare issues at work.	32
	2.7 Recognize the appropriate responsible persons to whom health and safety- and welfare-related matters should be reported.	33
	2.8 State the procedure for manual handling and lifting	35
3. Understand the basic electrical safety requirements.	3.1 Recognize how electric shock occurs.	36
	3.2 Recognize potential electrical hazards to be aware of on-site.	37
	3.3 Identify the different supply sources for electrical equipment and power tools.	37
	3.4 State why a reduced low voltage electrical supply is used for portable and handheld power tools on-site.	37
	3.5 Outline the user checks on portable electrical equipment before use.	39
	3.6 State the actions to be taken when portable electrical equipment is found to be damaged or faulty.	43
	3.7 State the need for safe isolation.	44
	3.8 Recognize the safe isolation procedure.	44
4. Know the safety requirements for using access equipment.	4.1 Identify the various types of access equipment available.	46
	4.2 Determine the most appropriate access equipment to gain access.	47
	4.3 Identify the safety checks to be carried out on access equipment.	47
	4.4 Describe safe practices and procedures for using access equipment.	47

Learning outcome	Assessment criteria – the learner can	Page number
5. Understand the importance of establishing a safe working environment. (i)	5.2 Outline the procedures for working in accordance with provided safety instructions.	52
	5.1 State how to produce a risk assessment and method statement.	52
	5.3 Describe the procedures to be taken to eliminate or minimize risk before deciding personal protective equipment is needed.	60
	5.4 State the purpose of personal protective equipment.	60
	5.5 Specify the appropriate protective clothing and equipment that is required for specific work tasks.	61
	5.6 Define the first-aid facilities that must be available in accordance with health and safety regulations.	63
	5.7 Clarify why first-aid supplies and equipment:	
	• must not be misused;	63
	• must be replaced when used.	
	5.8 Describe safe practices and procedures using equipment, tools and substances.	63
6. Understand the importance of establishing a safe working environment. (ii)	6.1 Recognize warning symbols of hazardous substances.	65
	6.2 Clarify what is meant by the term 'hazard'.	65
	6.3 Recognize hazards associated with the installation and maintenance of electrotechnical systems and equipment.	66
	6.4 Identify the main types of safety signs that may be used on-site.	67
	6.5 Identify hazardous situations in the workplace.	69
	6.6 Describe the practices and procedures for addressing hazards in the workplace.	69
	6.7 Define how combustion occurs.	74
	6.8 Identify the classes of fire.	74
	6.9 Identify the correct type of fire extinguisher for a particular fire.	75
	6.10 State the danger of being exposed to asbestos.	75
	6.11 Recognize situations where asbestos may be encountered.	76
	6.12 Outline the procedures for dealing with the suspected presence of asbestos in the workplace.	76

Electrical science and principles

EAL Electrical Installation Work – Level 2, 2nd Edition 978 0 367 19562 5
© 2019 Linsley. Published by Taylor & Francis. All rights reserved.
www.routledge.com/9780367195618

Learning outcomes

When you have completed this chapter you should:

1. Understand common units of measurement used in electrotechnical work.
2. Understand the principles of electrical circuits.
3. Understand electrical supply systems.
4. Understand micro-generation technologies.
5. Understand basic series and parallel electrical circuits.
6. Understand the principles of electromagnetism.
7. Understand fundamental mechanics.
8. Understand the principles of three phase circuits.
9. Understand the operating principles of electrical equipment.

DOI: 10.1201/9780429203176-2

Assessment criteria 1.1

Identify (SI) units of measurement for general quantities

Units

Very early units of measurement were based on the things easily available – the length of a stride, the distance from the nose to the outstretched hand, the weight of a stone and the time-lapse of one day. Over the years, new units were introduced and old ones modified. Different branches of science and engineering were working in isolation, using their own systems of measurement, resulting in an overwhelming variety of units.

In all branches of science and engineering there is a need for a practical system of units that everyone can use. In 1960, the General Conference of Weights and Measures agreed to an international system called the Système International d'Unités (since abbreviated to SI units).

Figure 2.1 Accurate measurements are essential when planning work.

SI units are based upon a small number of fundamental units from which all other units may be derived. Table 2.1 describes some of the basic units we will be using in our electrical studies.

Like all metric systems, SI units have the advantage that prefixes representing various multiples or sub-multiples may be used to increase or decrease the size of the unit by various powers of 10. Some of the more common prefixes and their symbols are shown in Table 2.2.

There is an expectation therefore that an electrician needs to convert from a multiple to a sub-multiple using these prefixes and because they occur as a multiple of three, this is called Engineering Notation.

For instance the rating of an RCD which is fitted to protect power sockets is normally represented by 30 milliamps or 30mA. In other words one amp divided into a thousand bits, but the bits are represented by the sub-multiple milli.

But equally we can also represent mA in amps, for instance 30mA is equivalent to 0.03A. Basically the decimal point (30.0) has been moved three times to the left, which appears to create a smaller number because it is representing a bigger unit 30mA = 0.03A.

The decimal point was moved three times because that is the difference between the prefixes (milli 10^{-3})

This is an important distinction and indeed a question you need to ask yourself when using prefixes and engineering notation.

- Am I converting from a small unit to a bigger one?

  Yes, then the number will appear to become smaller

- Am I converting from a big unit to a smaller one?

  Yes, then the number will appear to become bigger

Assessment criteria 1.2

State the SI or derived SI unit for electrical quantities

Table 2.1 Basic SI units

Quantity	Measure of	Basic unit	Symbol	Notes
Area	Length × length	Metre squared	m^2	
Current I	Electric current	Ampere	A	
Energy	Ability to do work	Joule	J	Joule is a very small unit 3.6×10^6 J × 1 kWh
Force	The effect on a body	Newton	N	
Frequency	Number of cycles	Hertz	Hz	Mains frequency is 50 Hz
Length	Distance	Metre	m	
Mass	Amount of material	Kilogram	Kg	One metric tonne = 1000 kg
Magnetic flux Φ	Magnetic energy	Weber	Wb	
Magnetic flux density B	Number of lines of magnetic flux	Tesla	T	
Potential or pressure	Voltage	Volt	V	
Period T	Time taken to complete one cycle	Second	s	The 50 Hz mains supply has a period of 20 ms
Power	Rate of doing work	Watt	W	
Resistance	Opposition to current flow	Ohm	Ω	
Resistivity	Resistance of a sample piece of material	Ohm metre	ρ	Resistivity of copper is $17.5 = 10^{-9} \Omega m$
Temperature	Hotness or coldness	Kelvin	K	0°C = 273 K. A change of 1 K is the same as 1°C
Time	Time	Second	s	60 s = 1 min 60 min = 1 h
Weight	Force exerted by a mass	Kilogram	kg	1000 kg = 1 tonne
Electric charge	Charge transported by a constant current of one ampere in one second:	Coulomb	C	Charge of approximately 6.241×10^{18} electrons

Note: A more detailed description may be found in this chapter.

Assessment criteria 1.3

Identify the common multiples and sub-multiples used within electrotechnical work

Table 2.2 Symbols and multiples for use with SI units

Prefix	Symbol	Multiplication factor		
Mega	M	$\times 10^6$	or	$\times 1,000,000$
Kilo	k	$\times 10^3$	or	$\times 1000$
Hecto	h	$\times 10^2$	or	$\times 100$
Deca	da	$\times 10$	or	$\times 10$
Deci	d	$\times 10^{-1}$	or	$\div 10$
Centi	c	$\times 10^{-2}$	or	$\div 100$
Milli	m	$\times 10^{-3}$	or	$\div 1000$
Micro	μ	$\times 10^{-6}$	or	$\div 1,000,000$
Nano	n	$\times 10^{-9}$	or	$\div 1,000,000,000$
Pico	r	$\times 10^{-12}$	or	$\div 1,000,000,000,000$

Assessment criteria 2.1

Transpose a basic formula

Transposition of formula

There are many different ways of transposing a formula but the simplest makes use of the **opposite rule**. For instance:

Rule number 1:

    The opposite of **addition** is **subtraction**

                +   –

Rule number 2:

    The opposite of **multiply** is **divide**

                ×   ÷

Rule number 3:

    The opposite of a **squaring** is **square root**

              x^2 √

By way of an example we can see this below:

Rule number 1

 $5 + 5 = 10$

$10 - 5 = 5$

(The opposite of addition is subtraction)

Top tip

If your maths skills are a little rusty, I recommend to my students a book called *Electrical Installation Calculations (Basic Level 2)* by Christopher Kitcher and A.J. Watkins.

Rule number 2

5 × 5 = 25

25 ÷ 5 = 5

(The opposite of multiply is divide)

In electrical installation we tend not to use the symbol ÷ when we are showing a divide sign. Therefore 1 ÷ 2, is effectively ½, but written in an equation it would

be: $\dfrac{1}{2}$

Rule number 3

Revision: (The square root of a number is a value that, when multiplied by itself, gives the number.

For example: 4 × 4 = 16, so the square root of 16 is 4.)

5^2 means 5 × 5 which equals 25.

The square root of 25 is 5.

(The opposite of squaring is square root)

The reason why the opposite rule works is that we can keep an equation or formula balanced by changing the opposite function.

Remember that whatever you do to one side you must then do the same **or the opposite**.

Figure 2.2 If you do the same or the opposite when transposing a formula it will balance out.

Let us work through a few examples in order to demonstrate solving certain values.

Example 1

a + c = d

However, what if I need to rearrange the formula to find **c**?

We do this by following three simple steps. Remember the aim is to get c by itself.

Step 1: focus on the letter you want to find and write this down.

c =

(continued)

Example 1 continued

Step 2: Look which letter or value is already on its own, or in other words which letter or value that is already on the other side of the = sign. In this case d

c = d

Step 3: Move the letter or value that is left away from the one you want to find.

In this case the letter remaining is +a, which becomes –a as it moves across

c = d – a

Example 2

Ohm's law states that **V = I × R.**

However, what if I need to rearrange the formula to find **I**?

Remember the aim is to get **I** by itself.

Step 1: focus on the letter you want to find and write this down.

I =

Step 2: Look which letter or value is already on its own, or in other words which letter or value that is already on the other side of the = sign. In this case V.

I = V

Step 3: Move the letter or value that is left away from the one you want to find.

In this case the letter remaining is R, which is shown on top (multiply), therefore the opposite of multiply is divide, then simply move the R to the bottom of the equation.

$$I = \frac{V}{R}$$

Example 3

$$a^2 + b^2 = c$$

Re-arrange the formula by transposing to find b

Step 1: focus on the letter you want to find and write this down. Just for now we write b^2 as we would if it was b.

b^2 =

Step 2: Look which letter or value is already on its own, or in other words which letter or value that is already on the other side of the = sign. In this case c

b^2 = c

Example 3

Step 3: Move the letter or value that is left away from the one you want to find. In this example a^2 becomes $-a^2$. Therefore our equation becomes:

$b^2 = c - a^2$.

Last part coming up, we need to find b not b^2. Therefore using the opposite rule remove the ^{2} from b, but keep the formula balanced by square rooting the other side.

$b = \sqrt{c - a^2}$.

If we now work back in order to find c, the first thing we need to do is get rid of the square root sign, and we do this by squaring b.

$b^2 = c - a^2$.

We then follow the 3 steps.

$c =$
$c = b^2$
$c = b^2 + a^2$

Assessment criteria 2.2

Determine electrical quantities using Ohm's law

Ohm's law

In 1826, Ohm published details of an experiment he had done to investigate the relationship between the current passing through and the potential difference between the ends of a wire. As a result of this experiment he arrived at a law, now known as **Ohm's law**, which says that the current passing through a conductor under constant temperature conditions is proportional to the potential difference across the conductor. This may be expressed mathematically as

$V = I \times R$ (Volts)

Transposing this formula, we also have

$I = \dfrac{V}{R}$ (Amps)   and $R = \dfrac{V}{I}$ (Ohms)

Definition

Every circuit offers some opposition to current flow, which we call the circuit *resistance,* measured in ohms (symbol Ω).

 Definition

Ohm's law, which says that the current passing through a conductor under constant temperature conditions is proportional to the potential difference across the conductor.

Example 1

An electric heater, when connected to a 230 V supply, was found to take a current of 4 A. Calculate the element resistance.

$R = \dfrac{V}{I}$

$\therefore R = \dfrac{230\ V}{4\ A} = 57.5\ \Omega$

 Definition

Ohm's law can be defined through a triangle:

$$\frac{V}{I \times R}$$

Example 2

The insulation resistance measured between line conductors on a 400 V supply was found to be 2 MΩ. Calculate the leakage current.

$$I = \frac{V}{R}$$

$$\therefore I = \frac{400\ V}{2 \times 10^6\ \Omega} = 200 \times 10^{-6}\ A = 200\ \mu A$$

Example 3

When a 4 Ω resistor was connected across the terminals of an unknown d.c. supply, a current of 3 A flowed. Calculate the supply voltage.

$$V = I \times R$$
$$\therefore V = 3A \times 4\Omega = 12V$$

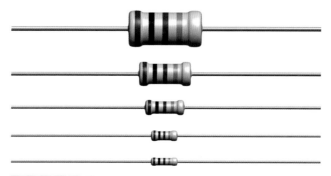

Figure 2.3 Resistors.

Assessment criteria 2.3

Calculate values of electrical power in basic circuits

Electrical energy

Power and energy are related. If we consider that energy is thought of as the amount of work that is being done or being converted. For instance when an electrical current flows in a circuit, heat is generated. The amount of heat that is generated depends directly on how much current is flowing (Amps), which in turn depends on how much voltage or electrical pressure there is (Volts).

This means that we can calculate the rate at which heat is generated through a formula for electrical Power: E = V × I, with the unit for Power being represented by the Watt.

This unit is named after the Scottish engineer James Watt, and can be thought of as the amount of energy used or generated per second.

Figure 2.4 These turbines convert wind into electrical energy

For instance if we consider the following situation, whereby if we need to calculate the amount of energy generated when a circuit is supplied with 50 V, and a current of 2 amperes flows for 2 minutes.

Using E = V × I, we can substitute the value for voltage and current, since they are directly responsible for the amount of energy used. However, the question is asking how much energy is expended over a specific amount of time, in this case 2 minutes. But if you remember the SI unit of time is seconds therefore we need to convert 2 minutes into seconds. 2 × 60 = 120 seconds.

Therefore to calculate how much energy is used over a specific amount of time our formula becomes E = V × I × t.

    E = 2 × 50 × 120

    E = 12000 or 12 KJ

These types of formula are very important for an electrician since all design calculations start off by calculating how much current an installation requires. Therefore: P = V × I should be thought of as the main formula for Power for an electrician.

Another way of remembering this formula is to use the power formula triangle

Figure 2.5 Saving energy is good for the environment, and it also reduces your energy bills

Simply cover the value required to extract the formula required.

$$I = \frac{P}{V}$$

Another two formulas for Power can be obtained by extracting values from Ohm's law as seen below:

If $V = I \times R$, by transposition we get $I = \frac{V}{R}$

Then substitute what I represents $\left(I = \frac{V}{R}\right)$ into $P = V \times I$

And you get $P = \frac{V \times V}{R}$

Which can be written as $P = \frac{V^2}{R}$

Examples

1 Calculate the current demanded by a 60 W lamp when they are connected to domestic mains supply of 230 V supply.

   Use the main formula for power for an electrician

   $P = V \times I$

   Transpose to find I

   $I = \frac{P}{V}$

   $I = \frac{60}{230}$

   $I = 0.26\,A$

2 A shower draws 40 A when supplied with 230 V. Work out the power consumed by the shower?

   $P = V \times I$

   $P = 230 \times 40$

   $P = 9200\,W$ or $9.2\,kW$

Have a go at the exercises below.

Exercise 1

1 If a resistor has 20 Volts across it, and 10 Amps through it, work out the circuit power?

2 A cable drops 4 V along its length. If the current flow is 25 A, how much power is dissipated?

3 If a resistor of 20 Ω, has 10 Amps flowing through it, work out the circuit power (hint, either work out the voltage using Ohm's law, or use the power triangle).

Assessment criteria 2.4

Calculate the values of electrical energy

Electrical energy (kW hours)

Although the joule (work done) is recognized as the unit for energy and is defined by the capacity for doing work, it is far too small to use when required to measure the energy used in a domestic setting.

Instead the kilowatt hour is used and it has a direct relation since:

1 kW (1000 W) has been used for a time of 1 hour (60 × 60 = 3600 seconds).

Given that:

1 Joule (J) = 1 watt (W) for one second (s)
1000 joules (J) = 1 kilowatt (kW) for one second.

In one hour there are 3600 seconds. Therefore it follows that:

3600 × 1000 J = 1 kW for one hour (kWh)

This is the measurement that appears on your electricity bill. If you have not seen an electricity bill ask your parents. Switching lights off in a room that is unoccupied saves electricity, which results in reduced bills as well as helping the environment.

Figure 2.6 Regularly examining your daily usage statistics can help you to save money in the long term.

Worked example

If a small house has the following items connected to the supply each day, calculate how much energy would be consumed over a seven day period:

- 12 × 60 W light fittings used for 4 hours
- 1 × 3 kW electric fire used for 2 hours
- 1 × 3 kW kettle for a total of 1 hour

Remembering that E = P × t and applying this to each load

Lights = (12 × 60) × 4 = 2880 W = 2.88 kWh

Fire = 3 × 2 = 6 kWh

Kettle = 3 × 1 = 3 kWh

Daily consumption = 2.88 + 6 + 3
                  = 11.88 kWh

Over 7 days 9.8 × 7 = 83.16 kWh

Assessment criteria 2.5

Determine the resistance of a conductor

Assessment criteria 2.6

Describe what is meant by resistance and resistivity in relation to electrical circuits

Resistivity

The resistance, or opposition, to current flow varies for different materials, each having a particular constant value. If we know the resistance of, say, 1m of a material, then the resistance of 5m will be five times the resistance of 1m. The **resistivity** (symbol ρ – the Greek letter 'rho') of a material is defined as the resistance of a sample of unit length and unit cross-section. Typical values are given in Table 2.3. Using the constants for a particular material we can calculate the resistance of any length and thickness of that material from the equation.

Table 2.3 gives the resistivity of silver as $16.4 \times 10^{-9}\,\Omega m$, which means that a cube of silver whose length, width and height all measure 1 m will have a resistance of $16.4 \times 10^{-9}\,\Omega$.

A handy rule to remember when converting millimetres to metres is to attach the magic number $\times 10^{-6}$. We obtain this number from the fact that 1m contains 1000mm. However, cable size is actually a cross-sectional area measured in m^2, so the difference between mm^2 and m^2 is shown through 6 decimal places or the magic number 10^{-6}. $1.5\,mm^2$ would therefore become $1.5 \times 10^{-6}\,m^2$.

A = area

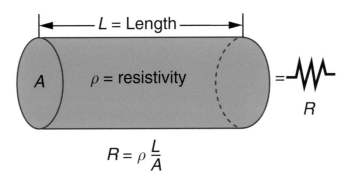

$$R = \rho \frac{L}{A}$$

Figure 2.7 Calculating the resistivity of an object.

Where
ρ = the resistivity constant for the material (Ωm)
l = the length of the material (m)
a = the cross-sectional area of the material (m^2)

Figure 2.8 Many safety critical tests depend on understanding how a conductor's length and cross sectional area affect its resistance.

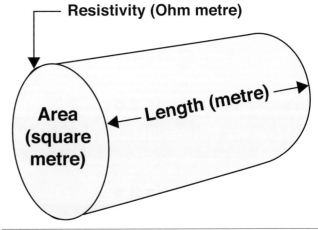

Figure 2.9 As the length of the conductor increases its resistance will increase. Resistivity however is based on conductor material and therefore remains constant.

Table 2.3 Resistivity values

Material	Resistivity
Silver	16.4×10^{-9}
Copper	17.5×10^{-9}
Aluminium	28.5×10^{-9}
Brass	75.0×10^{-9}
Iron	100.0×10^{-9}

It is vital that we emphasize the difference between resistance and resistivity.

- Resistance is the opposition to current flow and will include every element within the circuit, including the conductor and all the internal connections linking the protective device, busbar, switch and load.
- Resistivity is concerned with the conducting material only and remains constant for a given type of material measured in Ω/m.

The resistivity formula is very important in installation work since it impacts on many aspects. Looking at the formula again

$$R = \frac{\rho L}{A}$$

The length of the conductor is directly proportional to the overall resistance of the circuit.

$$R = \frac{\rho L}{A}$$

Directly proportional means that if the length increases so will the resistance.

If the length of a circuit is very long the circuit resistance will also increase. This means that more voltage will be dropped across the cable, depriving the load such as a lamp of voltage. This means that a lamp can appear dim, or motor circuits could run slower than expected. How do we compensate for this?

Again, looking at the formula the cross-sectional area of the conductor is indirectly proportional to the overall resistance of the circuit.

$$R = \frac{\rho L}{A}$$

Indirectly proportional means that if the area increases the resistance will decrease accordingly.

Therefore, when we encounter volt drop problems we counter this by increasing the cross-sectional area of a cable, in other words: select a bigger size.

To summarize the relationship between resistivity and resistance:

Increasing the length *increases* the resistance
Increasing the area *decreases* the resistance

Decreasing the length *decreases* the resistance
Decreasing the area *increases* the resistance

Doubling the length *doubles* the resistance
Doubling the area *halves* the resistance

Definition

Resistivity is concerned with conductor material.

Calculating resistivity

The general rule that you need to remember when working out any problem with regard to resistivity is to make sure that the units that you use are consistent, i.e. all the units in metres or millimetres. However, whilst cable length and resistivity tend to be represented in metres and Ω/m cable size is normally used and given in mm^2.

A handy rule to remember when converting millimetres to metres is to attach the magic number $\times 10^{-6}$. We obtain this number from the fact that 1 m contains 1000mm. However, cable size is actually a cross-sectional area measured in m^2, so the difference between mm and metres can be shown by 6 decimal places or the magic number 10^{-6}. $1.5\,mm^2$ would therefore become 1.5×10^{-6}.

Example 1

Calculate the resistance of 100 m of copper cable of $1.5\,mm^2$ cross-sectional area (csa) if the resistivity of copper is taken as $17.5 \times 10^{-9}\,\Omega m$.

$$R = \frac{\rho l}{a}\,(\Omega)$$

$$\therefore R = \frac{17.5 \times 10^{-9}\,\Omega\,m \times 100\,m}{1.5 \times 10^{-6}\,m^2} = 1.16\,\Omega$$

Example 2

Calculate the resistance of 100 m of aluminium cable of $1.5\,mm^2$ cross-sectional area if the resistivity of aluminium is taken as $28.5 \times 10^{-9}\,\Omega m$.

$$R = \frac{\rho l}{a}\,(\Omega)$$

$$\therefore R = \frac{28.5 \times 10^{-9}\,\Omega\,m \times 100\,m}{1.5 \times 10^{-6}\,m^2} = 1.9\,\Omega$$

The above examples show that the resistance of an aluminium cable is some 60% greater than a copper conductor of the same length and cross-section. Therefore, if an aluminium cable is to replace a copper cable, the conductor size must be increased to carry the rated current as given by the tables in Appendix 4 of the IET Regulations and Appendix 6 of the *On-Site Guide*.

The other factor which affects the resistance of a material is the temperature, and we will consider this later.

> **Key fact**
>
> *Resistance*: If the length is doubled, resistance will be doubled. If the cross-sectional area is doubled, the resistance will be halved.

Try this

Resistance

- Take two 100 m lengths of single cable (2 coils)

- Measure the resistance of 100 m of cable (1 coil)

- Value _____ Ω
- Join the two lengths together (200 m) and again measure the resistance
- Value _____ Ω
- Does this experiment prove resistance is proportional to length?
- If the resistance is doubled, it is proved (QED)! (QED, *quod erat demonstrandum*) is the Latin for 'which was to be demonstrated'.)

Assessment criteria 2.7

Outline the principles of an electrical circuit

Basic circuit theory

All matter is made up of atoms which arrange themselves in a regular framework within the material. The atom is made up of a central, positively charged nucleus, surrounded by negatively charged electrons. The electrical properties of a material depend largely upon how tightly these electrons are bound to the central nucleus.

A **conductor** is a material in which the electrons are loosely bound to the central nucleus and are, therefore, free to drift around the material at random from one atom to another, as shown in Fig. 2.10(a). Materials which are good conductors include copper, brass, aluminium and silver.

An **insulator** is a material in which the outer electrons are tightly bound to the nucleus, so that there are no free electrons to move around the material. Good insulating materials are PVC, rubber, glass and wood.

If a battery is attached to a conductor as shown in Fig. 2.10(b), the free electrons drift purposefully in one direction only. The free electrons close to the positive plate of the battery are attracted to it since unlike charges attract, and the free electrons near the negative plate will be repelled from it. For each electron entering the positive terminal of the battery, one will be ejected from the negative terminal, so the number of electrons in the conductor remains constant.

The drift of electrons within a conductor is known as an **electric current,** measured in amperes and given the symbol *I*.

 Definition

A *conductor* is a material in which the electrons are loosely bound to the central nucleus and are, therefore, free to drift around the material at random from one atom to another, as shown in Fig. 2.10(a). Materials which are good conductors include copper, brass, aluminium and silver.

 Definition

An *insulator* is a material in which the outer electrons are tightly bound to the nucleus, so that there are no free electrons to move around the material. Good insulating materials are PVC, rubber, glass and wood.

 Definition

The drift of electrons within a conductor is known as an *electric current*, measured in amperes and given the symbol *I*.

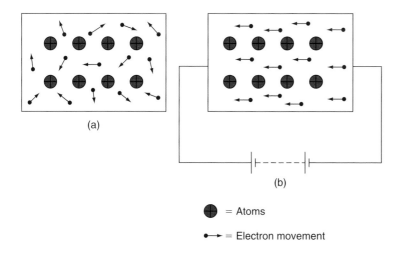

(a)

(b)

⊕ = Atoms

•→ = Electron movement

Figure 2.10 Atoms and electrons on a material.

Definition

The *potential difference* (p.d.) is the change in energy levels measured across the load terminals. This is also called the volt drop or terminal voltage, since e.m.f. and p.d. are both measured in volts.

Definition

Every circuit offers some opposition to current flow, which we call the circuit *resistance,* measured in ohms (symbol Ω).

Figure 2.11 Both conductors and insulators are used in electrical cable.

Definition

A *conductor* is a material, usually a metal, in which the electrons are loosely bound to the central nucleus. These electrons can easily become 'free electrons' which allows heat and electricity to pass easily through the material.

An *insulator* is a material, usually a non-metal, in which the electrons are very firmly bound to the nucleus and, therefore, will not allow heat or electricity to pass through it.

For a current to continue to flow, there must be a complete circuit for the electrons to move around. If the circuit is broken by opening a switch, for example, the electron flow and therefore the current will stop immediately.

To cause a current to flow continuously around a circuit, a driving force is required, just as a circulating pump is required to drive water around a central heating system. This driving force is the *electromotive force* (e.m.f.). Each time an electron passes through the source of e.m.f., more energy is provided to send it on its way around the circuit.

An e.m.f. is always associated with energy conversion, such as chemical to electrical in batteries and mechanical to electrical in generators. The energy introduced into the circuit by the e.m.f. is transferred to the load terminals by the circuit conductors.

The **potential difference** (p.d.) is the change in energy levels measured across the load terminals. This is also called the volt drop or terminal voltage, since e.m.f. and p.d. are both measured in volts. **Resistance** in every circuit offers some opposition to current flow, which we call the circuit *resistance*, measured in ohms (symbol Ω), to commemorate the famous German physicist Georg Simon Ohm, who was responsible for the analysis of electrical circuits.

Assessment criteria 2.8

Differentiate between materials which are good conductors and insulators

Properties of conductors and insulators

In Fig. 2.10 earlier in this chapter we looked at the atomic structure of materials. All materials are made up of atoms and electrons. What makes them different materials is the way in which the atoms and electrons are arranged and how strongly the electrons are attracted to the atoms.

A **conductor** is a material, usually a metal, in which the electrons are loosely bound to the central nucleus. These electrons can easily become 'free electrons' which allows heat and electricity to pass easily through the material.

An **insulator** is a material, usually a non-metal, in which the electrons are very firmly bound to the nucleus and, therefore, will not allow heat or electricity to pass through it.

Let us now define the terms and properties of some of the materials used in the electrical industry.

Ferrous A word used to describe all metals in which the main constituent is iron. The word 'ferrous' comes from the Latin word *ferrum*, meaning iron. Ferrous metals have magnetic properties. Cast iron, wrought iron and steel are all ferrous metals.

Non-ferrous Metals which *do not* contain iron are called non-ferrous. They are non-magnetic and resist rusting. Copper, aluminium, tin, lead, zinc and brass are examples of non-ferrous metals.

Alloy An alloy is a mixture of two or more metals. Brass is an alloy of copper and zinc, usually in the ratio 70–30% or 60–40%.

Corrosion The destruction of a metal by chemical action. Most corrosion takes place when a metal is in contact with moisture (see also mild steel and zinc).

Thermoplastic polymers These may be repeatedly warmed and cooled without appreciable changes occurring in the properties of the material. They are good insulators, but give off toxic fumes when burned. They have a flexible quality when operated up to a maximum temperature of 70°C but should not be flexed when the air temperature is near 0°C, otherwise they may crack.

Polyvinylchloride (PVC) used for cable insulation is a thermoplastic polymer.

Thermosetting polymers Once heated and formed, products made from thermosetting polymers are fixed rigidly. Plug tops, socket outlets and switch plates are made from this material.

Rubber is a tough elastic substance made from the sap of tropical plants. It is a good insulator, but degrades and becomes brittle when exposed to sunlight.

Synthetic rubber is manufactured, as opposed to being produced naturally. Synthetic or artificial rubber is carefully manufactured to have all the good qualities of natural rubber – flexibility, good insulation and suitability for use over a wide range of temperatures.

Silicon rubber Introducing organic compounds into synthetic rubber produces a good insulating material which is flexible over a wide range of temperatures and which retains its insulating properties even when burned. These properties make it ideal for cables used in fire alarm installations such as FP200 cables.

Magnesium oxide The conductors of mineral-insulated metal-sheathed (MICC) cables are insulated with compressed magnesium oxide, a white chalk-like substance which is heat resistant and a good insulator, and lasts for many years. The magnesium oxide insulation, copper conductors and sheath, often additionally manufactured with various external sheaths to provide further protection from corrosion and weather, produce a cable designed for long-life and high-temperature installations. However, the magnesium oxide is very hygroscopic, which means that it attracts moisture and, therefore, the cable must be terminated with a special moisture-excluding seal, as shown in Fig. 3.41.

Copper

Copper is extracted from an ore which is mined in South Africa, North America, Australia and Chile. For electrical purposes it is refined to about 98.8% pure copper, the impurities being extracted from the ore by smelting and electrolysis. It is a very good conductor, is non-magnetic and offers considerable resistance to atmospheric corrosion. Copper toughens with work, but may be annealed, or softened, by heating to dull red before quenching.

Copper forms the largest portion of the alloy brass, and is used in the manufacture of electrical cables, domestic heating systems, refrigerator tubes and vehicle radiators. An attractive soft reddish-brown metal, copper is easily worked and is also used to manufacture decorative articles and jewellery.

Aluminium

Aluminium is a grey-white metal obtained from the mineral bauxite which is found in the United States, Germany and the Russian Federation. It is a very good conductor, is non-magnetic, offers very good resistance to atmospheric corrosion and is notable for its extreme softness and lightness. It is used in the manufacture of power cables. The overhead cables of the National Grid are made of an aluminium conductor reinforced by a core of steel. Copper conductors would be too heavy to support themselves between the pylons.

Definition

Ferrous A word used to describe all metals in which the main constituent is iron.

Non-ferrous Metals which *do not* contain iron are called non-ferrous.

Alloy An alloy is a mixture of two or more metals.

Corrosion The destruction of a metal by chemical action.

Thermoplastic polymers These may be repeatedly warmed and cooled without appreciable changes occurring in the properties of the material.

Polyvinylchloride (PVC) used for cable insulation is a thermoplastic polymer.

Definition

Thermosetting polymers Once heated and formed, products made from thermosetting polymers are fixed rigidly. Plug tops, socket outlets and switch plates are made from this material.

Rubber is a tough elastic substance made from the sap of tropical plants.

Synthetic rubber is manufactured, as opposed to being produced naturally.

Silicon rubber Introducing organic compounds into synthetic rubber produces a good insulating material such as FP 200 cables.

Magnesium oxide The conductors of mineral-insulated metal-sheathed (MICC) cables are insulated with compressed magnesium oxide.

Key fact

Silver is the best conductor.

Lightness and resistance to corrosion make aluminium an ideal metal for the manufacture of cooking pots and food containers.

Aluminium alloys retain the corrosion resistance properties of pure aluminium with an increase in strength. The alloys are cast into cylinder heads and gearboxes for motor cars, and switch-boxes and luminaires for electrical installations. Special processes and fluxes have now been developed which allow aluminium to be welded and soldered.

Try this
Materials

List five 'good insulator' materials being used at your place of work.

Brass

Brass is a non-ferrous alloy of copper and zinc which is easily cast. Because it is harder than copper or aluminium it is easily machined. It is a good conductor and is highly resistant to corrosion. For these reasons it is often used in the electrical and plumbing trades. Taps, valves, pipes, electrical terminals, plug top pins and terminal glands for steel wire armour (SWA) and MI cables are some of the many applications.

Brass is an attractive yellow metal which is also used for decorative household articles and jewellery. The combined properties of being an attractive metal which is highly resistant to corrosion make it a popular metal for ships' furnishings.

Cast steel

Cast steel is also called tool steel or high-carbon steel. It is an alloy of iron and carbon which is melted in airtight crucibles and then poured into moulds to form ingots. These ingots are then rolled or pressed into various shapes from which the finished products are made. Cast steel can be hardened and tempered and is therefore ideal for manufacturing tools. Hammer heads, pliers, wire cutters, chisels, files and many machine parts are also made from cast steel.

Mild steel

Mild steel is also an alloy of iron and carbon but contains much less carbon than cast steel. It can be filed, drilled or sawn quite easily and may be bent when hot or cold, but repeated cold bending may cause it to fracture. In moist conditions corrosion takes place rapidly unless the metal is protected. Mild steel is the most widely used metal in the world, having considerable strength and rigidity without being brittle. Ships, bridges, girders, motor car bodies, bicycles, nails, screws, conduit, trunking, tray and SWA are all made of mild steel.

Try this
Materials

List five 'good conductor' materials being used at your place of work.

Zinc

Zinc is a non-ferrous metal which is used mainly to protect steel against corrosion and in making the alloy brass. Mild steel coated with zinc is sometimes called *galvanized steel*, and this coating considerably improves steel's resistance to corrosion. Conduit, trunking, tray, SWA, outside luminaires and electricity pylons are made of galvanized steel.

Assessment criteria 2.9

Define the sources of electromotive force

Assessment criteria 2.10

Clarify the main effects of electric currents

The three effects of an electric current

When an electric current flows in a circuit it can have one or more of the following three effects: **heating**, **magnetic** or **chemical**.

Definition

When an electric current flows in a circuit it can have one or more of the following three effects: *heating, magnetic* or *chemical*.

Heating effect

The movement of electrons within a conductor, which is the flow of an electric current, causes an increase in the temperature of the conductor. The amount of heat generated by this current flow depends upon the type and dimensions of the conductor and the quantity of current flowing. By changing these variables, a conductor may be operated hot and used as the heating element of a fire, or be operated cool and used as an electrical installation conductor.

The heating effect of an electric current is also the principle upon which a fuse gives protection to a circuit. The fuse element is made of a metal with a low melting point and forms a part of the electrical circuit. If an excessive current flows, the fuse element overheats and melts, breaking the circuit.

Other domestic items that operate through a thermal effect include: cookers, water heaters, soldering irons, toasters and electric fires, to name just a few.

Magnetic effect

Whenever a current flows in a conductor a magnetic field is set up around the conductor like an extension of the insulation. The magnetic field increases with the current and collapses if the current is switched off. A conductor carrying current and wound into a solenoid produces a magnetic field very similar to a permanent magnet, but has the advantage of being switched on and off by any switch which controls the circuit current.

The magnetic effect of an electric current is the principle upon which electric bells, relays, instruments, motors and generators work.

Chemical effect

When an electric current flows through a conducting liquid, the liquid is separated into its chemical parts. The conductors which make contact with the liquid are called the anode and cathode. The liquid itself is called the electrolyte, and the process is called *electrolysis*.

Electrolysis is an industrial process used in the refining of metals and electroplating. It was one of the earliest industrial applications of electric current. Most of the aluminium produced today is extracted from its ore by electrochemical methods. Electroplating serves a double purpose by protecting a base metal from atmospheric erosion and also giving it a more expensive and attractive appearance. Silver and nickel plating has long been used to enhance the appearance of cutlery, candlesticks and sporting trophies.

An anode and cathode of dissimilar metal placed in an electrolyte can react chemically and produce an e.m.f. When a load is connected across the anode and cathode, a current is drawn from this arrangement, which is called a cell. A battery is made up of a number of cells. It has many useful applications in providing portable electrical power, but electrochemical action can also be undesirable since it is the basis of electrochemical corrosion which rots our motor cars, industrial containers and bridges.

When a load such as a light bulb is connected across the battery, the circuit is complete and a chemical reaction occurs which breaks down the electrolyte and then deposits positive ions on one plate and negative electrons on the other. This then creates a flow of electrical energy in the device, in this case lighting up the lamp. Any voltage that is created is known as an e.m.f. (electromotive force).

Figure 2.12 It takes a chemical reaction to turn this light bulb on.

A battery is a sequence or collection of cells and there are two types of battery in general use: **primary** and **secondary** cells.

A primary cell can only be used once and has to be thrown way because it cannot be recharged. A secondary cell however can be recharged. There are two general types of secondary cell: **lead-acid** and **alkaline**.

Assessment criteria 2.11

State what is meant by the term 'voltage drop' in relation to electrical circuits

Volt drop

As explained previously, when dealing with the term resistivity earlier, the trouble with using excessive lengths of cable is that the overall resistance of the circuit will increase. Increased resistance means increased volt drop, which means that the load could be deprived of voltage.

To protect against this condition, the 18th Edition currently says that the maximum volt drop for lighting is 3% and for power it is 5%. This means that for a single-phase domestic supply:

$$\frac{3 \times 230}{100} = 6.9\,V \text{ (maximum dropped)}$$

$$\frac{5 \times 230}{100} = 11.5\,V \text{ (maximum dropped)}$$

If the volt drop was more than that specified above, then the cable size involved must be increased. Increasing the cross-sectional area of the cable, will decrease the resistance, which in turn will reduce the volt drop.

Assessment criteria 2.12

Identify how measurement instruments are connected into electrical circuits

Measuring volts and amps

The electrical contractor is required by the IET regulations to test all new installations and major extensions during erection and upon completion before being put into service. The contractor may also be called upon to test installations and equipment in order to identify and remove faults. These requirements imply the use of appropriate test instruments, and in order to take accurate readings, consideration should be given to the following points:

- Is the instrument suitable for this test?
- Has the correct scale been selected?
- Is the test instrument correctly connected to the circuit?

Many commercial instruments are capable of making more than one test or have a range of scales to choose from and are known as multimeters, which are designed to measure voltage, current or resistance. Before taking measurements, the appropriate volt, ampere or ohm scale should be selected.

To avoid damaging the instrument it is good practice first to switch to the highest value on a particular scale range. For example, if the 10 A scale is first selected and a reading of 2.5 A is displayed, we then know that a more appropriate scale would be the 3 A or 5 A range. This will give a more accurate reading which might be, say, 2.49 A.

When the multimeter is used as an ammeter to measure current it must be connected in series with the test circuit, as shown in Fig. 2.13(a).

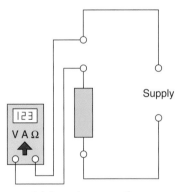

(a) Ammeter connection

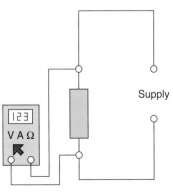

(b) Voltmeter connection

Figure 2.13 Using a multimeter (a) as an ammeter and (b) as a voltmeter.

This is because an ammeter has a very low input resistance since effectively it has to be connected in the circuit, therefore a low resistance will not affect or disturb the circuit. When used as a voltmeter, the multimeter must be connected in parallel with the component, as shown in Fig, 2.13(b). Again, there is a reason for this because a voltmeter has a very high input resistance, effectively being designed to measure across a component, which means that very little current will be disturbed. In fact, the higher the input resistance of a voltmeter the better, since it will make the meter more accurate. When using a meter as a voltmeter or ammeter there is an expectation regarding safety that the black lead is connected first, followed by the red positive terminal.

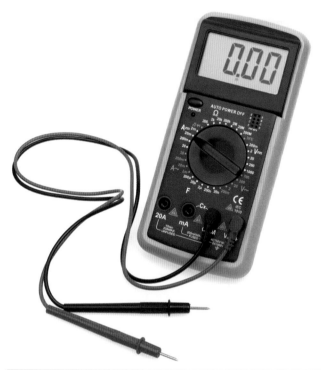

Figure 2.14 Devices such as this multimeter can be used to measure volts and amps.

In certain circumstances, connecting the red lead first could be dangerous since if the user, whilst still holding the black lead in their hand, touched a live part the circuit could be completed through the user. Disconnection would be in the reverse order, red lead disconnected first followed by the black lead.

Remember

A voltmeter measures potential difference which means that when you measure voltage you measure a difference between two points.

When using a commercial multi-range meter as an ohmmeter for testing electronic components, care must be exercised in identifying the positive terminal. The red terminal of the meter, identifying the positive input for testing voltage and current, usually becomes the negative terminal when the meter is used as an ohmmeter because of the way the internal battery is connected to the meter movement.

Check the meter manufacturer's handbook before using a multimeter to test electronic components.

Low resistance ohmmeter

Continuity of CPC or any other resistance reading is carried out with the main supply off, which is why it is known as dead testing. We do not need the supply to operate the meter because it has an internal battery which will send out a current.

- If the circuit is complete this current will return to the meter and the meter will give us a low reading (circuit complete).
- If the circuit is not complete this current cannot return to the meter and the meter will give us a very high value called infinity (circuit incomplete).

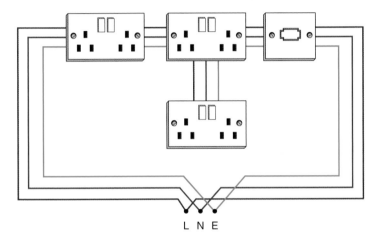

L N E

Figure 2.15 A low resistance ohmmeter uses an internal battery to push out a current. If the circuit is complete the current will return and the meter will indicate this through a low resistance reading.

We have to take away the resistance of the leads or use the *null* facility

Meter will give us a low value if the CPC circuit is complete

The power in a resistive load may be calculated from the readings of voltage and current since $P = VI$. This will give accurate calculations on both a.c. and d.c. supplies, but when measuring the power of an a.c. circuit which contains inductance or capacitance, a wattmeter must be used because the voltage and current will be out of phase. Fig. 2.17 clearly shows how an ammeter and voltmeter are connected in circuit.

A wattmeter is also closely linked to an energy meter (kWh meter), which measures the amount of energy used by domestic, commercial or industrial outlets. The unit for energy is the joule, but it is too small a value to realistically measure. Instead, it measures energy use over a given amount of time, and the unit selected is the kWh, which is seen on your energy bills by calculating how many kWh are normally used in a three-month period in relation to your tariff.

Figure 2.16 A voltmeter is connected in parallel which can be thought of as looking at the circuit from outside. Its resistance therefore needs to be high so that it does not disturb the circuit. An ammeter is actually connected in the circuit which means its resistance needs to be as low as possible so as not to disturb the circuit too much.

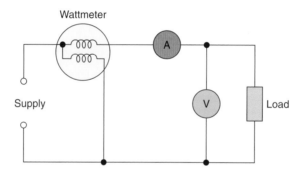

Figure 2.17 Wattmeter, ammeter and voltmeter correctly connected to a load.

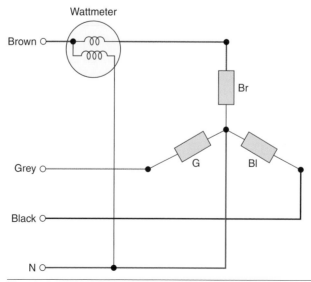

Figure 2.18 One-wattmeter measurement of power.

Tong tester

The tong tester, or clip-on ammeter, works on the same principle as the transformer. The laminated core of the transformer can be opened and passed over the busbar or single-core cable. In this way a measurement of the current being carried can be made without disconnection of the supply. The construction is shown in Fig. 2.19.

Figure 2.19 Tong tester or clip-on ammeter.

Assessment criteria 3.1

Identify how electricity is generated

Generation

Figure 2.20 shows a simple a.c. generator or alternator producing an a.c. waveform. We generate electricity in large, modern power-stations using the same basic principle of operation. However, in place of a single loop of wire, the power-station alternator has a three-phase winding and powerful electromagnets. The generated voltage is three identical sinusoidal waveforms each separated by 120°, as shown in Fig. 2.21. The prime mover is not, of course, a simple crank handle, but a steam turbine. Hot water is heated until it becomes superheated steam, which drives the vanes of a steam turbine which is connected to the alternator. The heat required to produce the steam may come from burning coal or oil or from a nuclear reactor. Whatever the primary source of energy, it is only being used to drive a turbine which is connected to an alternator, to generate electricity.

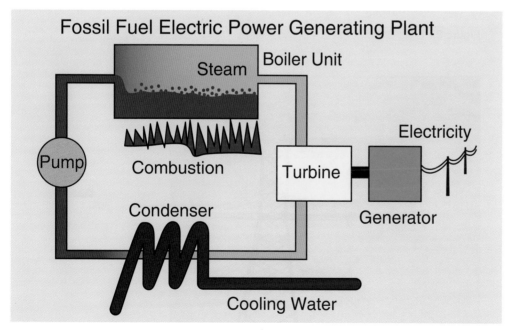

Figure 2.20 Fossil fuel electricity generating plant.

2

Assessment criteria 3.2

Identify the features of a generation, transmission and distribution system

Transmission

Electricity is generated in the power-station alternator at 25 kV. This electrical energy is fed into a transformer to be stepped up to a very high voltage for transmission on the National Grid network at 400 kV, 275 kV or 132 kV. These very high voltages are necessary because, for a given power, the current is greatly reduced, which means smaller grid conductors and the transmission losses are reduced.

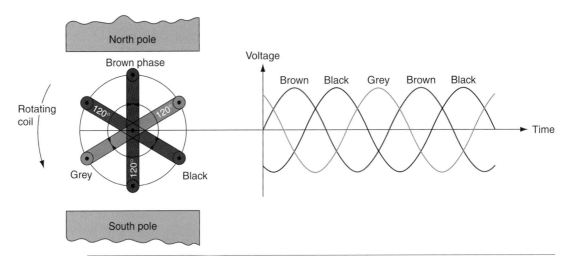

Figure 2.21 Generation of a three-phase voltage.

Figure 2.22 Transmission line steel pylon.

The National Grid network was established in 1934 and consists of over 5,000 miles of overhead aluminium conductors suspended from steel pylons which link together all the power-stations. Environmentalists say that these steel towers are ugly, but this method is about 10 times cheaper than the equivalent underground cable at these high voltages. Fig. 2.22 shows a transmission line steel pylon. Electricity is taken from the National Grid by appropriately located substations which eventually transform the voltage down to 11 kV at a local substation. At the local substation the neutral conductor is formed for single-phase domestic supplies and three-phase supplies to shops, offices and garages. These supplies are usually underground radial supplies from the local substation, but in rural areas we still see transformers and overhead lines suspended on wooden poles. Figures 2.23 and 2.24 give an overview of the system from power-station to consumer.

Assessment criteria 3.3

Identify the applications of supply systems

Distribution to the consumer

The electricity leaves the local substation and arrives at the consumer's mains intake position. The final connections are usually by simple underground radial feeders at 400 V/230 V. Underground cable distribution is preferred within a city, town or village because people find the overhead distribution, which we see in rural and remote areas, unsightly. Further, at these lower distribution voltages, the cost of underground cables is not prohibitive. The 400 V/230 V is derived from the 11 kV/400 V substation transformer by connecting the secondary winding in star, as shown in Fig. 2.23. The star point is earthed to an earth electrode sunk into the ground below the substation and from this point is taken the fourth conductor and the neutral. Loads connected between phases are fed at 400 V and those fed between one phase and neutral at 230 V. A three-phase 400 V supply is used for supplying small industrial and commercial loads such as garages, schools and blocks of flats. A single-phase 230 V supply is usually provided for individual domestic consumers.

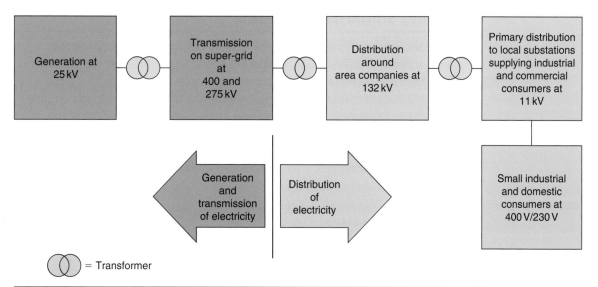

Figure 2.23 Generation, transmission and distribution of electrical energy.

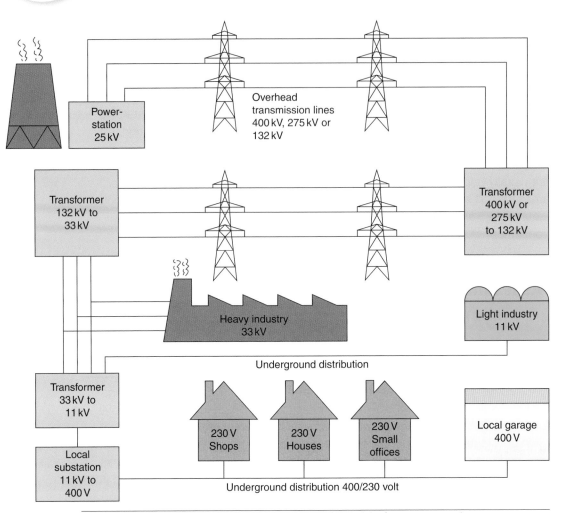

Figure 2.24 Simplified diagram of the distribution of electricity from power station to consumer.

At the mains intake position, the supplier will provide a sealed HBC fuse (BS 88–2: 2010) and a sealed energy meter to measure the consumer's electricity consumption. It is after this point that we reach the consumer's installation.

Balancing single-phase loads

A three-phase load such as a motor has equally balanced phases since the resistance of each phase winding will be the same. Therefore, the current taken by each phase will be equal. When connecting single-phase loads to a three-phase supply, care should be taken to distribute the single-phase loads equally across the three phases so that each phase carries approximately the same current. Equally distributing the single-phase loads across the three-phase supply is known as 'balancing' the load. A lighting load of 18 luminaires would be 'balanced' if six luminaires were connected to each of the three phases.

Assessment criteria 4.1

Outline the principles of electricity micro-generation technologies

Assessment criteria 4.2

Outline the principles of heat and co-generation micro-technologies

Environmental technology systems and renewable energy

Energy is vital to the modern industrial economy in the UK and Europe. We also need energy in almost every aspect of our lives, to heat and light our homes and offices, to enable us to travel on business or for pleasure, and to power our business and industrial machines.

In the past the UK has benefited from its fossil fuel resources of coal, oil and gas but respectable scientific sources indicate that the fossil fuel era is drawing to a close. Popular estimates suggest that gas and oil will reach peak production in the year 2060 with British coal reserves lasting only a little longer. Therefore we must look to different ways of generating electricity so that:

- the remaining fossil fuel is conserved;
- our CO_2 emissions are reduced to avoid the consequences of climate change;
- we ensure that our energy supplies are secure, and not dependent upon supplies from other countries.

Figure 2.25 We have been using renewable energy for a lot longer than you may think. It is thought that this windmill in Brill was originally erected in the 1680s, and there have been windmills there since at least the thirteenth century.

Following the introduction of the Climate Change Act in 2008 the UK and other Member States agreed an EU-wide target of 20% renewable energy by the year 2020 and 60% by 2050. Meeting these targets will mean basing much of the new energy infrastructure around renewable energy, particularly offshore wind power.

The 'energy hierarchy' states that organisations and individuals should address energy issues in the following order so as to achieve the agreed targets.

1 Reduce the need for energy: reducing energy demand is cost saving, reduces greenhouse gas emissions and contributes to the security of supply. Reducing the energy loss from buildings by better insulation and switching off equipment when not in use is one way of achieving this target.

2 Use energy more efficiently: use energy-efficient lamps and 'A'-rated equipment.

3 Use renewable energy: renewable energy refers to the energy that occurs naturally and repeatedly in the environment. This energy may come from wind, waves or water, the sun, or heat from the ground or air.

4 Any continuing use of fossil fuels should use clean and efficient technology. Power-stations generating electricity from coal and oil (fossil fuel) release a lot of CO_2 in the generating process. New-build power-stations must now be fitted with carbon capture filters to reduce the bad environmental effects.

Figure 2.26 The Hoover Dam generates electricity from a renewable source for millions of homes, and is also used to store much needed water for Las Vegas and the local area.

Funding for environmental technology systems

Renewable energy is no less reliable than energy generated from more traditional sources. Using renewable energy does not mean that you have to change your lifestyle or your domestic appliances. There has never been a better time to consider generating energy from renewable technology than now because grants and funding are available to help individuals and companies.

The Low Carbon Building Programme implemented by the Department of Energy and Climate Change (DECC) provides grants towards the installation

Key fact

Renewable energy is no less reliable than energy generated from more traditional sources.

of renewable technologies and is available to householders, public non-profitmaking organisations and commercial organisations across the UK.

The government's 'feed in tariff' pays a tax-free sum which is guaranteed for 25 years. It is called 'clean energy cashback' and has been introduced to promote the uptake of small-scale renewable and low carbon electricity generation technologies. The customer receives a generation tariff from the electricity supplier, whether or not any electricity generated is exported to the National Grid, and an additional export tariff when electricity is transported to the electricity grid through a smart meter.

From April 2010, clean energy generators were paid 41.3p for each kWh of electricity generated. Surplus energy fed back into the National Grid earned an extra 3p per unit. However, the scheme has been very popular and the feedback tariffs were soon reduced to 21p. From 1 August 2012 the fee is only 16p for each kWh. If you add to this the electricity bill savings, a normal householder could still make some savings. Savings vary according to energy use and type of system used. The Energy Saving Trust at www.energysavingtrust.org.uk (tel. 0300 123 1234) and www.britishgas.co.uk and Ofgem at www.ofgem.gov.uk/ provide an online calculator to determine the cost, size of system and CO_2 savings for PV systems.

Micro-generation technologies

Micro-generation is defined in The Energy Act 2004 Section 82 as the generation of heat energy up to 45 kW and electricity up to 50 kW.

Today, micro-generation systems generate relatively small amounts of energy at the site of a domestic or commercial building. However, it is estimated that by 2050, 30 to 40% of the UK's electricity demand could be met by installing micro-generation equipment to all types of buildings.

> **Key fact**
>
> The government's 'feed in tariff' pays a tax-free sum which is guaranteed for 25 years.

> **Key fact**
>
> Micro-generation technologies include small wind turbines, solar photovoltaic (PV) systems, small-scale hydro and micro CHP (combined heat and power) systems.

> **Key fact**
>
> Smart electricity meters are designed to be used in conjunction with micro-generators.

Figure 2.27 These solar panels harness the power of the sun to generate electricity, and can be seen on many houses nowadays.

In the USA, the EU and the UK buildings consume more than 70% of the nations' electricity and contribute almost 40% of the polluting CO_2 greenhouse gases. Any reductions which can be made to these figures will be good for the planet, and hence the great interest today in micro-generation systems. Micro-generation technologies include small wind turbines, solar photovoltaic (PV) systems, small-scale hydro and micro CHP (combined heat and power) systems. Micro-generators that produce electricity may be used as stand-alone systems, or may be run in parallel with the low-voltage distribution network, that is, the a.c. mains supply.

The April 2008 amendments to the Town and Country Planning Act 1990 now allow for the installation of micro-generation systems within the boundary of domestic premises without obtaining planning permission. However, size limitations have been set to reduce the impact upon neighbours. For example, solar panels attached to a building must not protrude more than 200 mm from the roof slope, and stand-alone panels must be no higher than 4 metres above ground level and no nearer than 5 metres from the property boundary. See the Electrical Safety Council site for advice on connecting micro-generation systems at www.esc.org.uk/bestpracticeguides.html.

Smart electricity meters

Smart electricity meters are designed to be used in conjunction with micro-generators. Electricity generated by the consumer's micro-generator can be sold back to the energy supplier using the 'smart' two-way meter.

The Department for Energy and Climate Change is planning to introduce smart meters into consumers' homes from 2012 and this is expected to run through until 2020 with the aim of helping consumers reduce their energy bills.

When introducing the proposal, Ed Miliband, the Energy and Climate Change Secretary, said 'the meters which most of us have in our homes were designed for a different age, before climate change. Now we need to get smarter with our energy. This is a big project affecting 26 million homes and several million businesses. The project will lead to extra work for electrical contractors through

Figure 2.28 An example of a smart electricity meter.

installing the meters on behalf of the utility companies and implementing more energy-efficient devices once customers can see how much energy they are using.'

Already available is the real time display (RTD) wireless monitor which enables consumers to see exactly how many units of electricity they are using through an easy-to-read, portable display unit. By seeing the immediate impact in pence per hour of replacing existing lamps with low-energy ones or switching off unnecessary devices throughout the home or office, consumers are naturally motivated to consider saving energy. RTD monitors use a clip-on sensor on the meter tails and include desktop software for PC and USB links. Let us now look at some of these micro-generation technologies.

Solar photovoltaic (PV) power supply systems

A **solar photovoltaic** (PV) system is a collection of PV cells known as a PV string, that forms a PV array and collectively is called a PV generator which turns sunlight directly into electricity. PV systems may be 'stand-alone' power supplies or be designed to operate in parallel with the public low-voltage distribution network, that is, the a.c. mains supply.

Stand-alone PV systems are typically a small PV panel of maybe 300 mm by 300 mm tilted to face the southern sky, where it receives the maximum amount of sunlight. They typically generate 12 to 15 volts and are used to charge battery supplies on boats, weather stations, road signs and any situation where electronic equipment is used in remote areas away from a reliable electrical supply.

The developing nations are beginning to see stand-alone PV systems as the way forward for electrification of rural areas beyond the National Grid rather than continuing with expensive diesel generators and polluting kerosene lamps.

The cost of PV generators is falling. The period 2009 to 2010 saw PV cells fall by 30% and, with new 'thin-film' cells being developed, the cost is expected to continue downwards. In the rural areas of developing nations they see PV systems linked to batteries bringing information technology, radio and television to community schools. This will give knowledge and information to the next generation which will help these countries to develop a better economy, a better way of life and to have a voice in the developed world.

Stand-alone systems are not connected to the electricity supply system and are therefore exempt from much of BS 7671, the IET Regulations. However, Regulation 134.1.1, 'good workmanship by skilled or instructed persons and proper materials shall be used in all electrical installations' will apply to any work done by an electrician who must also pay careful attention to the manufacturer's installation instructions.

Solar photovoltaic (PV) systems designed to operate in parallel with the public low-voltage distribution network are the type of micro-generator used on commercial and domestic buildings. The PV cells operate in exactly the same way as the stand-alone system described above, but will cover a much greater area. The PV cells are available in square panels which are clipped together and laid over the existing roof tiles as shown in Figure 2.29, or the PV cells may be manufactured to look just like the existing roof tiles which are integrated into the existing roof.

A solar PV system for a domestic three-bedroomed house will require approximately 15 to 20 square metres generating 2 to 3 kilowatts of power and this will save around 1,200 kg of CO_2 per year. On the positive side, a PV system for a three-bedroomed house will save around 1,200 kg of CO_2 per year.

These bigger micro-generation systems are designed to be connected to the power supply system and the installation must therefore comply with Section 712 of BS 7671: 2018. Section 712 contains the requirements for protective measures comprising automatic disconnection of the supply wiring systems, isolation, switching and control, earthing arrangements and labelling. In addition, the installation must meet the requirements of the Electricity Safety Quality and Continuity Regulations 2006. This is a mandatory requirement. However, where the output does not exceed 16 A per line, they are exempt from some of the requirements providing that:

- the equipment is type tested and approved by a recognized body;
- the installation complies with the requirements of BS 7671, the IET regulations;
- the PV equipment disconnects from the distributor's network in the event of a network fault;
- the distributor is advised of the installation before or at the time of commissioning.

Figure 2.29 PV system in a domestic situation.

Installations of less than 16 A per phase but up to 5 kilowatt peak (kWp) will also be required to meet the requirements of the Energy Network Association's Engineering Recommendation G83/1 for small-scale embedded generators in parallel with public low-voltage distribution networks. Installations generating more than 16 A must meet the requirements of G59/1 which requires approval from the distributor before any work commences.

Solar thermal hot water heating

Solar thermal hot water heating systems are recognized as a reliable way to use the energy of the sun to heat water. The technology is straightforward and solar thermal panels for a three-bedroomed house cost at the time of going to press between £3,000 and £6,000 for a 3 to 6 m² panel and they will save about 260 kg of CO_2 annually.

Definition

Solar thermal hot water heating systems are recognized as a reliable way to use the energy of the sun to heat water.

The solar panel comprises a series of tubes containing water that is pumped around the panel and a heat exchanger in the domestic water cylinder as shown in Figure 2.30. Solar energy heats up domestic hot water. A solar panel of about 4 m² will deliver about 1,000 kWh per year which is about half the annual water demand of a domestic dwelling. However, most of the heat energy is generated during the summer and so it is necessary to supplement the solar system with a boiler in the winter months. Figure 2.31 shows a photo of an installed solar hot water panel.

If you travel to Germany, you will see a lot of photovoltaic and solar thermal panels on the roofs there. In the UK, planning requirements for solar thermal and PV installations have already been made much easier. A website detailing planning requirements for solar and wind can be found at www.planningportal.gov.uk/uploads/hhghouseguide.html.

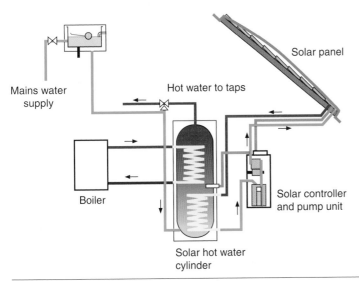

Figure 2.30 Solar-powered hot water system.

Figure 2.31 Solar hot water panel.

Wind energy generation

Modern large-scale wind machines are very different from the traditional windmill of the last century which gave no more power than a small car engine. Very large structures are needed to extract worthwhile amounts of energy from the wind. Modern large-scale wind generators are taller than electricity pylons, with a three-blade aeroplane-type propeller to catch the wind and turn the generator. If a wind turbine was laid down on the ground, it would be longer and wider than a football pitch. They are usually sited together in groups in what has become known as 'wind energy farms', as shown in Fig. 2.32.

Each modern grid-connected wind turbine generates about 600 kW of electricity. A wind energy farm of 20 generators will therefore generate 12 MW, a useful contribution to the National Grid, using a naturally occurring, renewable, non-polluting source of energy. The Department of Energy and Climate Change considers wind energy to be the most promising of the renewable energy sources.

Figure 2.32 Wind farm energy generators.

At time of print there are currently 5,085 turbines onshore in the UK producing 8,248 MW. There are 1,452 turbines offshore producing 5054 MW, a total of 13,302 MW. The Countryside Commission, the government's adviser on land use, has calculated that to achieve a target of generating 10% of the total electricity supply by wind power will require 40,000 generators of the present size. In 2017 5% of the UK's electricity was produced by wind.

Wind power is an endless renewable source of energy, is safer than nuclear power and provides none of the polluting emissions associated with fossil fuel. If there was such a thing as a morally pure form of energy, then wind energy would be it. However, wind farms are, by necessity, sited in some of the most beautiful landscapes in the UK. Building wind energy farms in these areas of outstanding natural beauty has outraged conservationists. Prince Charles has reluctantly joined the debate, saying that he was in favour of renewable energy sources but he believed that 'wind farms are a horrendous blot on the landscape'. He believes that if they are to be built at all they should be constructed well out to sea.

The next generation of wind farms will mostly be built offshore, where there is more space and more wind, but the proposed size of these turbines creates considerable engineering problems. From the sea bed foundations to the top of the turbine blade will measure up to a staggering 250 metres, three times the height of the Statue of Liberty. Each offshore turbine, generating between 5 and 7 MW, will weigh between 200 and 300 tonnes. When large wind forces are put onto that structure a very big cantilever effect will be created which creates a major engineering challenge.

The world's largest offshore wind energy farm built so far was opened in September 2010. The 100-turbine 'Thanet' wind farm just off the Kent coast will generate enough power to supply 200,000 homes. The Thanet project cost £780 million to build. The turbines are up to 380 feet high and cover an area as large as 4,000 football pitches. The Thanet project will not retain its title as the world's largest wind farm for long because in addition we now have the 'Greater Gabbard' wind farm, off the north-east coast with 140 turbines which opened in 2015. These projects bring Britain's total wind energy capacity above 5,000 megawatts for the first time and more are being built. The UK Government announced in March 2019 that they will double the present capacity of offshore wind farms around the UK, so that we can meet a new target of generating 30% of the total energy required from renewable generators by 2030.

The Department of Energy and Climate Change has calculated that 10,000 wind turbines could provide the energy equivalent of 8 million tonnes of coal per year and reduce CO_2 emissions. While this is a worthwhile saving of fossil fuel, opponents point out the obvious disadvantages of wind machines, among them the need to maintain the energy supply during periods of calm means that wind machines can only ever supplement a more conventional electricity supply. Small wind micro-generators can be used to make a useful contribution to a domestic property or a commercial building. They can stand alone and are about the size of a tall street lamp. A 12 m-high turbine costs about £24,000 and, with a good wind, can generate 10,000 kWh per year, enough for three small domestic homes. However, if you live in a village, town or city you are unlikely to obtain the local authority building and planning permissions to install a wind generator because your neighbours will object.

Small wind generators of the type shown in Fig. 2.33 typically generate between 1.5 A and 15 A in wind speeds of 10 to 40 mph.

Heat pumps

In applications where heat must be upgraded to a higher temperature so that it can be usefully employed, a **heat pump** must be used. Energy from a low temperature source such as the earth, the air, a lake or river is absorbed by a gas or liquid contained in pipes, which is then mechanically compressed by an electric pump to produce a temperature increase. The high-temperature energy is then transferred to a heat exchanger so that it might do useful work, such as providing heat to a building. For every 1 kWh of electricity used to power the heat pump compressor, approximately 3 to 4 kWh of heating is produced.

Definition

Wind power is an endless renewable source of energy, is safer than nuclear power and provides none of the polluting emissions associated with fossil fuel.

Key fact

The Department of Energy and Climate Change has calculated that 10,000 wind turbines could provide the energy equivalent of 8 million tonnes of coal per year and reduce CO_2 emissions.

Definition

In applications where heat must be upgraded to a higher temperature so that it can be usefully employed, a *heat pump* must be used.

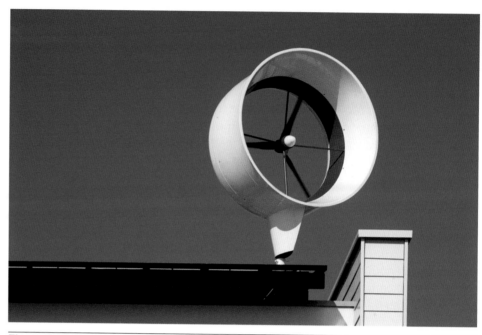

Figure 2.33 Small wind generator on a domestic property.

How a heat pump works

1 A large quantity of low-grade energy is absorbed from the environment and transferred to the refrigerant inside the heat pump (called the evaporator). This causes the refrigerant temperature to rise, causing it to change from a liquid to a gas.

2 The refrigerant is then compressed, using an electrically driven compressor, reducing its volume but causing its temperature to rise significantly.

3 A heat exchanger (condenser) extracts the heat from the refrigerant to heat the domestic hot water or heating system.

4 After giving up its heat energy, the refrigerant turns back into a liquid, and, after passing through an expansion valve, is once more ready to absorb energy from the environment and the cycle is repeated, as shown in Fig. 2.34.

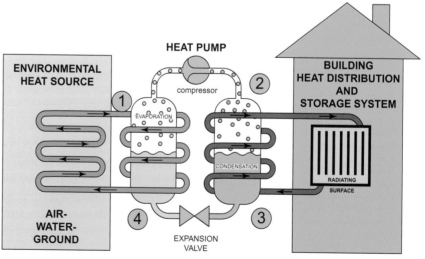

Figure 2.34 Heat pump working principle.

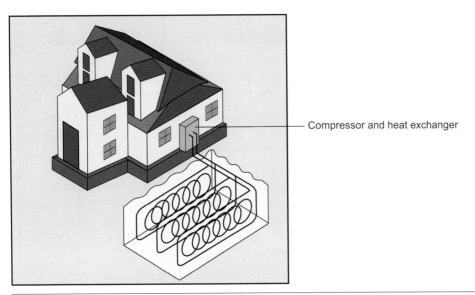

Figure 2.35 Ground source heating system.

A refrigerator works on this principle. Heat is taken out of the food cabinet, compressed and passed on to the heat exchanger or radiator at the back of the fridge. This warm air then radiates by air convection currents into the room.

Thus the heat from inside the cabinet is moved into the room, leaving the sealed refrigerator cabinet cold.

Heat pumps (ground source)

Ground source heat pumps extract heat from the ground by circulating a fluid through polythene pipes buried in the ground in trenches or in vertical boreholes, as shown in Fig. 2.35. The fluid in the pipes extracts heat from the ground and a heat exchanger within the pump extracts heat from the fluid. These systems are most effectively used to provide underfloor radiant heating or water heating. Calculations show that the length of pipe buried at a depth of 1.5 m required to produce 1.2 kW of heat will vary between 150 m in dry soil and 50 m in wet soil.

The average heat output can be taken as 28 watts per metre of pipe. A rule of thumb guideline is that the surface area required for the ground heat exchanger should be about 2.5 times the area of the building to be heated. This type of installation is only suitable for a new-build project and the ground heat exchanger will require considerable excavation and installation. The installer must seek local authority building control permissions before work commences.

Heat pumps (air source)

The performance and economics of heat pumps are largely determined by the temperature of the heat source and so we seek to use a high-temperature source. The heat sources used by heat pumps may be soil, the air, ground or surface water. Unfortunately all these sources follow the external temperature, being lower in winter when demand is highest. Normal atmosphere is an ideal heat source in that it can supply an almost unlimited amount of heat although unfortunately at varying temperatures, but relatively mild winter temperatures in the UK mean excellent levels of efficiency and performance throughout the year. For every 1 kWh of electricity used to power the heat pump compressor, between 3 and 4 kWh of heating energy is produced. They also have an advantage over ground source heat pumps with lower installation costs because

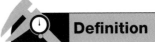

Definition

Ground source heat pumps extract heat from the ground by circulating a fluid through polythene pipes buried in the ground in trenches or in vertical boreholes.

Figure 2.36 Air source heat pump unit.

they do not require any groundwork. Figure 2.36 shows a commercial air source heat pump.

If the air heat pump is designed to provide full heating with an outside temperature of 2 to 4 °C, then the heat pump will provide approximately 80% of the total heating requirement with high performance and efficiency.

The point at which the output of a given heat pump meets the building heat demand is known as the 'balance point'. In the example described above, the 20% shortfall of heating capacity below the balance point must be provided by some supplementary heat. However, an air-to-air heat pump can also be operated in the reverse cycle which then acts as a cooling device, discharging cold air into the building during the summer months. So, here we have a system which could be used for air conditioning in a commercial building.

Hydroelectric power generation

The UK is a small island surrounded by water. Surely we could harness some of the energy contained in tides, waves, lakes and rivers? Many different schemes have been considered in the past 20 years and a dozen or more experimental schemes are being tested now.

Water power makes a useful contribution to the energy needs of Scotland, but the possibility of building similar hydroelectric schemes in England is unlikely chiefly due to the topographical nature of the country.

The Severn Estuary has a tidal range of 15 m, the largest in Europe, and is a reasonable shape for building a dam across the estuary. This would allow the basin to fill with water as the tide rose, and then allow the impounded water to flow out through electricity-generating turbines as the tide falls. However, such a tidal barrier might have disastrous ecological consequences for the many wildfowl and wading bird species by submerging the mudflats which now provide winter shelter for these birds. Therefore, the value of the power which might be produced must be balanced against the possible ecological consequences.

Figure 2.37 Lots of research currently focuses upon harnessing the tides to generate electricity – because they can generally be predicted they should in theory be able to provide a more reliable source of electricity than some other forms of micro-generation.

France has successfully operated a 240 MW tidal power-station at Rance in Brittany for the past 25 years.

Marine Current Turbines Ltd are carrying out research and development on submerged turbines which will rotate by exploiting the principle of flowing water in general and tidal streams in particular. The general principle is that an 11 m diameter water turbine is lowered into the sea down a steel column drilled in the seabed. The tidal movement of the water then rotates the turbine and generates electricity.

The prototype machines were submerged in the sea off Lynmouth in Devon. In May 2008, the world's first tidal turbine was installed in the Strangford Narrows in Northern Ireland, where it is now grid-connected and generating 1.2 MW. All the above technologies are geared to providing hydroelectric power connected to the national grid, but other micro-hydro schemes are at the planning and development stage.

Micro-hydro generation

The use of small hydropower (SHP) or micro-hydropower has grown over recent decades led by continuous technical developments, partly brought about in the UK by the 2010 coalition government's 'feed in tariff' where green electricity producers are paid a premium to produce electricity from renewable sources. The normal perception of hydropower is of huge dams, but there is a much bigger use of hydropower in smaller installations. Asia, and especially China, is set to become a leader in hydroelectric generation. Australia and New Zealand are focusing on small hydro plants. Canada, a country with a long tradition of hydropower, is developing small hydropower as a replacement for expensive diesel generation in remote off-grid communities.

Small hydropower schemes generate electricity by converting the power available in rivers, canals and streams. The object of a hydropower scheme is to convert the potential energy of a mass of water flowing in a stream with a certain fall, called the head, into electrical energy at the lower end of the stream where the powerhouse is located. The power generated is proportional to the flow, called the discharge, and to the head of water available. The fundamental asset of hydropower is that it is a clean and renewable energy source and the fuel used, water, is not consumed in the electricity-generating process.

In the Derbyshire Peak District along the fast-flowing River Goyt, there were once 16 textile mills driven by water wheels. The last textile mill closed in 2000 but the Old Torr Mill has been saved. Where once the water wheel stood, there is now a gigantic 12-tonne steel screw, 2.4 metres in diameter. The water now drives the Reverse Archimedian Screw, affectionately called 'Archie', to produce 130,000 kWh per year, enough electricity for 40 homes. The electricity-generating project is owned by the residents of New Mills in a sharing cooperative in which surplus electricity is sold back to the grid. Archie will produce 250 MWh of electricity each year and the installation cost was £300,000 in 2008. See Figure 2.38 and the Torrs Hyrdro New Mills website and Mann Power for more information and interesting video of fish swimming through the turbine (www.stockport-hydro.co.uk/index.php?page=history).

The type of turbine chosen for any hydro scheme will depend upon the discharge rate of the water and the head of water available. A Pelton wheel is a water turbine in which specially-shaped buckets attached to the periphery of the wheel are struck by a jet of water. The kinetic energy of the water turns the wheel which is coupled to the generator.

Key fact

Green electricity producers are paid a premium to produce electricity from renewable sources.

Key fact

The normal perception of hydropower is of huge dams, but there is a much bigger use of hydropower in smaller installations.

Definition

Small hydropower schemes generate electricity by converting the power available in rivers, canals and streams.

Figure 2.38 An example of an Archimedian Screw at the River Dart country park, Devon, similar to Archie. Credit: WRE Limited.

Key fact

The type of turbine chosen for any hydro scheme will depend upon the discharge rate of the water and the head of water available.

Definition

CHP is the simultaneous generation of usable heat and power in a single process. That is, heat is produced as a by-product of the power-generation process.

Definition

Biomass is derived from plant materials and animal waste. It can be used to generate heat and to produce fuel for transportation.

Definition

'Biomass renewables' means energy from crops.

Axial turbines comprise a large cylinder in which a propeller-type water turbine is fixed at its centre. The water moving through the cylinder causes the turbine blade to rotate and generate electricity.

A Francis turbine or Kaplan turbine is also an axial turbine but the pitch of the blades can be varied according to the load and discharge rate of the water.

Small water turbines will reach a mechanical efficiency at the coupling of 90%.

Up and down the country, riverside communities must be looking at the relics of our industrial past and wondering if they might provide a modern solution for clean, green, electrical energy. **However, despite the many successes and obvious potential, there are many barriers to using waterways for electricity generation in European countries**. It is very difficult in this country to obtain permission from the Waterways Commission to extract water from rivers, even though, once the water has passed through the turbine, it is put back into the river. Environmental pressure groups are opposed to micro-hydro generation because of its perceived local environmental impact on the river ecosystem and the disturbance to fishing. Therefore, once again, the value of the power produced would have to be balanced against the possible consequences.

Combined heat and power (CHP)

CHP is the simultaneous generation of usable heat and power in a single process. That is, heat is produced as a by-product of the power-generation process. A chemical manufacturing company close to where I live has a small power-station which meets some of their electricity requirements using the smart meter principle. Their 100 MW turbine is driven by high-pressure steam. When the steam condenses after giving up its energy to the turbine, a lot of very hot water remains which is then piped around the offices and some production plant buildings for space heating. Combining heat and power in this way can increase the overall efficiency of the fuel used because it is performing two operations. CHP can also use the heat from incinerating refuse to heat a nearby hospital, school or block of flats.

Biomass heating

Biomass is derived from plant materials and animal waste. It can be used to generate heat and to produce fuel for transportation. The biomass material may be straw and crop residues, crops grown especially for energy production, or rape seed oil and waste from a range of sources including food production. The nature of the fuel will determine the way that energy can best be recovered from it.

There is a great deal of scientific research being carried out at the moment into **'biomass renewables'**; that is, energy from crops. This area of research is at an early stage, but is expected to flourish in the next decade. The first renewable energy plant, which is to be located at Teesport on the River Tees in the northeast of England, has received approval from the Department for Energy and Climate Change for building to commence.

The facility will be one of the largest biomass plants to be built in the world and is scheduled to enter commercial operation in late 2020. Young trees will be grown as a crop to produce wood chips. The plant will use 2.5 million tonnes of wood chips each year to produce 300 MW of electrical energy. The plant will operate

24 hours a day, all year round, to meet some of the National Grid base load. The USA have over 200 biomass plants operating, which meets approximately 4% of the total USA energy consumption.

Water conservation

Conservation is the preservation of something important, especially of the natural environment. Available stored water is a scarce resource in England and Wales where there are only 1,400 cubic meters per person per year; very little compared with France, which has 3,100 cubic meters per person per year, Italy, which has 2,900 and Spain, 2,800. About half of the water used in an average home is used to shower, bathe and wash the laundry, and another third is used to flush the toilet. At a time when most domestic and commercial properties have water meters installed, it saves money to harvest and reuse water.

The City and Guilds has asked us to look at two methods of water conservation: rainwater harvesting and grey water recycling.

Definition

Conservation is the preservation of something important, especially of the natural environment.

Rainwater harvesting

Rainwater harvesting is the collection and storage of rainwater for future use. Rainwater has in the past been used for drinking, livestock and irrigation. It is now also being used to provide water for car cleaning and garden irrigation in domestic and commercial buildings.

Many gardeners already harvest rainwater for garden use by collecting run-off from a roof or greenhouse and storing it in a water butt or water barrel. However, a 200-litre water butt does not give much drought protection in spite of garden plants preferring rainwater to fluoridated tap water. To make a useful contribution, the rainwater storage tank should be between 2,000- and 7,000-litre capacity. The rainwater-collecting surfaces will be the roof of the building and any hard paved surfaces such as patios. Downpipes and drainage pipes then route the water to the storage tank situated, perhaps, under the lawn. An electric pump lifts the water from the storage tank to the point of use, possibly a dedicated outdoor tap. The water is then distributed through a hosepipe or sprinkler system to the garden in the normal way.

With a little extra investment, rainwater can be filtered and used inside the house to supply washing machines and WCs. Installing domestic pipes and interior plumbing can be added to existing homes although it is more straightforward in a new-build home.

With the move towards more sustainable homes, UK architects are becoming more likely to specify rainwater harvesting in their design to support alternatives to a mains water supply. In Germany, rainwater harvesting systems are now installed as standard in all new commercial buildings.

Definition

Rainwater harvesting is the collection and storage of rainwater for future use.

Wave energy

Waves are generally created from the strength of surface winds, which can create huge swells and surges. These swells can therefore generate the mechanical force required to create electricity by driving certain turbines strategically placed to tap into this natural resource. Given its abundance, wave energy is both renewable and environmentally friendly since it is both green and clean. It does have certain disadvantages since the system is wave dependent, and even when certain tidal patterns are known, in many locations, wave behaviour is erratic.

Figure 2.39 A PB40 PowerBuoy, built by Ocean Power Technologies, floats three-quarters of a mile off Marine Corps Base, Hawaii, drawing electrical power from the ocean's waves.

Assessment criteria 4.3

State the benefits of micro-generation

Micro-generation and co-generation technologies represent an investment not only in our future but also our children's future by safeguarding the continued reduction in carbon dioxide emissions, which in turn will combat climate change by lessening such destructive phenomena as:

• acid rain;
• depletion of the ozone layer;
• rising sea levels;
• rising surface temperatures.

Micro-generation uses clean and renewable resources that can also be adapted and implemented in rural locations which are not necessarily connected to the National Grid. Any domestic or industrial user generating their own supply of electricity would not be affected by raising energy prices. Government grants are also possible through initiatives such as the Green Deal, whereby things like purchasing a more efficient boiler can be subsidized. For the electrical installation industry there are also job opportunities, especially in the installation of solar PV, and despite the tariffs having been reduced by the government, the cost of a PV package is now far lower than it was in the past, which makes it a far more affordable and cost-effective option.

Assessment criteria 5.1

State the relationship between current, voltage and resistance in parallel and series d.c. circuits

Assessment criteria 5.2

Determine electrical quantities in series d.c. circuits

Resistors in series and in parallel

In an electrical circuit resistors may be connected in series, in parallel, or in various combinations of series and parallel connections.

Series-connected resistors

In any series circuit a current I will flow through all parts of the circuit as a result of the potential difference supplied by a battery V_T. Therefore, we say that in a series circuit the current is common throughout that circuit.

When the current flows through each resistor in the circuit, for example, R_1, R_2 and R_3 in Fig. 2.40, there will be a voltage drop across that resistor whose value will be determined by the values of I and R, since from Ohm's law $V = I \times R$. The sum of the individual voltage drops, for example, V_1, V_2 and V_3 in Fig. 2.40, will be equal to the total voltage V_T.

We can summarize these statements as follows. For any series circuit, I is common throughout the circuit and

$$V_T = V_1 + V_2 + V_3 \qquad \text{(Equation 1)}$$

Let us call the total circuit resistance R_T. From Ohm's law we know that $V = I \times R$ and therefore

Total voltage $V_T = I \times R_T$
Voltage drop across R_1 is $V_1 = I \times R_1$
Voltage drop across R_2 is $V_2 = I \times R_2$ $\qquad$ (Equation 2)
Voltage drop across R_3 is $V_3 = I \times R_3$

We are looking for an expression for the total resistance in any series circuit and, if we substitute equation (2) into equation (1), we have:

$$V_T = V_1 + V_2 + V_3$$
$$\therefore I \times R_T = I \times R_1 + I \times R_2 + I \times R_3$$

Now, since I is common to all terms in the equation, we can divide both sides of the equation by I. This will cancel out I to leave us with an expression for the circuit resistance:

$$R_T = R_1 + R_2 + R_3$$

Note that the derivation of this formula is given for information only. Craft students need only state the expression $R_T \times R_1 \times R_2 \times R_3$ for series connections.

Top tip

Series resistance
The total value of resistance in a series circuit is always greater than the largest individual value.

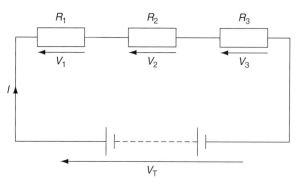

Figure 2.40 A series circuit.

Note that the derivation of this formula is given for information only. Craft students need only state the expression $R_T = R_1 + R_2 + R_3$ for series connections.

Assessment criteria 5.3

Determine electrical quantities in parallel d.c. circuits

Parallel-connected resistors

In any parallel circuit, as shown in Fig. 2.41, the same voltage acts across all branches of the circuit. The total current will divide when it reaches a resistor junction, part of it flowing in each resistor. The sum of the individual currents, for example, I_1, I_2 and I_3 in Fig. 2.41, will be equal to the total current I_T.

We can summarize these statements as follows. For any parallel circuit, V is common to all branches of the circuit and

$$I_T = I_1 + I_2 + I_3 \qquad \text{(Equation 3)}$$

Let us call the total resistance R_T.

From Ohm's law we know, that $I = \dfrac{V}{R}$, and therefore

the total current $I_T = \dfrac{V}{R_T}$

the current through R_1 is $I_1 = \dfrac{V}{R_1}$

the current through R_2 is $I_2 = \dfrac{V}{R_2}$ \qquad (Equation 4)

the current through R_3 is $I_3 = \dfrac{V}{R_3}$

We are looking for an expression for the equivalent resistance R_T in any *parallel* circuit and, if we substitute equations (4) into equation (3), we have:

$$I_T = I_1 + I_2 + I_3$$

$$\therefore \frac{V}{R_T} = \frac{V}{R_1} + \frac{V}{R_2} + \frac{V}{R_3}$$

Key fact

Resistors
In a series circuit, total resistance $R_T \times R_1 \times R_2 \times R_3$ ohms.

Top tip

Parallel resistance
The total value of resistance in a parallel circuit is always less than the smallest individual value.

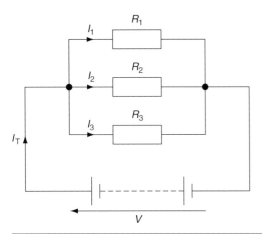

Figure 2.41 A parallel circuit.

Now, since V is common to all terms in the equation, we can divide both sides by V, leaving us with an expression for the circuit resistance:

$$\frac{1}{R_T} = \frac{1}{R_1} + \frac{1}{R_2} + \frac{1}{R_3}$$

Note that the derivation of this formula is given for information only. Craft students need only state the expression $1/R_T \times 1/R_1 \times 1/R_2 \times 1/R_3$ for parallel connections.

Example 1

Three $6\,\Omega$ resistors are connected (a) in series (see Fig. 2.42), and (b) in parallel (see Fig. 2.43), across a 12 V battery. For each method of connection, find the total resistance and the values of all currents and voltages.

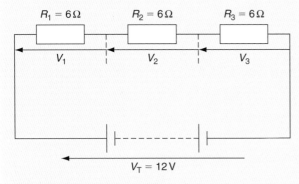

Figure 2.42 Resistors in series.

For any series connection

$$R_T = R_1 + R_2 + R_3 \text{ (see Figure 2.42)}$$
$$\therefore R_T = 6\,\Omega + 6\,\Omega + 6\,\Omega = 18\,\Omega$$

Total current $I_T = \dfrac{V_T}{R_T}$

$$\therefore I_T = \frac{12\,V}{18\,\Omega} = 0.67\,A$$

(continued)

Example 1 continued

The voltage drop across R_1 is

$$V_1 = I_T \times R_1$$
$$\therefore V_1 = 0.67 \text{ A} \times 6 \text{ } \Omega = 4 \text{ V}$$

The voltage drop across R_2 is

$$V_2 = I_T \times R_2$$
$$\therefore V_2 = 0.67 \text{ A} \times 6 \text{ } \Omega = 4 \text{ V}$$

The voltage drop across R_3 is

$$V_3 = I_T \times R_3$$
$$\therefore V_3 = 0.67 \text{ A} \times 6 \text{ } \Omega = 4 \text{ V}$$

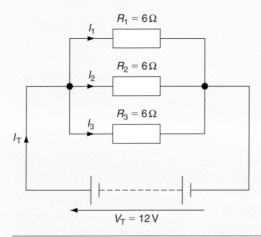

Figure 2.43 Resistors in parallel.

For any parallel connection

$$\frac{1}{R_T} = \frac{1}{R_1} + \frac{1}{R_2} + \frac{1}{R_3} \text{ (see Figure 2.43)}$$
$$\therefore \frac{1}{R_T} = \frac{1}{6 \text{ } \Omega} + \frac{1}{6 \text{ } \Omega} + \frac{1}{6 \text{ } \Omega}$$
$$\frac{1}{R_T} = \frac{1 + 1 + 1}{6 \text{ } \Omega} = \frac{3}{6 \text{ } \Omega}$$
$$R_T = \frac{6 \text{ } \Omega}{3} = 2 \text{ } \Omega$$

Total current $I_T = \dfrac{V_T}{R_T}$

$$\therefore I_T = \frac{12 \text{ V}}{2 \text{ } \Omega} = 6 \text{ A}$$

(continued)

Example 1 continued

The current flowing through R_1 is

$$I_1 = \frac{V_T}{R_1}$$

$$\therefore I_1 = \frac{12 \text{ V}}{6 \text{ } \Omega} = 2 \text{ A}$$

The current flowing through R_2 is

$$I_2 = \frac{V_T}{R_2}$$

$$\therefore I_2 = \frac{12 \text{ V}}{6 \text{ } \Omega} = 2 \text{ A}$$

The current flowing through R_3 is

$$I_3 = \frac{V_T}{R_3}$$

$$\therefore I_3 = \frac{12 \text{ V}}{6 \text{ } \Omega} = 2 \text{ A}$$

Series and parallel combinations

The most complex arrangement of series and parallel resistors can be simplified into a single equivalent resistor by combining the separate rules for series and parallel resistors.

Example 2

Resolve the circuit shown in Fig. 2.44 into a single resistor and calculate the potential difference across each resistor.

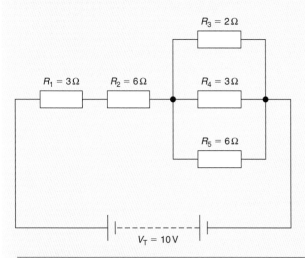

Figure 2.44 A series/parallel circuit.

(*continued*)

Example 2 continued

By inspection, the circuit contains a parallel group consisting of R_3, R_4 and R_5 and a series group consisting of R_1 and R_2 in series with the equivalent resistor for the parallel branch.

Consider the parallel group. We will label this group R_P. Then

$$= \frac{1}{R_3} + \frac{1}{R_4} + \frac{1}{R_5} \text{ (see Figure 2.44)}$$

$$= \frac{1}{2\,\Omega} + \frac{1}{3\,\Omega} + \frac{1}{6\,\Omega}$$

$$= \frac{3 + 2 + 1}{6\,\Omega} = \frac{6}{6\,\Omega}$$

$$= \frac{6\,\Omega}{6} = 1\,\Omega$$

Figure 2.44 may now be represented by the more simple equivalent shown in Fig. 2.45.

Since all resistors are now in series,

$$R_T = R_1 + R_2 + R_P$$
$$\therefore R_T = 3\,\Omega + 6\,\Omega + 1\,\Omega = 10\,\Omega$$

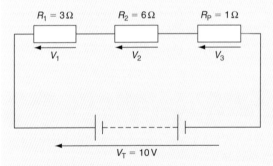

Figure 2.45 Equivalent series circuit.

Thus, the circuit may be represented by a single equivalent resistor of value $10\,\Omega$ as shown in Fig. 2.46. The total current flowing in the circuit may be found by using Ohm's law:

$$I_T = \frac{V_T}{R_T} = \frac{10\,\text{V}}{10\,\Omega} = 1\,\text{A}$$

The potential differences across the individual resistors are

$$V_1 = I_T \times R_1 = 1\,\text{A} \times 3\,\Omega = 3\,\text{V}$$
$$V_2 = I_T \times R_2 = 1\,\text{A} \times 6\,\Omega = 6\,\text{V}$$
$$V_P = I_T \times R_P = 1\,\text{A} \times 1\,\Omega = 1\,\text{V}$$

(continued)

Example 2 continued

Since the same voltage acts across all branches of a parallel circuit the same p.d. of 1 V will exist across each resistor in the parallel branch R_3, R_4 and R_5.

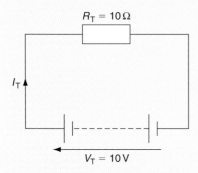

$R_T = 10\,\Omega$

I_T

$V_T = 10\,V$

Figure 2.46 Single equivalent resistor for Fig. 2.44.

Example 3

Determine the total resistance and the current flowing through each resistor for the circuit shown in Fig. 2.47.

By inspection, it can be seen that R_1 and R_2 are connected in series while R_3 is connected in parallel across R_1 and R_2. The circuit may be more easily understood if we redraw it as in Fig. 2.48.

For the series branch, the equivalent resistor can be found from

$$R_S = R_1 + R_2$$
$$\therefore R_S = 3\,\Omega + 3\,\Omega = 6\,\Omega$$

Figure 2.48 may now be represented by a more simple equivalent circuit, as in Fig. 2.49.

Since the resistors are now in parallel, the equivalent resistance may be found from

$$\frac{1}{R_T} = \frac{1}{R_S} + \frac{1}{R_3}$$

$$\therefore \frac{1}{R_T} = \frac{1}{6\,\Omega} + \frac{1}{6\,\Omega}$$

$$\frac{1}{R_T} = \frac{1+1}{6\,\Omega} = \frac{2}{6\,\Omega}$$

$$R_T = \frac{6\,\Omega}{2} = 3\,\Omega$$

The total current is

$$I_T = \frac{V_T}{R_T} = \frac{12\,V}{3\,\Omega} = 4\,A$$

(continued)

Example 3 continued

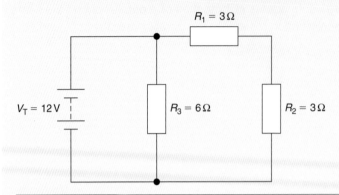

Figure 2.47 A series/parallel circuit for Example 3.

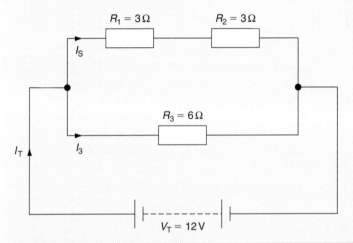

Figure 2.48 Equivalent circuit for Example 3.

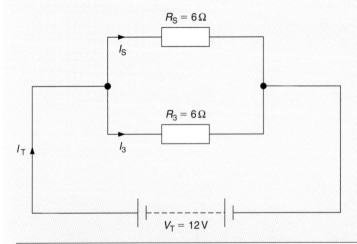

Figure 2.49 Simplified equivalent circuit for Example 3.

(continued)

Example 3 continued

Let us call the current flowing through resistor $R_3 I_3$

$$\therefore I_3 = \frac{V_T}{R_3} = \frac{12\,V}{6\,\Omega} = 2\,A$$

Let us call the current flowing through both resistors R_1 and R_2, as shown in Fig. 2.21, I_S

$$\therefore I_S = \frac{V_T}{R_S} = \frac{12\,V}{6\,\Omega} = 2\,A$$

Assessment criteria 6.1

Identify magnetic flux patterns of circuits

Magnetism

The Greeks knew as early as 600 BC that a certain form of iron ore, now known as magnetite or lodestone, had the property of attracting small pieces of iron. Later, during the Middle Ages, navigational compasses were made using the magnetic properties of lodestone. Small pieces of lodestone attached to wooden splints floating in a bowl of water always came to rest pointing in a north–south direction. The word lodestone is derived from an old English word meaning 'the way', and the word magnetism is derived from Magnesia, the place where magnetic ore was first discovered.

Iron, nickel and cobalt are the only elements which are attracted strongly by a magnet. These materials are said to be *ferromagnetic*. Copper, brass, wood, PVC and glass are not attracted by a magnet and are, therefore, described as *non-magnetic*.

Some basic rules of magnetism

1. Lines of magnetic flux have no physical existence, but they were introduced by Michael Faraday (1791–1867) as a way of explaining the magnetic energy existing in space or in a material. They help us to visualize and explain the magnetic effects. The symbol used for magnetic flux is the Greek letter Φ (phi) and the unit of magnetic flux is the weber (symbol Wb), pronounced 'veber', to commemorate the work of the German physicist Wilhelm Weber (1804–1891).
2. Lines of magnetic flux always form closed loops.
3. Lines of magnetic flux behave like stretched elastic bands, always trying to shorten themselves.
4. Lines of magnetic flux never cross over each other.
5. Lines of magnetic flux travel along a magnetic material and always emerge out of the 'north pole' end of the magnet.
6. Lines of magnetic flux pass through space and non-magnetic materials undisturbed.

Definition

The region of space through which the influence of a magnet can be detected is called the *magnetic field* of that magnet.

Definition

The places on a magnetic material where the lines of flux are concentrated are called the *magnetic poles*.

Definition

Like poles repel; unlike poles attract. These two statements are sometimes called the *first laws of magnetism* and are shown in Fig. 2.51.

7 The region of space through which the influence of a magnet can be detected is called the **magnetic field** of that magnet.

8 The number of lines of magnetic flux within a magnetic field is a measure of the flux density. Strong magnetic fields have a high-flux density. The symbol used for flux density is B, and the unit of flux density is the tesla (symbol T), to commemorate the work of the Croatian-born American physicist Nikola Tesla (1856–1943).

9 The places on a magnetic material where the lines of flux are concentrated are called the **magnetic poles**.

10 Like poles repel; unlike poles attract. These two statements are sometimes called the **'first laws of magnetism'** and are shown in Fig. 2.51.

Magnetic fields

If a permanent magnet is placed on a surface and covered by a piece of paper, iron filings can be shaken on to the paper from a dispenser. Gently tapping the paper then causes the filings to take up the shape of the magnetic field surrounding the permanent magnet. The magnetic fields around a permanent magnet are shown in Figs 2.50 and 2.51.

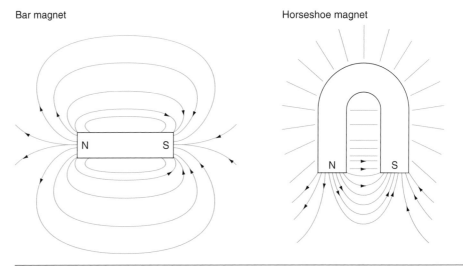

Bar magnet   Horseshoe magnet

Figure 2.50 Magnetic fields around a permanent magnet.

Electromagnetism

Electricity and magnetism have been inseparably connected since the experiments by Oersted and Faraday in the early nineteenth century. Electricity and magnetism are interconnected, in other words you cannot have one without the other. An electric current flowing in a conductor produces a magnetic field 'around' the conductor which is proportional to the current. Thus, a small current produces a weak magnetic field, while a large current will produce a strong magnetic field. The magnetic field 'spirals' around the conductor, as shown in Figure 2.52, and its direction can be determined by the 'dot' or 'cross' notation and the 'screw rule'.

To do this, we think of the current as being represented by a dart or arrow inside the conductor. The dot represents current coming towards us when we would see the point of the arrow or dart inside the conductor. The cross represents current going away from us when we would see the flights of the dart or arrow.

Unlike poles attract

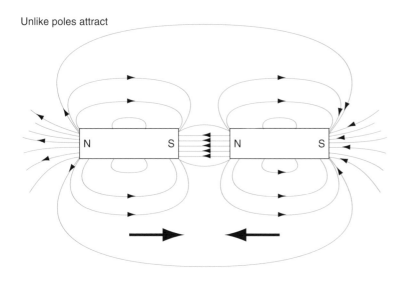

Like poles repel

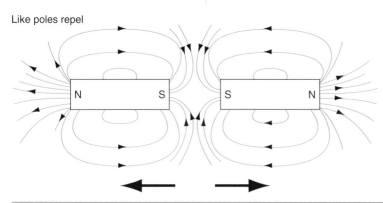

Figure 2.51 The first laws of magnetism.

Figure 2.52 A solenoid is an electromagnetic device.

Imagine a corkscrew or screw being turned so that it will move in the direction of the current. Therefore, if the current was coming out of the paper, as shown in Figure 2.53(a), the magnetic field would be spiralling anticlockwise around the conductor. If the current was going into the paper, as shown by Figure 2.53(b), the magnetic field would spiral clockwise. A current flowing in a coil of wire or solenoid establishes a magnetic field which is very similar to that of a bar magnet. Winding the coil around a soft iron core increases the flux density because the lines of magnetic flux concentrate on the magnetic material. The advantage of the electromagnet when compared with the permanent magnet is that the magnetism of the electromagnet can be switched on and off by a functional switch controlling the coil current. This effect is put to practical use in the electrical relay as used in a motor starter or alarm circuit.

Assessment criteria 6.2

Apply Fleming's right hand rule in relation to a basic alternator

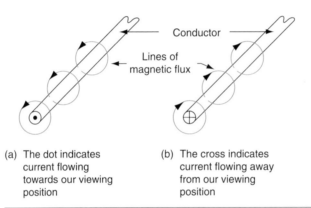

(a) The dot indicates current flowing towards our viewing position

(b) The cross indicates current flowing away from our viewing position

Figure 2.53 Magnetic fields around a current-carrying conductor.

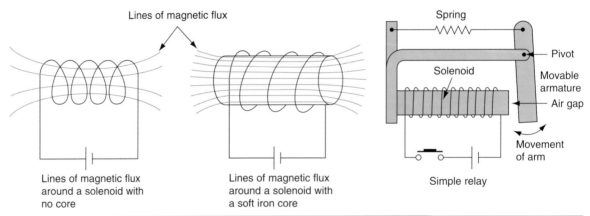

Lines of magnetic flux around a solenoid with no core

Lines of magnetic flux around a solenoid with a soft iron core

Simple relay

Figure 2.54 The solenoid and one practical application: the relay.

The relationship between electricity and magnetism can be seen through the work of John Ambrose Fleming, who came up with a method of working out how they interact in two distinct ways.

Fleming's Left Hand Rule (Motor Rule)

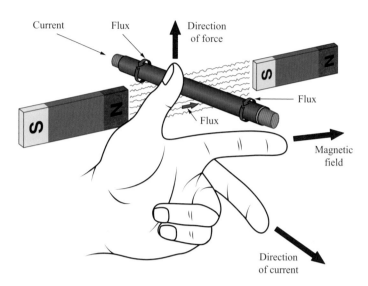

Figure 2.55 A motor accepts electrical energy and produces mechanical energy. This is shown through the formula F = B I L, whereby F is the force measured in newtons. As a learning aid remember: motors drive on the left hand side.

Fleming's Right Hand Rule (Generator Rule)

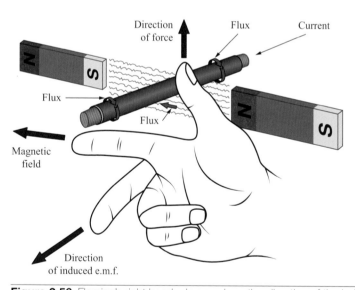

Figure 2.56 Fleming's right hand rule can show the direction of the induced e.m.f.

When a current is passed through a conductor, a magnetic field is generated, however if the same conductor happens to be placed within another magnetic field, then the reaction of both fields will cause the conductor to move. This is the motor principle and Fleming used his left hand to show how electricity, magnetism and motion are linked.

Then literally on the other hand, if we reverse the process and move a conductor across a magnetic field, we will create or generate electricity and this process is known as the generator effect. This time Fleming used his right hand to indicate how we can determine the direction of this induced e.m.f.

Assessment criteria 6.3

Identify applications of electromagnetism

The electrical relay

A **relay** is an electromagnetic switch operated by a solenoid. We looked at the action of a solenoid in Figure 2.54. The solenoid in a relay operates a number of switch contacts as it moves under the electromagnetic forces. Relays can be used to switch circuits on or off at a distance remotely. The energizing circuit, the solenoid, is completely separate to the switch contacts and, therefore, the relay can switch high-voltage, high-power circuits from a low-voltage switching circuit. This gives the relay many applications in motor control circuits, electronics and instrumentation systems. Figure 2.57 shows a simple relay.

Transformers

A **transformer** is an electrical machine which is used to change the value of an alternating voltage. They vary in size from miniature units used in electronics to huge power transformers used in power-stations. A transformer will only work when an alternating voltage is connected. It will not normally work from a d.c. supply such as a battery.

Definition

A *relay* is an electromagnetic switch operated by a solenoid.

Definition

A *transformer* is an electrical machine which is used to change the value of an alternating voltage.

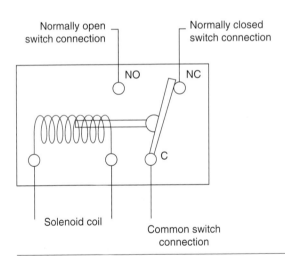

Figure 2.57 A simple relay.

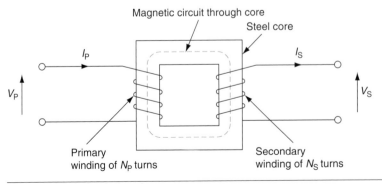

Figure 2.58 A simple transformer.

A transformer, as shown in Fig. 2.55, consists of two coils, called the primary and secondary coils, or windings, which are insulated from each other and wound on to the same steel or iron core.

Electrical machines

Electrical machines are energy converters. If the machine input is mechanical energy and the output electrical energy then that machine is a generator, as shown in Figure 2.60(a). Alternatively, if the machine input is electrical energy and the output mechanical energy then the machine is a motor, as shown in Figure 2.60(b).

An electrical machine may be used as a motor or a generator, although in practice the machine will work more efficiently when operated in the mode for which it was designed.

Figure 2.59 Electric cars use electrical energy to power electrical motors, which in turn create mechanical motion.

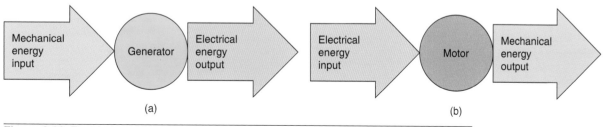

(a)   (b)

Figure 2.60 Electrical machines as energy converters.

Simple d.c. generator or dynamo

A d.c. generator uses a commutator where contact with the external circuit is made through carbon brushes which make contact with the commutator copper segment as shown in Fig 2.61. The action of the commutator is to reverse the generated e.m.f. every half-cycle, rather like an automatic change-over switch. What is generated in essence is a d.c. output.

However, this simple arrangement produces a very bumpy d.c. output. In a practical machine, the commutators would contain many windings interspaced with insulation pieces made from Mica, which in turn would produce a smoother d.c. output similar to the unidirectional battery supply shown in Figure 2.62.

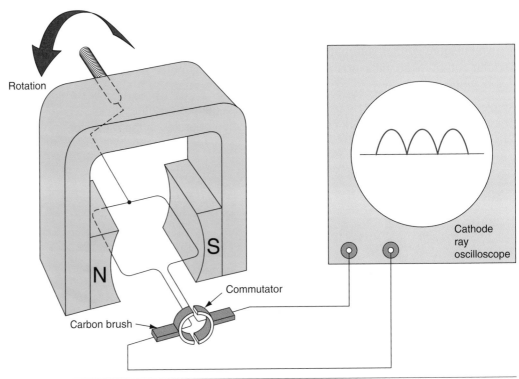

Figure 2.61 Simple d.c. generator or dynamo.

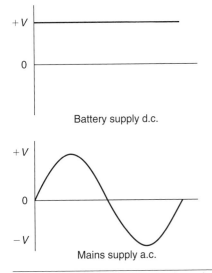

Figure 2.62 Unidirectional and alternating supply.

Figure 2.63 Electrical motor coil and electronic circuit.

Alternating current theory

Commercial quantities of electricity for industry, commerce and domestic use are generated as a.c. in large power-stations and distributed around the United Kingdom on the National Grid to the end user. The d.c. electricity has many applications where portability or an emergency stand-by supply is important but for large quantities of power it has to be an a.c. supply because it is so easy to change the voltage levels using a transformer.

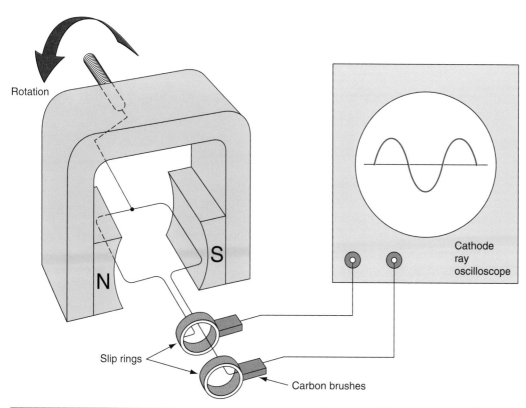

Figure 2.64 Simple a.c. generator or alternator.

Assessment criteria 6.5

Determine sinusoidal quantities

Most electrical equipment makes use of alternating current supplies, and for this reason a knowledge of alternating waveforms is necessary for all practising electricians.

When a coil of wire is rotated inside a magnetic field as shown in Figure 2.65, a voltage is induced in the coil. The induced voltage follows a mathematical law known as the sinusoidal law and, therefore, we can say that a sine wave has been generated.

Such a waveform has the characteristics displayed in Figure 2.67. All generators therefore produce alternating current and after each revolution it will have completed a 360° cycle.

A sine wave is therefore a different representation of a circle but is shown laterally broken down in stages and degrees in Fig 2.67.

Figure 2.65 One complete cycle of a sinusoidal wave is equivalent to 360°

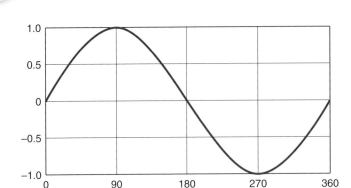

Figure 2.66 Sinusoidal wave is sometimes shortened to sine wave. 90° and 270° are equivalent to the wave's positive and negative peaks respectively

In the United Kingdom, we generate electricity at a frequency of 50 Hz and the time taken to complete each cycle is given by

$$T = \frac{1}{f}$$

$$\therefore T = \frac{1}{50\,\text{Hz}} = 0.02\,\text{s}$$

An alternating waveform is constantly changing from zero to a maximum, first in one direction, then in the opposite direction, and so the instantaneous values of the generated voltage are always changing. A useful description of the electrical effects of an a.c. waveform can be given by the maximum, average and r.m.s. values of the waveform.

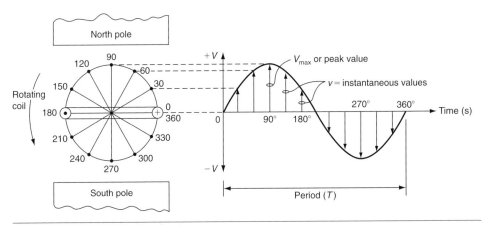

Figure 2.67 Characteristics of a sine wave.

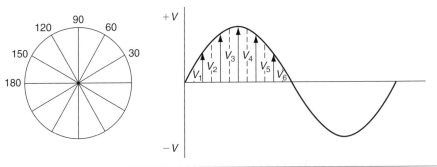

Figure 2.68 Sinusoidal waveform showing instantaneous values of voltage.

The maximum or peak value is the greatest instantaneous value reached by the generated waveform. Cable and equipment insulation levels must be equal to or greater than this value.

Normally, when calculating averages simply add up all the numbers and then divide by how many numbers there are.

Calculate the average of: 4, 11, 9

- Add the numbers: **4 + 11 + 9 = 24**
- Divide by *how many* numbers (there are three sets of numbers): **24 ÷ 3 = 8**

The average is 8.

Unfortunately we cannot use the same mathematical method when calculating the average value of an alternating wave since mathematically the positive half cancels out the negative half. Therefore, mathematically the average value of a sine wave is zero.

To find the average value therefore we use plotted instantaneous values over one half-cycle as they change from zero to a maximum and can be found from the following formula which is applied to the sinusoidal waveform shown in Fig. 2.68.

$$V_{av} = \frac{V_1 + V_2 + V_3 + V_4 + V_5 + V_6}{6} = 0.637 V_{max}$$

This calculates as 0.637 of the peak or maximum value.

Energy loss over a specific time in a resistive circuit is not directly proportional to the voltage and current, which is why in order to work out the useful element of a sine wave, we use RMS, which stands for root mean square.

It is calculated in the same format as the average value except we also take the square root.

RMS works out as 70.7% of a sine wave.

All instruments measure RMS, therefore 230V mains voltage is actually 230V RMS, the useful part or the same heating effect as a d.c. voltage.

The value can be found from the following formula applied to the sinusoidal waveform shown in Figure 2.68.

$$V_{r.m.s.} = \sqrt{\frac{V_1^2 + V_2^2 + V_3^2 + V_4^2 + V_5^2 + V_6^2}{6}}$$
$$= 0.7071 V_{max}$$

For any sinusoidal waveform the RMS value is equal to 0.707 x the maximum value.

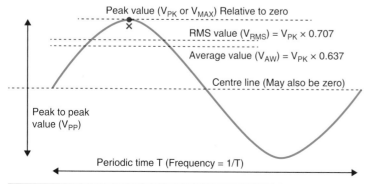

Figure 2.69 RMS is a useful bit of a sine wave or the equivalant to a direct current supply. It is roughly 3/4 of the peak (maximum) value or 70.7% when shown as a percentage

Example

The sinusoidal waveform applied to a particular circuit has a maximum value of 325.3 V. Calculate the average and RMS value of the waveform.

$$\text{Average value } V_{av} = 0.637 \times V_{max}$$
$$\therefore V_{av} = 0.637 \times 325.3 = 207.2 \text{ V}$$
$$\text{RMS value } V_{RMS} = 0.7071 \times V_{max}$$
$$V_{RMS} = 0.7071 \times 325.3 = 230 \text{ V}$$

When we say that the main supply to a domestic property is 230 V, we really mean 230 V_{RMS}. Such a waveform has an average value of about 207.2 V and a maximum value of almost 325.3 V but because the RMS value gives the d.c. equivalent value we almost always give the RMS value without identifying it as such.

Assessment criteria 6.6

State the reasons for using alternating current transmission and distribution

One of the reasons for using alternating supplies for the electricity mains supply is because we can very easily change the voltage levels by using a transformer which will only work on an a.c. supply.

The generated alternating supply at the power-station is transformed up to 132,000 V, or more, for efficient transmission along the national grid conductors. This is because the power into a transformer must equal the power out, consequently boosting the voltage up will effectively reduce the circuit current.

Reducing the current means that smaller conductors can be used and losses due to heat and volt drop are also reduced.

Assessment criteria 7.1

Distinguish what is meant by mass and weight

Basic mechanics and machines

Mechanics is the scientific study of 'machines', where a machine is defined as a device which transmits motion or force from one place to another. An engine is one particular type of machine, an energy-transforming machine, converting fuel energy into a more directly useful form of work.

Most modern machines can be traced back to the five basic machines described by the Greek inventor Hero of Alexandria who lived at about the time of Christ. The machines described by him were the wedge, the screw, the wheel and axle, the pulley, and the lever. Originally they were used for simple purposes, to raise water and to move objects which man alone could not lift, but today

Definition

Mechanics is the scientific study of 'machines', where a machine is defined as a device which transmits motion or force from one place to another.

their principles are of fundamental importance to our scientific understanding of mechanics. Let us now consider some fundamental mechanical principles and calculations.

Mass

Mass is a measure of the amount of material in a substance, such as metal, plastic, wood, brick or tissue, which is collectively known as a body. The mass of a body remains constant and can easily be found by comparing it on a set of balance scales with a set of standard masses. The SI unit of mass is the kilogram (kg).

Definition

Mass is a measure of the amount of material in a substance, such as metal, plastic, wood, brick or tissue, which is collectively known as a body. The mass of a body remains constant and can easily be found by comparing it on a set of balance scales with a set of standard masses. The SI unit of mass is the kilogram (kg).

Weight

Weight is a measure of the force which a body exerts on anything which supports it. Normally it exerts this force because it is being attracted towards the Earth by the force of gravity.

For scientific purposes the weight of a body is *not* constant, because gravitational force varies from the Equator to the Poles; in space a body would be 'weightless' but here on Earth under the influence of gravity a 1 kg mass would have a weight of approximately 9.81 N (see also the definition of 'force').

Definition

Weight is a measure of the force which a body exerts on anything which supports it. Normally it exerts this force because it is being attracted towards the Earth by the force of gravity.

Speed

The feeling of speed is something with which we are all familiar. If we travel in a motor vehicle we know that an increase in speed would, excluding accidents, allow us to arrive at our destination more quickly. Therefore, **speed** is concerned with distance travelled and time taken. Suppose we were to travel a distance of 30 miles in one hour; our speed would be an average of 30 miles/h:

$$\text{Speed} = \frac{\text{Distance (m)}}{\text{Time (s)}}$$

Definition

Speed is concerned with distance travelled and time taken.

Velocity

In everyday conversation we often use the word **velocity** to mean the same as speed, and indeed the units are the same. However, for scientific purposes this is not acceptable since velocity is also concerned with direction. Velocity is speed in a given direction. For example, the speed of an aircraft might be 200 miles/h, but its velocity would be 200 miles/h in, say, a westerly direction. Speed is a scalar quantity, while velocity is a vector quantity.

$$\text{Velocity} = \frac{\text{Distance (m)}}{\text{Time (s)}}$$

Definition

In everyday conversation we often use the word *velocity* to mean the same as speed, and indeed the units are the same. However, for scientific purposes this is not acceptable since velocity is also concerned with direction.

Acceleration

When an aircraft takes off, it starts from rest and increases its velocity until it can fly. This change in velocity is called its acceleration. By definition, **acceleration** is the rate of change in velocity with time.

$$\text{Acceleration} = \frac{\text{Velocity}}{\text{Time}} = (\text{m/s}^2)$$

Definition

By definition, *acceleration* is the rate of change in velocity with time.

2

Example 1

If an aircraft accelerates from a velocity of 15 m/s to 35 m/s in 4 s, calculate its average acceleration.

Average velocity = 35 m/s – 15 m/s = 20 m/s

$$\text{Average acceleration} = \frac{\text{Velocity}}{\text{Time}} = \frac{20}{4} = 5 \text{ m/s}^2$$

Thus, the average acceleration is 5 m/s².

Assessment criteria 7.2

Outline the principles of basic mechanics as they apply to levers, gears and pulleys

Levers and turning force

Definition

A *lever* allows a heavy load to be lifted or moved by a small effort.

Definition

A *lever* is any rigid body which pivots or rotates about a fixed axis or fulcrum.

A **lever** allows a heavy load to be lifted or moved by a small effort. Every time we open a door, turn on a tap or tighten a nut with a spanner, we exert a lever-action turning force. A **lever** is any rigid body which pivots or rotates about a fixed axis or fulcrum. The simplest form of lever is the crowbar, which is useful because it enables a person to lift a load at one end which is greater than the effort applied through his or her arm muscles at the other end. In this way the crowbar is said to provide a 'mechanical advantage'. A washbasin tap and a spanner both provide a mechanical advantage through the simple lever action. The mechanical advantage of a simple lever is dependent upon the length of lever on either side of the fulcrum. Applying the principle of turning forces to a lever, we obtain the formula:

Load force × Distance from fulcrum = Effort force × Distance from fulcrum

Figure 2.70 An example of a lever in use.

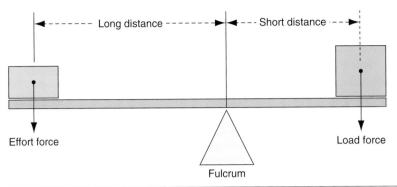

Figure 2.71 Turning forces of a simple lever.

Example

Calculate the effort required to raise a load of 500 kg when the effort is applied at a distance of five times the load distance from the fulcrum (assume the acceleration due to gravity to be 10 m/s²).

Load force = Mass × Acceleration (N)

Load force = 500 kg × 10 m/s² = 5000 N

Load force × Distance from fulcrum = Effort force × Distance from fulcrum

5000 N × 1 m = Effort force × 5 m

$$\therefore \text{Effort force} = \frac{5000\,\text{N} \times 1\,\text{m}}{5\,\text{m}} = 1000\,\text{N}$$

Thus an effort force of 1000 N can overcome a load force of 5000 N using the mechanical advantage of this simple lever.

This formula can perhaps better be understood by referring to Fig. 2.71. A small effort at a long distance from the fulcrum can balance a large load at a short distance from the fulcrum. Thus a 'turning force' or 'turning moment' depends upon the distance from the fulcrum and the magnitude of the force.

Levers are grouped into three classes which depend upon the position of the fulcrum or pivot point. Examples of the three classes are shown in Fig. 2.72.

Class 1 lever – the fulcrum is in the middle, between the force and the load. A crowbar and a sack truck are good examples of a class 1 lever. A sack truck is shown in Chapter 1 at Fig 1.24 and in Fig 2.72 on the next page. A pair of pliers or side cutters is an example of two class 1 levers working together.

Class 2 lever – the load is in the middle, between the force and the fulcrum. A wheelbarrow is a good example of a class 2 lever.

Class 3 lever – the force is in the middle, between the load and the fulcrum. You will see many examples of this class 3 lever toning muscles in the gym. A builder's shovel is another example of a class 3 lever.

Simple machines

Our physical abilities in the field of lifting and moving heavy objects are limited. However, over the centuries we have used our superior intelligence to design tools, mechanisms and machines which have overcome this physical inadequacy.

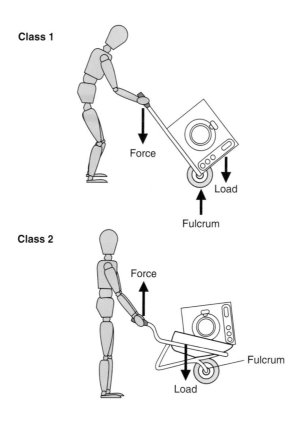

Class 1

Force

Load

Fulcrum

Class 2

Force

Fulcrum

Load

Class 3

Force

Fulcrum

Load

Figure 2.72 Classes of lever.

Gears and pulleys

A pulley is a mechanical advantage device that uses wheels, has a groove contained within its outer edge and turns freely on an axle attached to a stationary or moveable position. The pulley has a rope which is wrapped around a wheel-and-axle system which permits a user to lift heavy loads by matching or reducing the effort required regarding the weight or load in question. The greater the number of pulleys involved, the greater the number of rope parts, which basically means that the user requires less effort.

This is explained through the diagram below, since the user of the single fixed pulley (number 1) requires an effort of 100 N to operate it, but in comparison, pulley number 4 only requires an effort of 25 N. Basically, the effort required is

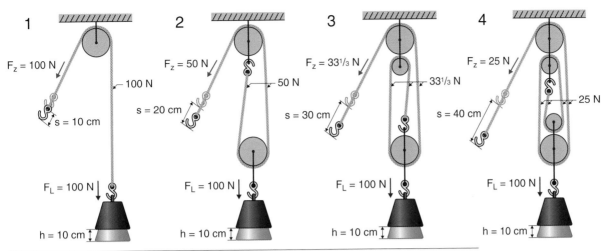

Figure 2.73 Notice that pulley system 4 only requires a quarter of the effort to operate.

shared across the pulleys and the four rope parts. Mechanical advantage = load/effort, 100/25 = 4.

Pulleys can also be used to transfer power when they are connected directly to or through a belt. This sort of belt drive uses at least two pulleys: a driver element which is connected to a drive shaft and the driven element.

Another system which uses a driving and driven wheel but is connected directly through the use of cogs which mesh with each other, is a *gear system*. The mechanical advantage can be calculated through the actual gear ratio in terms of the numbers of teeth on the gears and the actual speed of the wheels.

For example if the driven has 500 teeth in comparison to the driver which has 100, then the gear ratio would be

$$\text{Gear ratio} = \frac{\text{number of teeth on the driven}}{\text{number of teeth on the driver}}$$

$$\text{Gear ratio} = \frac{500}{100}$$

Gear ratio = 5.

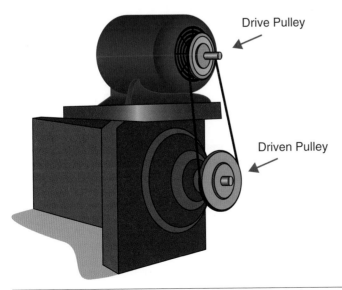

Figure 2.74 Pulleys can be used to provide mechanical advantage.

Figure 2.75 A cable car is a practical example of a pulley system.

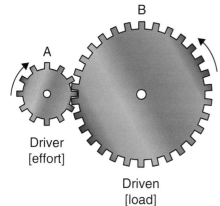

Figure 2.76 By revolving at a quicker rate, the smaller cog drives a much bigger system. This is similar to an electrical relay which uses a small electrical signal to control a big electrical supply.

If the driven load was revolving around 1000 rpm, then we can determine the speed of the driver.

$$\text{Gear ratio} = \frac{\text{input speed}}{\text{output speed}}$$

Input speed = Gear ratio × output speed
Input speed = 5 x 1000 rpm
Input speed = 5000 rpm.

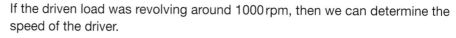

Assessment criteria 7.3

Determine mechanical quantities

Force

The presence of a **force** can only be detected by its effect on a body. A force may cause a stationary object to move or bring a moving body to rest. For example, a number of people pushing a broken-down motor car exert a force which propels it forward, but applying the motor car brakes applies a force on the brake drums which slows down or stops the vehicle. Gravitational force causes objects to fall to the ground. The apple fell from the tree on to Isaac Newton's head as a result of gravitational force. The standard rate of acceleration due to gravity is accepted as 9.81 m/s². Therefore, an apple weighing 1 kg will exert a force of 9.81 N, since

Force = Mass × Acceleration (N)

The SI unit of force is the newton, symbol N, to commemorate the great English scientist Sir Isaac Newton (1642–1727).

Definition

The presence of a *force* can only be detected by its effect on a body. A force may cause a stationary object to move or bring a moving body to rest.

Example 1

A 50 kg bag of cement falls from a forklift truck while being lifted on to a storage shelf. Determine the force with which the bag will strike the ground:

Force = Mass × Acceleration (N)

$50 \, \text{kg} \times 9.81 \, \text{m/s}^2 = 490.5 \, \text{N}$

A force can manifest itself in many different ways. Let us consider a few examples:

- 'inertial force' is the force required to get things moving, to change direction or stop, like the motor car discussed above;
- 'cohesive or adhesive force' is the force required to hold things together;
- 'tensile force' is the force pulling things apart;
- 'compressive force' is the force pushing things together;
- 'friction force' is the force which resists or prevents the movement of two surfaces in contact;
- 'shearing force' is the force which moves one face of a material over another;
- 'centripetal force' is the force acting towards the centre when a mass attached to a string is rotated in a circular path;
- 'centrifugal force' is the force acting away from the centre, the opposite to centripetal force;
- 'gravitational force' is the force acting towards the centre of the Earth due to the effect of gravity;
- 'magnetic force' is the force created by a magnetic field;
- 'electrical force' is the force created by an electrical field.

Pressure or stress

To move a broken-down motor car I might exert a force on the back of the car to propel it forward. My hands would apply a pressure on the body panel at the point of contact with the car. **Pressure** or **stress** is a measure of the force per unit area.

$$\text{Pressure or stress} = \frac{\text{Force}}{\text{Area}} \, (\text{N/m}^2)$$

Example 2

A young woman of mass 60 kg puts all her weight on to the heel of one shoe which has an area of 1 cm². Calculate the pressure exerted by the shoe on the floor (assuming the acceleration due to gravity to be 9.81 m/s²).

$$\text{Pressure} = \frac{\text{Force}}{\text{Area}} \, (\text{N/m}^2)$$

$$\text{Pressure} = \frac{60 \, \text{kg} \times 9.81 \, \text{m/s}^2}{1 \times 10^{-4} \, \text{m}^2} = 5886 \, \text{kN/m}^2$$

Definition

- 'Inertial force' is the force required to get things moving, to change direction or stop, like the motor car discussed above.
- 'Cohesive or adhesive force' is the force required to hold things together.
- 'Tensile force' is the force pulling things apart.
- 'Compressive force' is the force pushing things together.
- 'Friction force' is the force which resists or prevents the movement of two surfaces in contact.
- 'Shearing force' is the force which moves one face of a material over another.
- 'Centripetal force' is the force acting towards the centre when a mass attached to a string is rotated in a circular path.
- 'Centrifugal force' is the force acting away from the centre, the opposite to centripetal force.
- 'Gravitational force' is the force acting towards the centre of the earth due to the effect of gravity.
- 'Magnetic force' is the force created by a magnetic field.
- 'Electrical force' is the force created by an electrical field.

Definition

Pressure or *stress* is a measure of the force per unit area.

Example 3

A small circus elephant of mass 1 tonne (1000 kg) puts all its weight on to one foot which has a surface area of 400 cm². Calculate the pressure exerted by the elephant's foot on the floor, assuming the acceleration due to gravity to be 9.81 m/s².

$$\text{Pressure} = \frac{\text{Force}}{\text{Area}} \, (\text{N/m}^2)$$

$$\text{Pressure} = \frac{1000 \, \text{kg} \times 9.81 \, \text{m/s}^2}{400 \times 10^{-4} \, \text{m}^2} = 245.3 \, \text{kN/m}^2$$

These two examples show that the young woman exerts 24 times more pressure on the ground than the elephant. This is because her mass exerts a force over a much smaller area than the elephant's foot and is the reason why many wooden dance floors are damaged by high-heeled shoes.

Work done

Suppose a broken-down motor car was to be pushed along a road. Work would be done on the car by applying the force necessary to move it along the road. Heavy breathing and perspiration would be evidence of the work done.

By definition, **work done** is dependent upon the force applied times the distance moved in the direction of the force.

Work done = Force × Distance moved in the direction of the force (J)

The SI unit of work done is the newton metre or joule (symbol J). The joule is the preferred unit and it commemorates an English physicist, James Prescott Joule (1818–1889).

Example 4

A building hoist lifts ten 50 kg bags of cement through a vertical distance of 30 m to the top of a high-rise building. Calculate the work done by the hoist, assuming the acceleration due to gravity to be 9.81 m/s².

$$\text{Work done} = \text{Force} \times \text{Distance moved (J)}$$
$$\text{but force} = \text{Mass} \times \text{Acceleration (N)}$$
$$\therefore \text{Work done} = \text{Mass} \times \text{Acceleration} \times \text{Distance moved (J)}$$
$$\text{Work done} = 10 \times 50 \, \text{kg} \times 9.81 \, \text{m/s}^2 \times 30 \, \text{m}$$
$$\text{Work done} = 147.15 \, \text{kJ}$$

Assessment criteria 7.4

Outline the main principle of kinetic and potential energy

Potential energy is energy that is stored, and can be defined as energy that acts on a body and in particular the position of a body. An example of potential energy can be seen below, with a car parked on top of a hill.

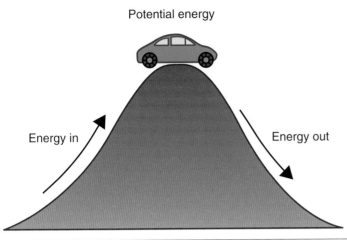

Potential energy

Energy in

Energy out

Figure 2.77 *Potential energy – a car positioned at the top of a hill, that is, energy in motion.*

Kinetic energy is energy that possesses motion.

When the car moves, the energy is transferred to movement, in other words potential energy becomes kinetic energy, that is, energy in motion.

A battery is a good example of electrical potential and kinetic energy. Any battery with charged cells that isn't connected to a circuit has potential energy. Once a battery is connected to a circuit, electrical current will flow, which means that the potential energy stored in the battery has now changed to one of motion with regard to the flow of electricity (kinetic energy).

Power

If one motor car can cover the distance between two points more quickly than another car, we say that the faster car is more powerful. It can do a given amount of work more quickly. By definition, **power** is the rate of doing work.

$$\text{Power} = \frac{\text{Work done}}{\text{Time taken}} (W)$$

The SI unit of power, both electrical and mechanical, is the watt (symbol W). This commemorates the name of James Watt (1736–1819), the inventor of the steam-engine.

Definition

By definition, *power* is the rate of doing work.

Example 1

A building hoist lifts ten 50 kg bags of cement to the top of a 30 m-high building. Calculate the rating (power) of the motor to perform this task in 60 s if the acceleration due to gravity is taken as 9.81 m/s².

$$\text{Power} = \frac{\text{Work done}}{\text{Time taken}} (W)$$

but work done = Force × Distance moved (J)
and force = Mass × Acceleration (N)

(continued)

Example 1 continued

By substitution,

$$\text{Power} = \frac{\text{Mass} \times \text{Acceleration} \times \text{Distance moved}}{\text{Time taken}} (W)$$

$$\text{Power} = \frac{10 \times 50\,\text{kg} \times 9.81\,\text{m/s}^2 \times 30\,\text{m}}{60\,\text{s}}$$

$$\text{Power} = 2452.5\,\text{W}$$

The rating of the building hoist motor will be 2.45 kW.

Example 2

A hydroelectric power-station pump motor working continuously during a seven-hour period raises 856 tonnes of water through a vertical distance of 60 m. Determine the rating (power) of the motor, assuming the acceleration due to gravity is 9.81 m/s².

From Example 1,

$$\text{Power} = \frac{\text{Mass} \times \text{Acceleration} \times \text{Distance moved}}{\text{Time taken}} (W)$$

$$\text{Power} = \frac{856 \times 1000\,\text{kg} \times 9.81\,\text{m/s}^2 \times 60\,\text{m}}{7 \times 60 \times 60\,\text{s}}$$

$$\text{Power} = 20{,}000\,\text{W}$$

The rating of the pump motor is 20 kW.

Example 3

An electric hoist motor raises a load of 500 kg at a velocity of 2 m/s. Calculate the rating (power) of the motor if the acceleration due to gravity is 9.81 m/s².

$$\text{Power} = \frac{\text{Mass} \times \text{Acceleration} \times \text{Distance moved}}{\text{Time taken}} (W)$$

$$\text{but Velocity} = \frac{\text{Distance}}{\text{Time}} (\text{m/s})$$

$$\therefore \text{Power} = \text{Mass} \times \text{Acceleration} \times \text{Velocity}$$
$$\text{Power} = 500\,\text{kg} \times 9.81\,\text{m/s}^2 \times 2\,\text{m/s}$$
$$\text{Power} = 9810\,\text{W}.$$

The rating of the hoist motor is 9.81 kW.

Assessment criteria 7.5

Calculate values of electrical efficiency

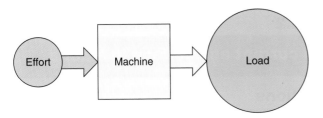

Figure 2.78 Simple machine concept.

By definition, a **machine** is an assembly of parts, some fixed, others movable, by which motion and force are transmitted. With the aid of a machine we are able to magnify the effort exerted at the input and lift or move large loads at the output.

Efficiency of any machine

In any machine the power available at the output is less than that which is put in because losses occur in the machine. The losses may result from friction in the bearings, wind resistance to moving parts, heat, noise or vibration.

The ratio of the output power to the input power is known as the **efficiency** of the machine. The symbol for efficiency is the Greek letter 'eta' (η). In general,

$$\eta = \frac{\text{Power output}}{\text{Power input}}$$

Since efficiency is usually expressed as a percentage we modify the general formula as follows:

$$\eta = \frac{\text{Power output}}{\text{Power input}} \times 100$$

Example

A transformer feeds the 9.81 kW motor driving the mechanical hoist of the previous example. The input power to the transformer was found to be 10.9 kW. Find the efficiency of the transformer.

$$\eta = \frac{\text{Power output}}{\text{Power input}} \times 100$$

$$\eta = \frac{9.81\,\text{kW}}{10.9\,\text{kW}} \times 100 = 90\%$$

Thus the transformer is 90% efficient. Note that efficiency has no units, but is simply expressed as a percentage.

Star and delta connections

The three-phase windings of an a.c. generator may be star connected or delta connected as shown in Fig. 2.79. The important relationship between phase and line currents and voltages is also shown. The square root of 3 ($\sqrt{3}$) is simply a constant for three-phase circuits, and has a value of 1.732. The delta connection is used for electrical power transmission because only three conductors are required. Delta connection is also used to connect the windings of most three-phase motors because the phase windings are perfectly balanced and, therefore, do not require a neutral connection.

Making a star connection at the local substation has the advantage that two voltages become available – a line voltage of 400 V between any two phases, and a phase voltage of 230 V between line and neutral which is connected to the star point.

In any star-connected system currents flow along the lines (I_L), through the load and return by the neutral conductor connected to the star point. In a *balanced* three-phase system all currents have the same value and when they are added up by phasor addition, we find the resultant current is zero. Therefore, no current flows in the neutral and the star point is at zero volts. The star point of the distribution transformer is earthed because earth is also at zero potential.

A star-connected system is also called a three-phase four-wire system and allows us to connect single-phase loads to a three-phase system.

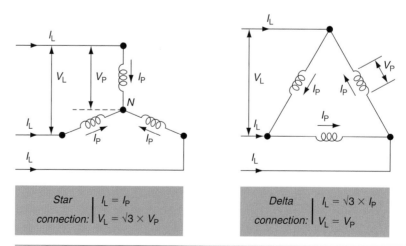

| Star connection: | $I_L = I_P$ $V_L = \sqrt{3} \times V_P$ | Delta connection: | $I_L = \sqrt{3} \times I_P$ $V_L = V_P$ |

Figure 2.79 Star and delta connections.

Three-phase power

We know from our single-phase alternating current theory earlier in this chapter that power can be found from the following formula:

$$\text{Power} = VI \cos\phi \text{ (W)}$$

In any balanced three-phase system, the total power is equal to three times the power in any one phase.

$$\therefore \text{ Total three-phase power} = 3V_P I_P \cos\phi \text{ (W)} \qquad \text{(Equation 1)}$$

Now for a star connection:

$$V_P = \frac{V_L}{\sqrt{3}} \quad \text{and} \quad I_L = I_P \qquad \text{(Equation 2)}$$

Substituting Equation (2) into Equation (1), we have:

$$\text{Total three-phase power} = \sqrt{3}\, V_L I_L \cos\phi \text{ (W)}$$

Now consider a delta connection:

$$V_P = V_L \quad \text{and} \quad I_P = \frac{I_L}{\sqrt{3}} \qquad \text{(Equation 3)}$$

Substituting Equation (3) into Equation (1) we have, for any balanced three-phase load,

$$\text{Total three-phase power} = \sqrt{3}\, V_L I_L \cos\phi \text{ (W)}$$

So, a general equation for three-phase power is:

$$\text{Power} = \sqrt{3}\, V_L I_L \cos\phi$$

Example 1

A balanced star-connected three-phase load of $10\,\Omega$ per phase is supplied from a 400 V, 50 Hz mains supply at unity power factor. Calculate (a) the phase voltage, (b) the line current and (c) the total power consumed.

For a star connection

$$V_L = \sqrt{3}\, V_P \text{ and } I_L = I_P$$

For (a)

$$V_P = \frac{V_L}{\sqrt{3}} \text{ (V)}$$

$$V_P = \frac{400\,\text{V}}{1.732} = 230.9\,\text{V}$$

For (b)

$$I_L = I_P = \frac{V_P}{R_P} \text{ (A)}$$

$$I_L = I_P = \frac{230.9\,\text{V}}{10\,\Omega} = 23.09\,\text{A}$$

For (c)

$$\text{Power} = \sqrt{3}\, V_L I_L \cos\phi \text{ (W)}$$

$$\therefore \text{Power} = 1.732 \times 400\,\text{V} \times 23.09\,\text{A} \times 1 = 16\,\text{kW}$$

Example 2

A 20 kW, 400 V balanced delta-connected load has a power factor of 0.8. Calculate (a) the line current and (b) the phase current.

We have that:

$$\text{Three-phase power} = \sqrt{3}\, V_L\, I_L\, \cos\phi \;\text{(W)}$$

For (a)

$$I_L = \frac{\text{Power}}{\sqrt{3}\, V_L\, \cos\phi} \;\text{(A)}$$

$$\therefore I_L = \frac{20{,}000\,\text{W}}{1.732 \times 400\,\text{V} \times 0.8}$$

$$I_L = 36.08 \;\text{(A)}$$

For delta connection

$$I_L = \sqrt{3}\, I_P \;\text{(A)}$$

Thus, for (b)

$$I_P = I_L / \sqrt{3} \;\text{(A)}$$

$$\therefore I_P = \frac{36.08\,\text{A}}{1.732} = 20.83\,\text{A}$$

Example 3

Three identical loads each having a resistance of 30 Ω and inductive reactance of 40 Ω are connected first in star and then in delta to a 400 V three-phase supply. Calculate the phase currents and line currents for each connection.

For each load

$$Z = \sqrt{R^2 + X_L^2} \;(\Omega)$$

$$\therefore Z = \sqrt{30^2 + 40^2}$$

$$Z = \sqrt{2500} = 50\,\Omega$$

For star connection

$$V_L = \sqrt{3}\, V_P \quad \text{and} \quad I_L = I_P$$

$$V_P = \frac{V_L}{\sqrt{3}}\;\text{(V)}$$

$$\therefore V_P = \frac{400\,\text{V}}{1.732} = 230.9\,\text{V}$$

$$I_P = \frac{V_P}{Z_P}\;\text{(A)}$$

$$\therefore I_P = \frac{230.9\,\text{V}}{50\,\Omega} = 4.62\,\text{A}$$

$$I_P = I_L$$

(continued)

Example 3 continued

Therefore phase and line currents are both equal to 4.62 A.

For delta connection

$$V_L = V_P \quad \text{and} \quad I_L = \sqrt{3}\, I_P$$

$$V_L = V_P = 400\,V$$

$$I_P = V_P / Z_P \; (A)$$

$$\therefore I = \frac{400\,V}{50\,\Omega} = 8\,A$$

$$I_L = \sqrt{3}\, I_P \; (A)$$

$$\therefore I_L = 1.732 \times 8\,A = 13.86\,A$$

Assessment criteria 8.3

Recognize the reasons for the neutral conductor

Assessment criteria 8.4

Identify the advantages of distributing loads evenly over the three lines

Balancing single-phase loads

As has already been discussed, the reason behind balancing three-phase loads is to ensure that the same current is drawn by each phase. However, in industrial and commercial buildings this is more difficult but every attempt should be made to balance the load. The neutral cable allows unbalanced loads to operate safely.

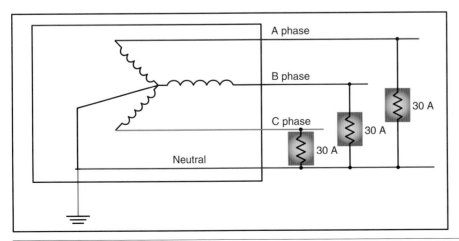

Figure 2.80 An example of a perfectly balanced star-connected system since there is 30A flowing in all three phases. This means no current will flow in the neutral conductor.

Summary of three-phase supplies:

- Star-connected systems give us two sets of voltages: 230 V single phase domestic (line to neutral) and 400 V three-phase (line-to-line) industrial supply.
- Delta systems are used when machines for instance need to draw a large volume of current.
- In star systems the line and phase currents are the same.
- In delta systems the line and phase voltage are the same.
- The neutral cable also allows unbalanced three-phase supplies to operate safely, since it accommodates any out of balance current component.

Assessment criteria 9.1

Outline the essential operational characteristics of electrical equipment

Direct current motors

All electric motors work on the principle that when a current-carrying conductor is placed in a magnetic field it will experience a force. An electric motor uses this magnetic force to turn the shaft of the electric motor. Let us try to understand this action. If a current-carrying conductor is placed into the field of a permanent magnet, as shown in Fig. 2.81(c), a force F will be exerted on the conductor to push it out of the magnetic field.

To understand the force, let us consider each magnetic field acting alone. Figure 2.81(a) shows the magnetic field due to the current-carrying conductor only. Figure 2.81(b) shows the magnetic field due to the permanent magnet in which is placed the conductor carrying no current. Figure 2.81(c) shows the effect of the combined magnetic fields which are distorted and, because lines of magnetic flux never cross but behave like stretched elastic bands, always trying to find the shorter distance between a north and south pole, the force F is exerted on the conductor, pushing it out of the permanent magnetic field.

Assessment criteria 9.2

Apply Fleming's left hand rule

This is the basic motor principle, and the force F is dependent upon the strength of the magnetic field B, the magnitude of the current flowing in the conductor I and the length of conductor within the magnetic field l. The following equation expresses this relationship:

$$F = BIl \ (N)$$

where B is in tesla, l is in metres, I is in amperes and F is in newtons.

We can use Fleming's left hand rule to define the direction of the force. Fleming's rules were discussed earlier in this chapter on page 147.

Example

A coil, which is made up of a conductor some 15 m in length, lies at right-angles to a magnetic field of strength 5 T. Calculate the force on the conductor when 15 A flows in the coil.

$$F = BIl \text{ (N)}$$

$$F = 5\,\text{T} \times 15\,\text{m} \times 15\,\text{A} = 1125\,\text{N}$$

Practical d.c. motors

Practical motors are constructed as shown in Fig. 2.82. All d.c. motors contain a field winding wound on pole pieces attached to a steel yoke. The armature winding rotates between the poles and is connected to the commutator. Contact with the external circuit is made through carbon brushes rubbing on the commutator segments. Direct current motors are classified by the way in which the field and armature windings are connected, which may be in series or in parallel.

Series motor

The field and armature windings are connected in series and consequently share the same current. The series motor has the characteristics of a high starting torque but a speed which varies with load. Figure 2.83 shows series motor

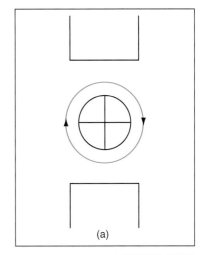

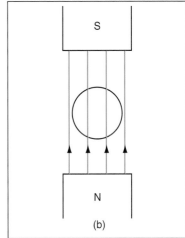

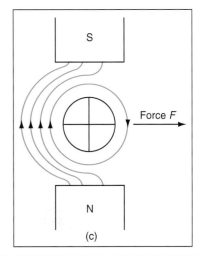

Figure 2.81 Force on a conductor in a magnetic field.

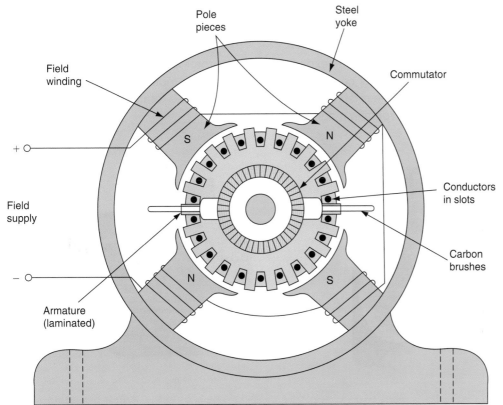

Figure 2.82 Showing d.c. machine construction.

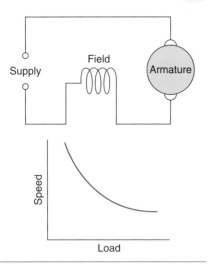

Figure 2.83 Series motor connections and characteristics.

connections and characteristics. For this reason the motor is only suitable for direct coupling to a load, except in very small motors, such as vacuum cleaners and hand drills, and is ideally suited for applications where the machine must start on load, such as electric trains, cranes and hoists.

Reversal of rotation may be achieved by reversing the connections of either the field or armature windings but not both. This characteristic means that the machine will run on both a.c. or d.c. and is, therefore, sometimes referred to as a 'universal' motor.

Three-phase a.c. motors

If a three-phase supply is connected to three separate windings equally distributed around the stationary part or stator of an electrical machine, an alternating current circulates in the coils and establishes a magnetic flux. The magnetic field established by the three-phase currents travels around the stator, establishing a rotating magnetic flux, creating magnetic forces on the rotor which turns the shaft on the motor.

Three-phase induction motor

When a three-phase supply is connected to insulated coils set into slots in the inner surface of the stator or stationary part of an induction motor, as shown in Fig. 2.85(a), a rotating magnetic flux is produced. The rotating magnetic flux cuts the conductors of the rotor and induces an e.m.f. in the rotor conductors by Faraday's law, which states that when a conductor cuts or is cut by a magnetic field an e.m.f. is induced in that conductor, the magnitude of which is proportional to the rate at which the conductor cuts or is cut by the magnetic flux. This induced e.m.f. causes rotor currents to flow and establish a magnetic flux which reacts with the stator flux and causes a force to be exerted on the rotor conductors, turning the rotor, as shown in Fig. 2.85(b).

The turning force or torque experienced by the rotor is produced by inducing an e.m.f. into the rotor conductors due to the *relative* motion between the conductors and the rotating field. The torque produces rotation in the same direction as the rotating magnetic field.

(a)

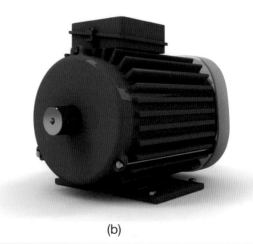

(b)

Figure 2.84 (a) The internal workings of an induction motor; (b) The external shell of an induction motor.

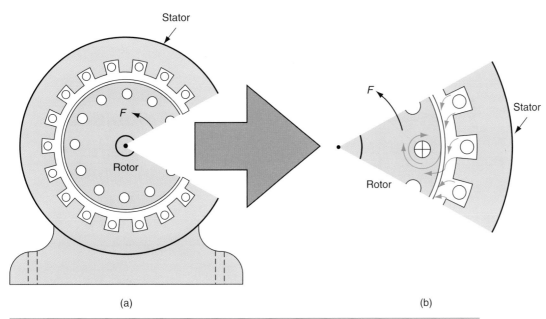

Figure 2.85 Segment taken out of an induction motor to show turning force: (a) construction of an induction motor, and (b) production of torque by magnetic fields.

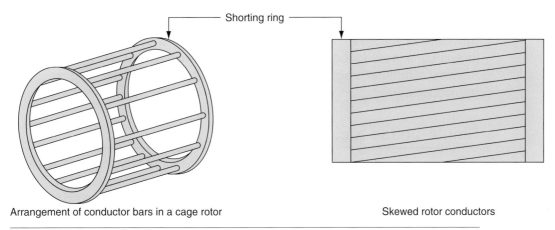

Arrangement of conductor bars in a cage rotor

Skewed rotor conductors

Figure 2.86 Construction of a cage rotor.

Rotor construction

There are two types of induction motor rotor – the wound rotor and the cage rotor. The cage rotor consists of a laminated cylinder of silicon steel with copper or aluminium bars slotted in holes around the circumference and short-circuited at each end of the cylinder, as shown in Fig. 2.86. In small motors the rotor is cast in aluminium. Better starting and quieter running are achieved if the bars are slightly skewed. This type of rotor is extremely robust and since there are no external connections there is no need for slip rings or brushes. A machine fitted with a cage rotor does suffer from a low starting torque and a machine must be chosen which has a higher starting torque than the load, as shown by curve (b) in Fig. 2.87. A machine with the characteristic shown by curve (a) in Fig. 2.87 would not start since the load torque is greater than the machine-starting torque. Alternatively the load may be connected after the motor has been run up to full speed.

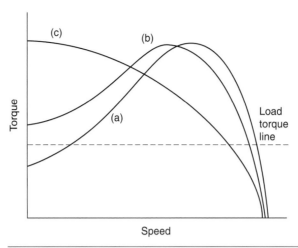

Figure 2.87 Various speed–torque characteristics for an induction motor.

The wound rotor consists of a laminated cylinder of silicon steel with copper coils embedded in slots around the circumference. The windings may be connected in star or delta and the end connections brought out to slip rings mounted on the shaft. Connection by carbon brushes can then be made to an external resistance to improve starting.

The cage induction motor has a small starting torque and should be used with light loads or started with the load disconnected. The speed is almost constant. Its applications are for constant speed machines such as fans and pumps. Reversal of rotation is achieved by reversing any two of the stator winding connections.

Single-phase a.c. motors

A single-phase a.c. supply produces a pulsating magnetic field, not the rotating magnetic field produced by a three-phase supply. All a.c. motors require a rotating field to start. Therefore, single-phase a.c. motors have two windings which are electrically separated by about 90°. The two windings are known as the start and run windings. The magnetic fields produced by currents flowing through these out-of-phase windings create the rotating field and turning force required to start the motor. Once rotation is established, the pulsating field in the run winding is sufficient to maintain rotation and the start winding is disconnected by a centrifugal switch which operates when the motor has reached about 80% of the full load speed.

A cage rotor is used on single-phase a.c. motors, the turning force being produced in the way described previously for three-phase induction motors and shown in Fig. 2.85. Because both windings carry currents which are out of phase with each other, the motor is known as a 'split-phase' motor. The phase displacement between the currents in the windings is achieved in one of two ways:

- by connecting a capacitor in series with the start winding, as shown in Fig. 2.88(a), which gives a 90° phase difference between the currents in the start and run windings;
- by designing the start winding to have a high resistance and the run winding a high inductance, once again creating a 90° phase shift between the currents in each winding, as shown in Fig. 2.88(b).

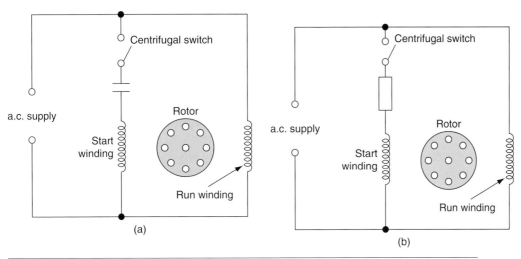

Figure 2.88 Circuit diagram of: (a) capacitor split-phase motors, and (b) resistance split-phase motors.

When the motor is first switched on, the centrifugal switch is closed and the magnetic fields from the two coils produce the turning force required to run the rotor up to full speed. When the motor reaches about 80% of full speed, the centrifugal switch clicks open and the machine continues to run on the magnetic flux created by the run winding only.

Split-phase motors are constant speed machines with a low starting torque and are used on light loads such as fans, pumps, refrigerators and washing machines. Reversal of rotation may be achieved by reversing the connections to the start or run windings, but not both.

Shaded pole motors

The shaded pole motor is a simple, robust, single-phase motor, which is suitable for very small machines with a rating of less than about 50 W. Figure 2.98 shows a shaded pole motor. It has a cage rotor and the moving field is produced by enclosing one side of each stator pole in a solid copper or brass ring, called a shading ring, which displaces the magnetic field and creates an artificial phase shift.

Shaded pole motors are constant speed machines with a very low starting torque and are used on very light loads such as oven fans, record turntable motors and electric fan heaters. Reversal of rotation is theoretically possible by moving the shading rings to the opposite side of the stator pole face. However, in practice this is often not a simple process, but the motors are symmetrical and it is sometimes easier to reverse the rotor by removing the fixing bolts and reversing the whole motor.

Discharge lamps

Discharge lamps do not produce light by means of an incandescent filament but by the excitation of a gas or metallic vapour contained within a glass envelope. A voltage applied to two terminals or electrodes sealed into the end of a glass tube containing a gas or metallic vapour will excite the contents and produce light directly. Fluorescent tubes and CFLs operate on this principle.

 Definition

Discharge lamps do not produce light by means of an incandescent filament but by the excitation of a gas or metallic vapour contained within a glass envelope.

Definition

A *luminaire* is equipment which supports an electric lamp and distributes or filters the light created by the lamp.

Fluorescent luminaires

A **luminaire** is equipment which supports an electric lamp and distributes or filters the light created by the lamp. It is essentially the 'light fitting'.

A lamp is a device for converting electrical energy into light energy. There are many types of lamps. General lighting service (GLS) lamps and tungsten halogen lamps use a very hot wire filament to create the light and so they also become very hot in use. Fluorescent tubes operate on the 'discharge' principle; that is, the excitation of a gas within a glass tube. They are cooler in operation and very efficient in converting electricity into light. They form the basic principle of most energy-efficient lamps.

Fluorescent lamps are linear arc tubes, internally coated with a fluorescent powder, containing a little low-pressure mercury vapour and argon gas. The lamp construction is shown in Fig. 2.89.

Passing a current through the electrodes of the tube produces a cloud of electrons that ionize the mercury vapour and the argon in the tube, producing invisible ultraviolet light and some blue light. The fluorescent powder on the inside of the glass tube is very sensitive to ultraviolet rays and converts this radiation into visible light.

Fluorescent luminaires require a simple electrical circuit to initiate the ionization of the gas in the tube and a device to control the current once the arc is struck and the lamp is illuminated. Such a circuit is shown in Fig. 2.90.

A typical application for a fluorescent luminaire is in suspended ceiling lighting modules used in many commercial buildings. Energy-efficient lamps use electricity much more efficiently.

The choke has two functions:

- it creates a high voltage strike when the starter switch opens;
- it lowers the circuit current once the light is lit due to its high impedance.

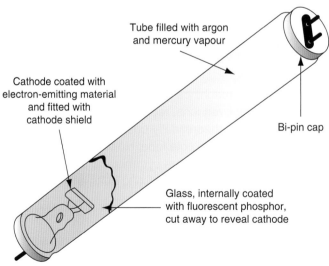

Figure 2.89 Fluorescent lamp construction.

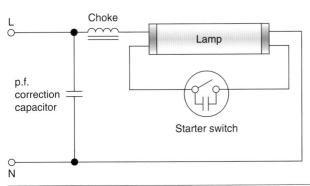

Figure 2.90 Fluorescent lamp circuit arrangement.

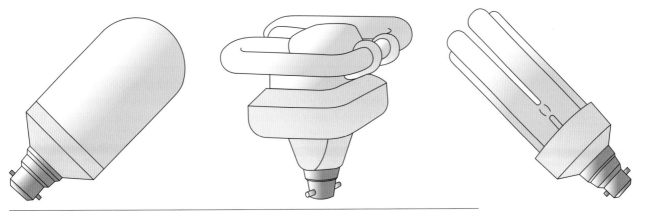

Figure 2.91 Energy-efficient lamps.

Compact fluorescent lamps

CFLs are miniature fluorescent lamps designed to replace ordinary GLS lamps. They are available in a variety of shapes and sizes so that they can be fitted into existing light fittings. Figure 2.91 shows three typical shapes. The 'stick' type give most of their light output radially while the flat 'double D' type give most of their light output above and below.

Definition

CFLs are miniature fluorescent lamps designed to replace ordinary GLS lamps.

LED lamps

Light emitting diode lamps have been around for more than fifty years as small signal and indicator lamps, but recent scientific research and new technology have enabled the super efficient LED lamps that we now know, to be created. LED lamps have come such a long way in recent years and are now probably the first choice for energy efficient lamp and luminaire replacement.

A light emitting diode (LED) is a very small, semi-conductor source of illumination. It is a P-N junction diode that emits photons of light when activated, called electroeluminence.

LED lamps have many advantages over GLS and compact fluorescent lamps (CFLs) including very low energy consumption, greater robustness, smaller size and longer life. They are available in a range of shapes and sizes with Edison Screw, Bayonet Cap, Bi-Pin, GU 10 and GX 53 connections so that they may easily replace existing lamps.

When installing LED Lamps and luminaires, the installer must take account of the colour of the light emitted by the LED lamp for a particular application. The colour of the light emitted by an LED lamp is measured in Kelvin, a temperature measurement. The higher the number, the whiter is the light output. A light output of 3,000K and above gives a bright white light, suitable for installation in contemporary kitchens and dentist surgeries. A 2,700K lamp gives a much warmer light, suitable for table lamps with shades in a domestic situation. A 2,700K LED lamp almost replicates the soft yellow/white light output of a GLS filament lamp but consumes much less energy.

Semi conductors, P-N junctions and LEDs are discussed in Chapter 6 of *Electrical Installation Work* (EAL) Edition, Level 3.

Definition

A transformer is an electrical machine which is used to change the value of an alternating voltage.

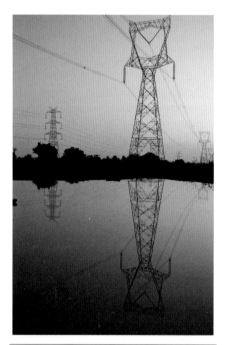

Figure 2.92 Transmission tower.

Assessment criteria 9.3

Describe the operating principle of transformers

Transformers

A **transformer** is an electrical machine which is used to change the value of an alternating voltage. They vary in size from miniature units used in electronics to extremely efficient huge power transformers used in power-stations (Fig. 2.92.)

A transformer will only work on an alternating supply, it will not normally work from a d.c. supply such as a battery.

The iron core of the transformer is not solid but made up of very thin sheets called laminations, to improve efficiency.

- An alternating voltage applied to the primary winding establishes an alternating magnetic flux in the core.
- The magnetic flux in the core causes a voltage to be induced in the secondary winding of the transformers.
- The voltage in both the primary and secondary windings is proportional to the number of turns.
- This means that if you increase the number of secondary turns you will increase the output voltage. This has an application in power distribution.
- Alternatively, reducing the number of secondary turns will reduce the output voltage. This is useful for low-voltage supplies such as domestic bell transformers. Because it has no moving parts, a transformer can have a very high efficiency. Large power transformers, used on electrical distribution systems, can have an efficiency of better than 90%.

Large power transformers need cooling to take the heat generated away from the core. This is often achieved by totally immersing the core and windings in insulating oil. A sketch of an oil-filled transformer can be seen in Figure 2.93.

When an alternating current is passed through a single coil, the coil will produce a second voltage, one that will act against the current that produces it.

This process is called self-induction and the second voltage is called a back e.m.f. (electro motive force).

A transformer works on the principle that both coils are linked through magnetism and this is called mutual induction. This is shown in Figure 2.94, which consists of two coils, called the primary and secondary coils, or windings, which are insulated from each other and wound on to the same steel or iron core.

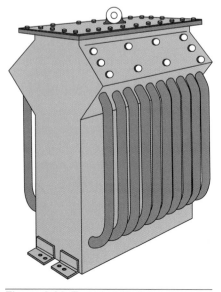

Figure 2.93 Typical oil-filled power transformer.

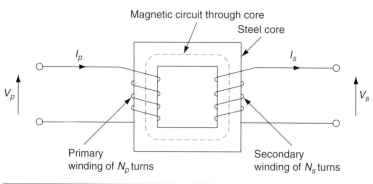

Figure 2.94 A simple transformer.

Assessment criteria 9.4

Determine by calculation and measurement: primary and secondary voltages, primary and secondary current

An alternating voltage applied to the primary winding produces an alternating current, which sets up an alternating magnetic flux throughout the core. This magnetic flux induces an e.m.f. in the secondary winding, as described by Faraday's law, which says that when a conductor is cut by a magnetic field, an e.m.f. is induced in that conductor. Since both windings are linked by the same magnetic flux, the induced e.m.f. per turn will be the same for both windings. Therefore, the e.m.f. in both windings is proportional to the number of turns. In symbols:

$$\frac{V_p}{N_p} = \frac{V_s}{N_s}$$

where

V_p = the primary voltage

V_s = the secondary voltage

Figure 2.95 A 3ph transformer.

N_p = the number of primary turns

N_s = the number of secondary turns

Moving the terms around, we have a general expression for a transformer:

$$\frac{V_p}{V_s} = \frac{N_p}{N_s} = IS/IP$$

Try this

Maths

Using the general equation for a transformer given above, follow this maths carefully, step by step, in the following example.

Example

A 230 V to 12 V emergency lighting transformer is constructed with 800 primary turns. Calculate the number of secondary turns required. Collecting the information given in the question into a usable form, we have:

$$V_p = 230 \text{ V}$$
$$V_s = 12 \text{ V}$$
$$N_p = 800$$

From the general equation:

$$\frac{V_p}{V_s} = \frac{N_p}{N_s}$$

the equation for the secondary turn is:

$$N_s = \frac{N_p V_s}{V_p}$$

$$\therefore N_s = \frac{800 \times 12V}{230V} = 42 \text{ turns}$$

42 turns are required on the secondary winding of this transformer to give a secondary voltage of 12 V.

Definition

Step down transformers are used to reduce the output voltage, often for safety reasons.

Definition

Step up transformers are used to increase the output voltage. The electricity generated in a power-station is stepped up for distribution on the National Grid network.

Assessment criteria 9.5

Identify transformer types

Types of transformer

Step down transformers are used to reduce the output voltage, often for safety reasons. Figure 2.96 shows a step down transformer where the primary winding has twice as many turns as the secondary winding. The turns ratio is 2:1 and, therefore, the secondary voltage is halved.

Step up transformers are used to increase the output voltage. The electricity generated in a power-station is stepped up for distribution on the National Grid

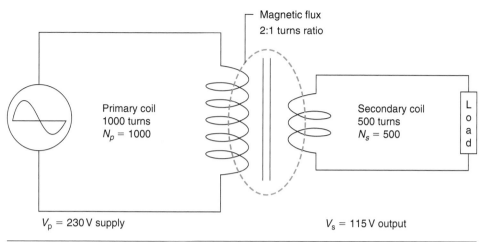

Figure 2.96 A step down transformer.

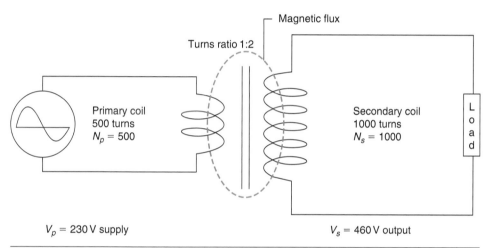

Figure 2.97 A step up transformer.

network. Figure 2.97 shows a step up transformer where the primary winding has only half the number of turns as the secondary winding. The turns ratio is 1:2 and, therefore, the secondary voltage is doubled.

Instrument transformers are used in industry and commerce so that large currents and voltage can be measured by small electrical instruments.

A current transformer (CT) has the large load currents connected to the primary winding of the transformer and the ammeter connected to the secondary winding. The ammeter is calibrated to take account of the turns ratio of the transformer, so that the ammeter displays the actual current being taken by the load when the ammeter is actually only taking a small proportion of the load current.

A voltage transformer (VT) has the main supply voltage connected to the primary winding of the transformer and the voltmeter connected to the secondary winding. The voltmeter is calibrated to take account of the turns ratio of the transformer, so that the voltmeter displays the actual supply voltage.

Isolating transformers – such as separated extra-low voltage (SELV) transformers. If the primary winding and the secondary winding of a double wound transformer have a separate connection to earth, then the output

Key fact

Transformers
- Transformers are very efficient (more than 80%) because they do not have any moving parts.

Top tip

The primary and secondary coil of a transformer are linked by magnetism only. This means that the user is electrically isolated from the supply.

of the transformer is effectively isolated from the input since the only connection between the primary and secondary windings is the magnetic flux in the core.

Assessment criteria 9.6

Outline applications of transformers

Transformers therefore are extensively used, ranging from the distribution of very high voltages to very small through the supply of electronic applications. A very important use is that of providing reduced voltage systems such as through autotransformers, for example, in which the autotransformer employs only one coil. Various voltages can then be tapped off such as 110V which is especially important when supplying hand held equipment on constructions sites. Damp conditions will need further precautions such as the use of SELV, which typically is run through a 12V supply, bearing in mind, of course, that equipment supplied from a SELV source may be installed in a bathroom or shower room, provided that it is suitably enclosed and protected from the ingress of moisture. This includes equipment such as water heaters, pumps for showers and whirlpool baths.

Although applications such as shaver sockets do not meet the SELV voltage requirements, it safeguards the user through the use of safety isolating transformers. The transformer operates through magnetism only, therefore by using a turns ratio of 1:1, the user is electrically isolated from the incoming supply.

Try this

Have you seen any transformers in action? Were they big or small – what were they being used for? Have you been close up to a transmission tower, perhaps when you were walking in the countryside? Make a note in the margin.

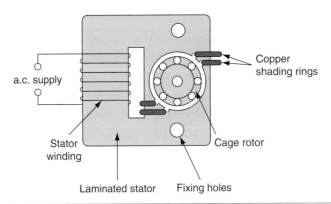

Figure 2.98 Shaded pole motor.

There are more motors operating from single-phase supplies than all other types of motor added together. Most of them operate as very small motors in domestic and business machines where single-phase supplies are most common.

Test your knowledge

When you have completed the questions, check out the answers at the back of the book.

Note: more than one multiple-choice answer may be correct.

Learning outcome 1

1   The SI unit of mass is the:
    a.   kilogram or kg
    b.   pound or lb
    c.   metre or m
    d.   millimetre or mm

2   The prefix mega applies to:
    a.   100
    b.   1000
    c.   1000,000
    d.   10

3   The SI unit of energy is the:
    a.   ohm
    b.   joule
    c.   watt
    d.   coulomb

4   Area is represented by:
    a.   kg
    b. m^2
    c. m^3
    d.   m

5   The SI unit of resistance is the:
    a.   ohm
    b.   joule
    c.   watt
    d.   coulomb

6   The SI unit of electrical charge is the:
    a.   ohm
    b.   joule
    c.   watt
    d.   coulomb

7   The watt is a measure of:
   a.   potential difference
   b.   power
   c.   force
   d.   coulomb

8   The volt is a measure of:
   a.   potential difference
   b.   power
   c.   voltage
   d.   coulomb

9   The newton is a measure of:
   a.   potential difference
   b.   power
   c.   force
   d.   volt

10   The prefix milli applies to:
   a. 10^{-2}
   b. 10^{-3}
   c. 10^{-4}
   d. 10^{-6}

Learning outcome 2

11   Rearrange the formula $R = \rho\, L/A$ to find ρ.
   a. $\rho = R/LA$
   b. $\rho = RA/L$
   c. $\rho = RL/A$
   d. $\rho = A/RL$

12   A shower unit draws 20 A from a domestic 230 V supply. Calculate the amount of power used.
   a.   4.6 kW
   b.   11.5 kW
   c.   46 kW
   d.   0.08 kW

13   The resistance of a kettle element which takes 12 A from a 230 V mains supply is:
   a.   2.88 Ω
   b.   5.00 Ω
   c.   12.24 Ω
   d.   19.16 Ω

14   A 12 Ω filament lamp was found to be taking a current of 2 A at full brilliance. The voltage across the lamp under these conditions is:
   a.   6 V
   b.   12 V
   c.   24 V
   d.   48 V

15 Calculate the energy generated when a circuit is supplied with 100 V, and a current of 2 amperes flows for 6 minutes?
   a. 72 KJ
   b. 1.2 KJ
   c. 12 KJ
   d. 72000 J

16 An electron:
   a. is negatively charged
   b. is positively charged
   c. has no charge
   d. is balanced

17 230 V is a measure of:
   a. potential difference
   b. amps
   c. watts
   d. ohms

18 Domestic wiring uses what kind of insulation material?
   a. Mica
   b. PVC
   c. Mineral oxide
   d. Glass

19 Good conductor materials include:
   a. copper
   b. PVC
   c. brass
   d. wood

20 Good insulator materials include:
   a. copper
   b. PVC
   c. mica
   d. mineral oxide

21 A good conductor material:
   a. has lots of free electrons
   b. has no free electrons
   c. may be made of copper
   d. may be made of plastic

22 A good insulator material:
   a. has lots of free electrons
   b. has few free electrons
   c. may be made of copper
   d. may be made of plastic

23 Which of the following would be considered a load:
    a. motor
    b. lamp
    c. circuit breaker
    d. switch

24 Water heaters use which of the following to operate?
    a. Heat
    b. Magnetism
    c. Chemical
    d. Light

25 If the voltage across a lamp is less than the supply. What causes this?
    a. Current drop
    b. Power drop
    c. Resistance drop
    d. Volt drop

26 A solenoid operates through:
    a. magnetism
    b. heat
    c. chemical reaction
    d. light

27 What kind of meter is connected in series?
    a. Volt meter
    b. Ammeter
    c. Frequency meter
    d. Watt meter

Learning outcome 3

28 Which of the following uses steam to generate electricity?
    a. Oil
    b. Solar thermal
    c. Coal
    d. Nuclear

29 Alternating current is used in the distribution and transmission of electricity because:
    a. it is cheaper than d.c.
    b. transformers do not work on d.c.
    c. transformers only work with a.c.
    d. it is safer than d.c

30 A large industrial outlet needs which of the following supplies?
    a. 400 V three phase and neutral
    b. 400 V single phase
    c. 230 V single phase
    d. 230 V three phase and neutral

31 A domestic setting is associated with:
   a. 400 V three phase and neutral
   b. 400 V single phase
   c. 230 V single phase
   d. 230 V three phase and neutral

32 Electricity is typically generated in large commercial power-stations at:
   a. 230 V
   b. 400 V
   c. 25 kV
   d. 132 kV

33 Transmission of electricity on the National Grid is associated with:
   a. 230 V
   b. 400 V
   c. 25 kV
   d. 132 kV

34 Electricity is transmitted at very high voltages because for the same amount of power the:
   a. current is reduced
   b. current is increased
   c. losses are reduced
   d. losses are increased

35 If any given power station fails, the National Grid will:
   a. fail
   b. shut down temporarily
   c. use standby power
   d. divert power from other sources

Learning outcome 4

36 If a person has a stream flowing through their land, which of the following could they take advantage of?
   a. Micro-hydro
   b. Micro-turbine
   c. Air source ground pump
   d. Ground source ground source

37 Which of the following uses the energy from the sun in order to produce hot water?
   a. Ground source heat pump
   b. Air source heat pump
   c. Solar thermal
   d. Photovoltaic

38 Micro-generation technology:
   a. can reduce the effect of changes in energy prices to households
   b. can increase the effect of changes in energy prices to households
   c. can reduce the emission of carbon dioxide
   d. is extremely inefficient

Learning outcome 5

39   A 20V battery causes 2 amps of current to flow in a circuit when two
     resistors R1 & R2 are connected to it. The resistors are connected in series
     with R1 being valued at 8 Ω. Work out the value of R2.

   a.   20Ω

   b.   12Ω

   c.   5Ω

   d.   2Ω

40   Three resistors are all valued at 6Ω and are connected in parallel. Work out
     the total resistance.

   a.   3Ω

   b.   6Ω

   c.   18Ω

   d.   2Ω

41   Resistors of 6Ω and 3Ω are connected in series. The combined resistance
     value will be:

   a.   2.0Ω

   b.   3.6Ω

   c.   6.3Ω

   d.   9.0Ω

42   Resistors of 6Ω and 3Ω are connected in parallel. The combined resistance
     value will be:

   a.   2.0Ω

   b.   3.6Ω

   c.   6.3Ω

   d.   9.0Ω

43   Resistors of 60Ω, 40Ω and 20Ω are connected in parallel. The total
     resistance values will be:

   a.   10.9Ω

   b.   20.04Ω

   c.   60.0Ω

   d.   120Ω

44   Resistors of 60Ω, 40Ω and 20Ω are connected in series. The total
     resistance values will be:

   a.   10.9Ω

   b.   20.0Ω

   c.   60.0Ω

   d.   120Ω

45   Two identical resistors are connected in series across a 24V battery. The
     volt drop across each resistor will be:

   a.   24V

   b.   12V

   c.   6V

   d.   48V

46   Two identical resistors are connected in parallel across a 24 V battery. The volt drop across each resistor will be:
    a.   24 V
    b.   12 V
    c.   6 V
    d.   48 V

47   In a series circuit:
    a.   the current is common to all resistors
    b.   the voltage is common to all resistors
    c.   Rt = R1 + R2
    d. $\dfrac{1}{R_T} = \dfrac{1}{R_1} + \dfrac{1}{R_2}$

48   In a parallel circuit:
    a.   the current is common to all resistors
    b.   the voltage is common to all resistors
    c.   Rt = R1 + R2
    d. $\dfrac{1}{R_T} = \dfrac{1}{R_1} + \dfrac{1}{R_2}$

Learning outcome 6

49   Lines of magnetic field flow from:
    a.   north to north
    b.   south to south
    c.   north to south
    d.   south to north

50   Fleming's right hand rule is associated with:
    a.   a generator
    b.   a motor
    c.   a relay
    d.   an inductor

51   An electromagnetic device which controls a large electrical supply by switching a smaller powered circuit is known as:
    a.   a transformer
    b.   a diode
    c.   a relay
    d.   an inductor

52   An electromagnetic device which uses a solenoid to operate a number of contacts is one definition of:
    a.   a transformer
    b.   a motor
    c.   a relay
    d.   an inductive coil

53   If a conductor within an alternator cuts the magnetic field at 45 degrees the output will be:
 a. maximum
 b. minimum
 c. 50%
 d. 25%

54   A relay operates through:
 a. self-induction
 b. electromagnetism
 c. magnetic induction
 d. mutual induction

55   Work out the frequency of a sine wave if its periodic time is 20 ms:
 a. 50 Hz
 b. 100 Hz
 c. 60 Hz
 d. 20 Hz

56   Work out the periodic time of a sine wave which has a frequency of 1 Khz:
 a. 1 mS
 b. 10 mS
 c. 100 mS
 d. 1 S

57   Work out the periodic time of a sine wave which has a frequency of 10 Mhz:
 a. 0.1 μS
 b. 100 nS
 c. 100,000 pS
 d. 100 S

58   RMS can be defined or seen as:
 a. the useful aspect of a sine wave
 b. the equivalent d.c. value
 c. 230 V
 d. average value

59   The principle of electrical transmission is:
 a. voltage is high, current is low, conductors are small, more efficient due to less heat loss
 b. voltage is low, current is high, conductors are small, more efficient due to less heat loss
 c. voltage is high, current is low, conductors are large, more efficient due to less heat loss
 d. voltage is high, current is low, conductors are small, more efficient due to more heat loss

Learning outcome 7

60   Which of the following may be defined as 'a measure of the amount of material in a substance'?
 a. Acceleration
 b. Force

c.   Mass

d.   Weight

61   Who defined the formula Force = Mass x acceleration?

a.   Fahrenheit

b.   Newton

c.   Bell

d.   Watt

62   Which of the following may be defined as 'the rate of doing work'?

a.   Acceleration

b.   Work done

c.   Power

d.   Velocity

63   Which of the following may be defined as 'the capacity for doing work'?

a.   Energy

b.   Work done

c.   Power

d.   Velocity

64   Which of the following may be defined as 'the speed in a given direction'?

a.   Acceleration

b.   Work done

c.   Power

d.   Velocity

Learning outcome 8

65   In a three-phase system, what is the function of the neutral cable?

a.   Produces 230 V

b.   Carries unbalanced currents

c.   Produces 400 V

d.   Carries balanced currents

66   The benefits of balancing three-phase loads and including a neutral cable include:

a.   providing a path for potentially dangerous circulating currents

b.   neutral currents are kept small

c.   neutral cable can be smaller

d.   all line conductors can be excessively large

67   Which configuration gives us two values of voltage?

a.   Star

b.   Delta

c.   Star two-phase

d.   Delta four-wire

68   Measuring across two lines of a star-connected three-phase four-wire supply system from the local substation would have a voltage of:

a.   230 V

b.   400 V

c.   25 kV

d.   132 kV

69   A load connected to phase and neutral of a star-connected three-phase four-wire supply system from the local substation would have a voltage of:
   a.   230 V
   b.   400 V
   c.   25 kV
   d.   132 kV

70   The phase voltage of a star-connected load is 100 V. The line voltage will be:
   a.   57.73 V
   b.   100 V
   c.   173.2 V
   d.   230 V

71   The phase voltage of a delta-connected load is 100 V. The line voltage will be:
   a.   57.73 V
   b.   100 V
   c.   173.2 V
   d.   230 V

72   The phase current of a star-connected load is 100 A. The line current will be:
   a.   57.73 A
   b.   100 A
   c.   173.2 A
   d.   230 A

73   The phase current of a delta-connected load is 100 A. The line current will be:
   a.   57.73 A
   b.   100 A
   c.   173.2 A
   d.   230 A

Learning outcome 9

74   A fluorescent luminaire emits light due to electricity passing through?
   a.   The starter
   b.   The filament
   c.   The gas
   d.   The choke

75   Applications for a series d.c. motor include:
   a.   an electric train
   b.   a microwave oven
   c.   a central heating pump
   d.   an electric hand drill

76   One application for a single phase induction motor is:
   a.   an electric train
   b.   a microwave oven
   c.   a central heating pump
   d.   an electric hand drill

77 One application for a shaded pole a.c. motor is:
   a. an electric train
   b. a microwave oven
   c. a central heating pump
   d. an electric hand drill

78 Shaded pole a.c. motors create motion through:
   a. a magnetic ring
   b. an iron ring
   c. a brass ring
   d. a copper ring

79 Fleming's left hand rule is associated with:
   a. a generator
   b. a motor
   c. a relay
   d. an inductor

80 A safety isolation transformer?
   a. requires earthing
   b. does not require earthing
   c. is used with shaver sockets
   d. only has one winding

81 A transformer is associated with:
   a. self-induction
   b. electromagnetism
   c. magnetic induction
   d. mutual induction

82 If a transformer has no losses, what does this mean?
   a. power in equals power out
   b. power in is less than power out
   c. power out is less than power in
   d. power out is double power in

83 An autotransformer has how many windings?
   a. 1
   b. 2
   c. 3
   d. 4

84 State the units of resistance and current.

85 Describe, with the aid of a simple diagram, how the atoms and electrons behave in a material said to be a good conductor of electricity.

86 Describe, with the aid of a simple diagram, how the atoms and electrons behave in a material said to be a good insulator.

87 List five materials which are used as good conductors in the electrical industry.

88 List five materials which are used as good insulators in the electrical industry.

89 Sketch a simple circuit of two resistors connected in series across a battery and explain how the current flows in this circuit.

90 Sketch a simple circuit of two resistors connected in parallel across a battery and explain how the current flows in this circuit.

91 Sketch a simple circuit to show how a voltmeter and ammeter would be connected into the circuit to measure total voltage and total current.

92 Sketch the construction of a simple transformer and label the primary and secondary windings. Why is the metal core of the transformer laminated? How do we cool a big power transformer?

93 List five practical applications for a transformer – for example, a shaver socket.

94 Describe the three effects of an electric current.

95 Sketch the magnetic flux patterns:
   a.   around a simple bar magnet
   b.   a horseshoe magnet
   c.   explain the action and state one application for a solenoid.

96 Briefly describe what we mean by 'a turning force' and give five practical examples of this effect.

97 Briefly define what we mean by a 'simple machine' and give five examples.

98 Briefly describe what we mean by 'the efficiency of a machine'.

99 Sketch the construction of a simple alternator and label all the parts.

100 State how an e.m.f. is induced in an alternator. Sketch and name the shape of the generated e.m.f.

101 Calculate or state the average r.m.s. and maximum value of the domestic a.c. mains supply and show these values on a sketch of the mains supply.

102 Use a sketch with notes of explanation to describe how a force is applied to a conductor in a magnetic circuit and how this principle is applied to an electric motor.

103 Use a sketch with notes of explanation to show how a turning force is applied to the rotor and, therefore, the drive shaft of an electric motor.

Unit Elec2/08

Chapter 2 checklist

Learning outcome	Assessment criteria – the learner can		Page number
1. Understand common units of measurement used in electrotechnical work.	1.1	Identify (SI) units of measurement for general quantities.	92
	1.2.	State the SI or derived SI unit for electrical quantities.	93
	1.3	Identify the common multiples and sub-multiples used within electrotechnical work.	94
2. Understand the principles of electrical circuits.	2.1	Transpose a basic formula.	94
	2.2	Determine electrical quantities using Ohm's law.	97
	2.3	Calculate values of electrical power in basic circuits.	98
	2.4	Calculate the values of electrical energy.	101
	2.5	Determine the resistance of a conductor.	101
	2.6	Describe what is meant by resistance and resistivity in relation to electrical circuits.	101
	2.7	Outline the principles of an electrical circuit.	105
	2.8	Differentiate between materials which are good: • Conductors • Insulators.	106
	2.9	Define the sources of electromotive force.	109
	2.10	Clarify the main effects of electric currents.	109
	2.11	State what is meant by the term 'voltage drop' in relation to electrical circuits.	111
	2.12	Identify how measurement instruments are connected into electrical circuits.	111
3. Understand electrical supply systems.	3.1	Identify how electricity is generated.	115
	3.2	Identify the features of a generation, transmission and distribution system.	116
	3.3	Identify the applications of supply systems.	117
4. Understand micro-generation technologies.	4.1	Outline the principles of electricity micro-generation technologies.	118
	4.2	Outline the principles of heat and co-generation micro-technologies.	119
	4.3	State the benefits of micro-generation.	134
5. Understand basic series and parallel electrical circuits.	5.1	State the relationship between current, voltage and resistance in parallel and series d.c. circuits.	135
	5.2	Determine electrical quantities in series d.c. circuits.	135
	5.3	Determine electrical quantities in parallel d.c. circuits.	136
6. Understand the principles of electro-magnetism.	6.1	Identify magnetic flux patterns of circuits.	143
	6.2	Apply Fleming's right hand rule in relation to a basic alternator.	146
	6.3	Identify applications of electromagnetism.	148
	6.4	Describe the basic principles of generating an alternating current.	149
	6.5	Determine sinusoidal quantities.	151
	6.6	State the reasons for using alternating current transmission and distribution.	154

Learning outcome	Assessment criteria – the learner can		Page number
7. Understand fundamental mechanics. (This Learning outcome is assessed by assignment only).	7.1	Distinguish what is meant by mass and weight.	154
	7.2	Outline the principles of basic mechanics as they apply to: • Levers • Gears • Pulleys	156
	7.3	Determine mechanical quantities.	160
	7.4	Outline the main principle of kinetic and potential energy.	162
	7.5	Calculate values of electrical efficiency.	165
8. Understand the principles of three phase circuits.	8.1	Distinguish between current and voltage in a star.	166
	8.2	Distinguish between current and voltage in a delta.	166
	8.3	Recognize the reasons for the neutral conductor.	169
	8.4	Identify the advantages of distributing loads evenly over the three lines.	169
9. Understand the operating principles of electrical equipment.	9.1	Outline the essential operational characteristics of electrical equipment.	170
	9.2	Apply Fleming's left hand rule.	170
	9.3	Describe the operating principle of transformers.	178
	9.4	Determine by calculation and measurement: • Primary and secondary voltages • Primary and secondary current.	179
	9.5	Identify transformer types	180
	9.6	Outline applications of transformers.	182

EAL Unit Elec2/04

CHAPTER **3**

Electrical installation theory and technology

EAL Electrical Installation Work – Level 2, 2nd Edition 978 0 367 19562 5
© 2019 Linsley. Published by Taylor & Francis. All rights reserved.
www.routledge.com/9780367195618

Learning outcomes

When you have completed this chapter you should:

1. Know the legislation, regulations and guidance that apply to electrical installation work.
2. Know the technical information used in electrical work.
3. Understand the properties, applications and limitations of different wiring systems.
4. Know the general layout of equipment at the service position.
5. Understand standard lighting circuits.
6. Understand standard ring and radial final circuits.
7. Know the basic requirements for circuits.
8. Know the importance of earthing and bonding for protection.
9. Know the principles of overcurrent protection.
10. Know the principles of circuit design.

DOI: 10.1201/9780429203176-3

Learners are reminded that unit ELEC2-01 covers statutory and non-statutory documents and can look back at this chapter for these outcomes.

This unit provides the learner an opportunity to acquire the knowledge underpinning how electrical systems are selected and how they operate, together with how regulations and requirements affect electrical installations.

Assessment criteria 1.1

Recognize the legal status of documents used in the electrical industry

Communications

When we talk about good communications we are talking about transferring information from one person to another both quickly and accurately. We do this by talking to other people, looking at drawings and plans and discussing these with colleagues from the same company and with other professionals who have an interest in the same project. The technical information used within our industry comes from many sources. The IET Regulations (BS 7671) are non statutory but enable electricians to comply with the Electricity at Work Regulations 1989. They are the 'electrician's bible' and form the basis of all our electrical design calculations and installation methods. British Standards, European Harmonised Standards and codes of practice provide detailed information for every sector of the electrical industry, influencing all design and build considerations.

Appendix 1 of BS 7671 lists the BS and BS EN numbers referred to throughout the regulation, giving a brief description of each.

Assessment criteria 1.2

Define the implications of not complying with regulations, documents and guidance

Approved codes of practice

Approved codes of practice (ACOP) give guidance on complying with the specific area each code is written for. While failure to comply with any code of practice is not an offence in law, courts will look for evidence that alternative ways to comply with legislation have been used if the ACOP has not been followed. It is therefore accepted that following ACOPs is best practice.

The HSE website lists electrical standards and approved codes of practice.

Assessment criteria 2.1

Identify the purpose of different sources of technical information used in electrical work

Sources of technical information

The equipment and accessories available to use in a specific situation can often be found in the very comprehensive manufacturers' catalogues and the catalogues of the major wholesalers that service the electrical industries.

All of this technical information may be distributed and retrieved by using:

- conventional drawings and diagrams which we will look at in more detail below;
- sketch drawing to illustrate an idea or the shape of, say, a bracket to hold a piece of electrical equipment;
- the Internet, for downloading British Standards and codes of practice;
- the Internet, also for viewing health and safety information from the health and safety executive at: www.hse.gov.uk;
- CDs, DVDs, USB memory sticks and email for communicating and storing information electronically;
- tablets and android devices for communicating with other busy professionals, information, say, about a project you are working on together.

If you are working at your company office with access to online computers, then technical information is only a fingertip or mouse click away. However, a construction site can be a hostile environment for a laptop and so a hard copy of any data is often preferable when on-site.

We will look at the types of drawings and diagrams which we use within our industry to communicate technical information between colleagues and other professionals. The type of diagram to be used in any particular situation is the one which most clearly communicates the desired information.

Figure 3.1 Computers can be very useful for reference, but be careful if using them on-site.

Types of technical information

Technical information is communicated to electrical personnel in lots of different ways. It comes in the form of:

- Specifications: these are details of the client's requirements, usually drawn up by an architect. For example, the specification may give information about the type of wiring system to be employed or detail the type of luminaires or other equipment to be used.
- Manufacturer's data: if certain equipment is specified, let's say a particular type of luminaire or other piece of equipment, then the manufacturer's data

sheet will give specific instructions for its assembly and fixing requirements. It is always good practice to read the data sheet before fitting the equipment. A copy of the data sheet should also be placed in the job file for the client to receive when the job is completed.

- Reports and schedules: a report is the written detail of something that has happened or the answer to a particular question asked by another professional person or the client. A schedule gives information about a programme or timetable of work: it might be a list or a chart giving details of when certain events will take place. For example, when the electricians will start to do the 'first fix' and how many days it will take.

- User instructions: give information about the operation of a piece of equipment. Manufacturers of equipment provide 'user instructions' and a copy should be placed in the job file for the client to receive when the project is handed over.

- Job sheets and time sheets: give 'on-site' information. Job sheets give information about what is to be done and are usually issued by a manager to an electrician. Time sheets are a record of where an individual worker has been spending his or her time, which job and for how long. This information is used to make up individual wages and to allocate company costs to a particular job. We will look at these again later under the subheading 'on-site documentation'.

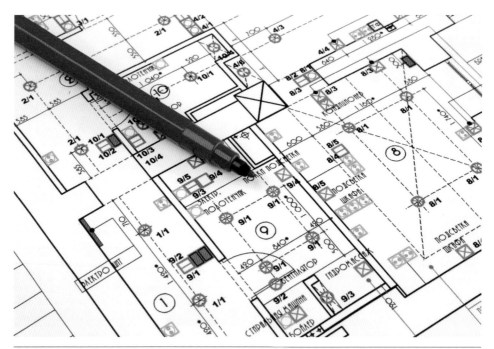

Figure 3.2 It is important that everyone involved in a job has access to the same information, one of the reasons specifications are so important.

Those who need or use technical information

Technical information is required by many of the professionals involved in any electrical activity, so who are the key people?

- The operative: in our case this will be the skilled electricians actually on-site, doing the job for the electrical company.

- The supervisor: he or she may have overall responsibility for a number of electricians on-site and will need the 'big picture'.

- The contractor: the main contractor takes on the responsibility of the whole project for the client. The main contractor may take on a subcontractor to carry out some part of the whole project. On a large construction site the electrical contractor is usually the subcontractor.
- The site agent: he or she will be responsible for the smooth running of the whole project and for bringing the contract to a conclusion on schedule and within budget. The site agent may be nominated by the architect.
- The customer or client: they also are the people ordering the work to be done. They will pay the final bill that pays everyone's wages.

Assessment criteria 2.2

Identify the different diagrams and drawings used in electrical work

Drawings and diagrams

So that the electrician will know where to install the sockets, lights and equipment they will probably be provided with a site plan or layout drawing by the architect or main contractor.

Site plans or layout drawings

These are scale drawings based upon the architect's site plan of the building and show the position of the electrical equipment which is to be installed. The electrical equipment is identified by a graphical symbol. The standard symbols used by the electrical contracting industry are those recommended by the British Standard EN 60617, Graphical Symbols for Electrical Power, Telecommunications and Electronic Diagrams. The inside rear cover of the IET *On-Site Guide* shows some of the symbols in most common usage and some have been given in Figure 3.3.

Assessment criteria 2.3

Identify graphical symbols used in diagrams and drawings

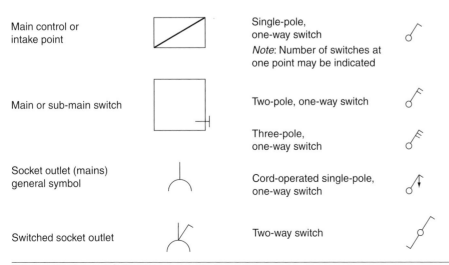

Figure 3.3 Some BS EN 60617 electrical installation symbols

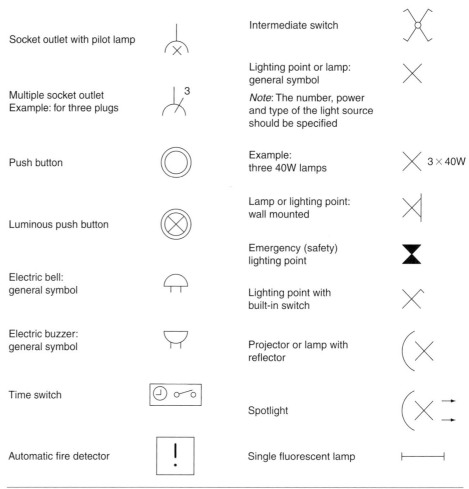

Figure 3.3 *(continued)*

Assessment criteria 2.4

Convert measurements from a scale drawing

The site plan or layout drawing will be drawn to a scale, smaller than the actual size of the building, so to find the actual measurement you must measure the distance on the drawing and multiply by the scale.

For example, if the site plan is drawn to a scale of 1:100, then 10 mm on the site plan represents 1 m measured in the building (10 mm × 100 = 1000 mm or 1 m).

Common scales used in the construction industry include 1:200, 1:100, 1:50, 1:20.

Scale	Measured on drawing	Actual size
1:200	1 mm	200 mm /20 cm/0.2 m
1:100	1 mm	100 mm/10 cm/0.1 m
1:50	1 mm	50 mm/5 cm/0.05 m
1:20	1 mm	20 mm/2 cm/0.02 m

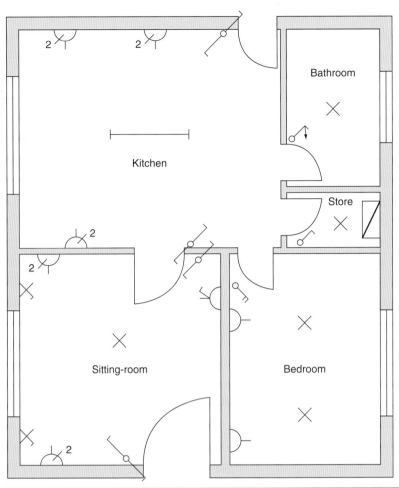

Figure 3.4 Layout drawing or site plan of a small electrical installation.

The layout drawing or site plan of a small domestic extension is shown in Figure 3.4. It can be seen that the mains intake position, probably a consumer unit, is situated in the storeroom which also contains one light controlled by a switch at the door. The bathroom contains one lighting point controlled by a one-way pull switch at the door. The kitchen has two doors and a switch is installed at each door to control the fluorescent luminaire. There are also three double sockets situated around the kitchen. The sitting-room has a two-way switch at each door controlling the centre lighting point. Two wall lights with built-in switches are to be wired, one at each side of the window. Two double sockets and one switched socket are also to be installed in the sitting-room. The bedroom has two lighting points controlled independently by two one-way switches at the door. The wiring diagrams and installation procedures for all these circuits will be looked at in more detail later in this chapter.

Try this

Drawings

The next time you are on-site ask your supervisor to show you the site plans. Ask them:

- how does the scale work?
- to help you put names to the equipment represented by British Standard symbols.

Figure 3.5 Being able to interpret drawings takes time but is an essential skill within the construction industry.

As-fitted drawings

When the installation is completed, a set of drawings should be produced which indicate the final positions of all the electrical equipment. As the building and electrical installation progresses, it is sometimes necessary to modify the positions of equipment indicated on the layout drawing because, for example, the position of a doorway has been changed. The layout drawings or site plans indicate the original intentions for the position of equipment, while the 'as-fitted' drawing indicates the actual positions of equipment upon completion of the contract or to identify a specific component in a piece of equipment.

Try this

Drawings

Take a moment to clarify the difference between:

- layout drawings; and
- as-fitted drawings.

Detail drawings and assembly drawings

These are additional drawings produced by the architect to clarify some point of detail. For example, a drawing might be produced to give a fuller description of a suspended ceiling arrangement or the assembly arrangements of the metalwork for the suspended ceiling.

Location drawings

Location drawings identify the place where something is located. It might be the position of the manhole covers giving access to the drains. It might be the position of all water stop taps or the position of the emergency lighting fittings. This type of information may be placed on a blank copy of the architect's site plan or on a supplementary drawing.

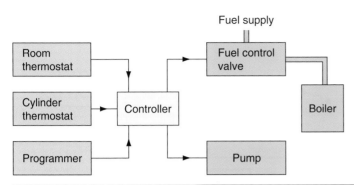

Figure 3.6 Block diagram – space-heating control system (Honeywell Y. Plan).

Distribution cable route plans

On large installations there may be more than one position for the electrical supplies. Distribution cables may radiate from the site of the electrical mains intake position to other sub-mains positions. The site of the sub-mains and the route taken by the distribution cables may be shown on a blank copy of the architect's site plan or on the electrician's 'as-fitted' drawings.

Block diagrams

A block diagram is a very simple diagram in which the various items or pieces of equipment are represented by a square or rectangular box. The purpose of the block diagram is to show how the components of the circuit relate to each other and, therefore, the individual circuit connections are not shown. Figure 3.6 shows the block diagram of a space-heating control system.

Wiring diagrams

A wiring diagram or connection diagram shows the detailed connections and relative positions between components or items of equipment. They do not indicate how a piece of equipment or circuit works and may not show the internal workings of a component. The purpose of a wiring diagram is to help someone with the actual connection of the circuit conductors. Figure 3.7 shows the wiring diagram for a space-heating control system and Figures 3.8 to 3.10 the wiring diagrams for a one-way, two-way and intermediate switch control of a light.

Circuit diagrams

A circuit diagram shows most clearly how a circuit works. All the essential parts and connections are represented by their graphical symbols. The purpose of a circuit diagram is to help our understanding of the circuit. It will be laid out as clearly as possible, without regard to the physical layout of the actual components and, therefore, it may not indicate the most convenient way to wire the circuit. Figure 3.11 shows the circuit diagram of our same space-heating control system.

Schematic diagrams

A schematic diagram is a diagram in outline of, for example, a motor starter circuit. It uses graphical symbols to indicate the interrelationship of the electrical elements in a circuit and the process of operation can be followed working from left to right. These help us to understand the working operation of the circuit but are not helpful in showing us how to wire the components. An electrical schematic diagram looks very like a circuit diagram. Figure 3.12 shows a schematic diagram.

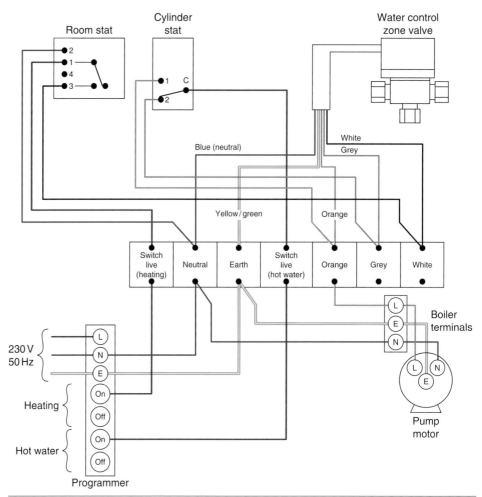

Figure 3.7 Wiring diagram – space-heating control system (Honeywell Y. Plan).

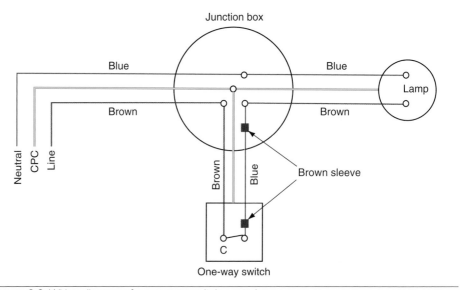

Figure 3.8 Wiring diagram of a one-way switch control.

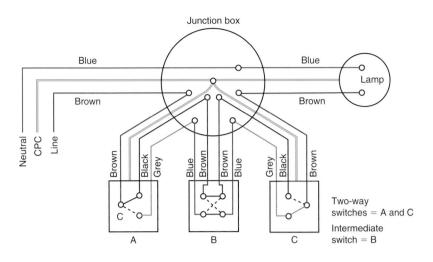

Figure 3.9 Wiring diagram of intermediate switch control.

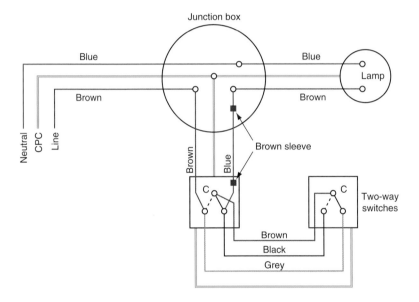

Figure 3.10 Wiring diagram of one-way converted to two-way switch control.

Freehand working diagrams

Freehand working drawings or sketches are another important way in which we communicate our ideas. A freehand sketch may be done as an initial draft of an idea before a full working drawing is made. It is often much easier to produce a sketch of your ideas or intentions than to describe them or produce a list of instructions.

To convey the message or information clearly it is better to make your sketch large rather than too small. It should also contain all the dimensions necessary to indicate clearly the size of the finished object depicted by the sketch.

All drawings and communications must be aimed at satisfying the client's wishes for the project. It is the client who will pay the final bill which, in turn, pays your wages. The detailed arrangements of what must be done to meet the client's wishes are contained in the client's specification documents and all your company's efforts must be directed at meeting the whole specification, but no more.

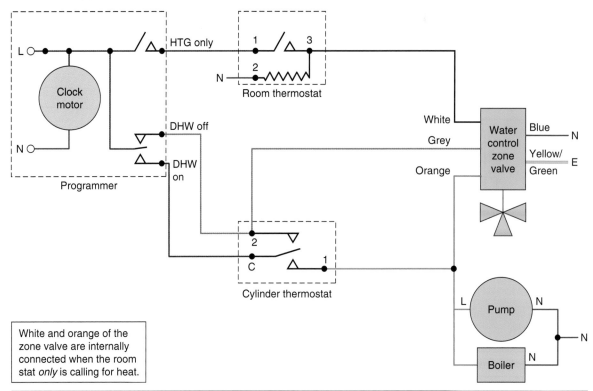

Figure 3.11 Circuit diagram – space heating control system.

White and orange of the zone valve are internally connected when the room stat *only* is calling for heat.

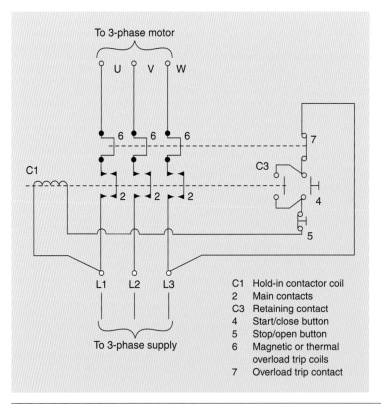

C1   Hold-in contactor coil
2   Main contacts
C3   Retaining contact
4   Start/close button
5   Stop/open button
6   Magnetic or thermal overload trip coils
7   Overload trip contact

Figure 3.12 Schematic diagram – DOL motor starter.

Assessment criteria 3.1

State the properties, applications, advantages and limitations of different cable types

Assessment criteria 3.2

State the features, applications, advantages and limitations of different containment systems

Wiring systems and enclosures

The final choice of a wiring system must rest with those designing the installation and those ordering the work, but whatever system is employed, good workmanship by skilled or instructed persons and the use of proper materials is essential for compliance with the regulations (IET Regulation 134.1.1). The necessary skills can be acquired by an electrical trainee who has the correct attitude and dedication to their craft.

Most cables can be considered to be constructed in three parts: the conductor, which must be of a suitable cross-section to carry the load current; the insulation, which has a colour or number code for identification; and the outer sheath, which may contain some means of providing protection from mechanical damage.

The conductors of a cable are made of either copper or aluminium and may be stranded or solid. Solid conductors are only used in fixed wiring installations and may be shaped in larger cables. Stranded conductors are more flexible and conductor sizes from 4.0 to $25\,mm^2$ contain seven strands. A $10\,mm^2$ conductor, for example, has seven $1.35\,mm$ diameter strands which collectively make up the $10\,mm^2$ cross-sectional area of the cable. Conductors above $25\,mm^2$ have more than seven strands, depending upon the size of the cable. Flexible cords have multiple strands of very fine wire conductors, as fine as one strand of human hair. This gives the cable its very flexible quality.

New wiring colours

Over 25 years ago, the United Kingdom agreed to adopt the European colour code for flexible cords, that is, brown for live or line conductor, blue for the neutral conductor and green combined with yellow for earth conductors.

However, no similar harmonization was proposed for non-flexible cables used for fixed wiring. These were to remain as red for live or line conductor, black for the neutral conductor and green combined with yellow for earth conductors.

On 31 March 2004, the IET published Amendment No. 2 to BS 7671: 2001 which specified new cable core colours for all fixed wiring in UK electrical installations.

These new core colours will 'harmonize' the United Kingdom with practice in mainland Europe.

Fixed cable core colours up to 2006

* Single-phase supplies red line conductors, black neutral conductors, and green combined with yellow for earth conductors;

Figure 3.13 UK cable colours match up with those used in mainland Europe.

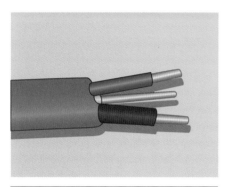

Figure 3.14 PVC insulated and sheathed cable – the cheapest and most widespread type of cable.

- Three-phase supplies red, yellow and blue line conductors, black neutral conductors and green combined with yellow for earth conductors.

These core colours could not be used after 31 March 2006.

New (harmonized) fixed cable core colours

Single-phase supplies brown line conductors, blue neutral conductors and green combined with yellow for earth conductors (just like flexible cables).

Three-phase supplies brown, black and grey line conductors, blue neutral conductors and green combined with yellow for earth conductors.

Cable core colours used from 31 March 2004 onwards

Extensions or alterations to existing single-phase installations do not require marking at the interface between the old and new fixed wiring colours. However, a warning notice must be fixed at the consumer unit or distribution fuse board which states:

> Caution – this installation has wiring colours to two versions of BS 7671. Great care should be taken before undertaking extensions, alterations or repair that all conductors are correctly identified.

Alterations to three-phase installations must be marked at the interface L1, L2, L3 for the lines and N for the neutral. Both new and old cables must be marked.

These markings are preferred to coloured tape and a caution notice is again required at the distribution board. Appendix 7 of BS 7671: 2018 deals with harmonized cable core colours.

PVC insulated and sheathed cables

Domestic and commercial installations use this cable, which may be clipped direct to a surface, sunk in plaster, or installed in conduit or trunking. It is the simplest and least expensive cable. Figure 3.15 shows a sketch of a twin and earth cable.

The conductors are covered with a colour-coded PVC insulation and then contained singly or with others in a PVC outer sheath.

It is referred to in BS 7671 as '70 °C thermoplastic insulated and sheathed flat cable with protective conductor'.

PVC/SWA cable

PVC insulated steel wire armour cables are used for wiring underground between buildings, for main supplies to dwellings, rising sub-mains and industrial installations. They are used where some mechanical protection of the cable conductors is required.

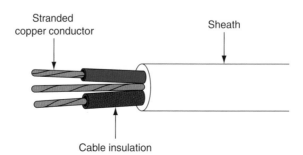

Stranded copper conductor

Sheath

Cable insulation

Figure 3.15 A twin and earth PVC insulated and sheathed cable.

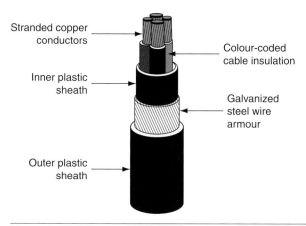

Figure 3.16 A four-core PVC/SWA cable.

The conductors are covered with colour-coded PVC insulation and then contained either singly or with others in a PVC sheath (see Figure 3.16). Around this sheath is placed an armour protection of steel wires twisted along the length of the cable, and a final PVC sheath covering the steel wires protects them from corrosion. The armour sheath also provides the circuit protective conductor (CPC) and the cable is simply terminated using a compression gland.

Fire performance cables

There are two different types of cable on the market, which have improved fire performance compared with standard cables. Those that react in a fire, give low smoke emission and reduced flame propagation, and those that are fire resistant, which will continue to operate during a fire when subjected to a specified flame source for a period of time.

Cables for fire alarm applications are now split into 'standard', where survival time of the cable is 30 minutes, and 'enhanced', where the survival time increases to 120 minutes.

Fixings, terminals and accessories must offer the same level of protection against fire as the cable you use.

Mineral insulated cable and FP200 cable are the most commonly used examples of these cables.

MI cable

A mineral insulated (MI) cable has a seamless copper sheath which makes it waterproof and fire- and corrosion-resistant. These characteristics often make it the only cable choice for hazardous or high-temperature installations such as oil refineries and chemical works, boiler houses and furnaces, petrol pump and fire alarm installations.

The cable has a small overall diameter when compared to alternative cables and may be supplied as bare copper or with a PVC over-sheath. It is colour coded orange for general electrical wiring, white for emergency lighting or red for fire alarm wiring. The copper outer sheath provides the CPC, and the cable is terminated with a pot and sealed with compound and a compression gland (see Figure 3.17).

The copper conductors are embedded in a white powder, magnesium oxide, which is non-ageing and non-combustible, but which is hygroscopic, which means that it readily absorbs moisture from the surrounding air, unless

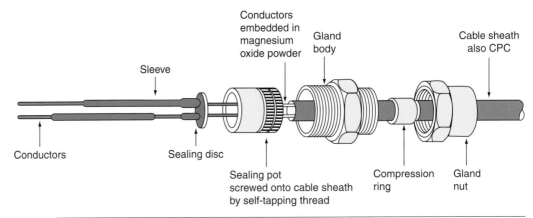

Figure 3.17 MI cable with terminating seal and gland.

adequately terminated. The termination of an MI cable is a complicated process requiring the electrician to demonstrate a high level of practical skill and expertise for the termination to be successful.

The marking on mineral insulated cables and accessories indicates the size and number of the conductors and its rating of light or heavy gauge. As an example, a mineral insulated cable sealing pot marked 3L1.5 would be constructed of three cores each of 1.5 mm² c.s.a. and light gauge (up to 500 V). Heavy gauge cable is rated up to 750 V.

FP 200 cable

FP 200 cable is similar in appearance to an MI cable in that it is a circular tube, or the shape of a pencil, and is available with a red or white sheath. However, it is much simpler to use and terminate than an MI cable.

The cable is available with either solid or stranded conductors that are insulated with 'insudite', a fire-resistant insulation material. The conductors are then screened by wrapping an aluminium tape around the insulated conductors, that is, between the insulated conductors and the outer sheath. This aluminium tape screen is applied metal side down and in contact with the bare CPC.

The sheath is circular and made of a robust thermoplastic low-smoke, zero halogen material.

FP 200 is available in 2, 3, 4, 7, 12 and 19 cores with a conductor size range from 1.0 to 4.0 mm².

The cable is as easy to use as a PVC insulated and sheathed cable. No special terminations are required; the cable may be terminated through a grommet into a knock-out box or terminated through a simple compression gland.

The cable is a fire-resistant cable, primarily intended for use in fire alarms and emergency lighting installations, or it may be embedded in plaster.

High-voltage overhead cables

Suspended from cable towers or pylons, overhead cables must be light, flexible and strong. The cable is constructed of stranded aluminium conductors formed around a core of steel-stranded conductors (see Figure 3.18). The aluminium conductors carry the current and the steel core provides the tensile strength required to suspend the cable between pylons. The cable is not insulated since it is placed out of reach and insulation would only add to the weight of the cable.

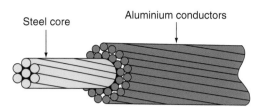

Figure 3.18 123 kV overhead cable construction.

Optical fibre cables

The introduction of fibre-optic cable systems and digital transmissions will undoubtedly affect future cabling arrangements and the work of the electrician.

Networks based on the digital technology currently being used so successfully by the telecommunications industry are very likely to become the long-term standard for computer systems. Fibre-optic systems dramatically reduce the number of cables required for control and communications systems, and this will in turn reduce the physical room required for these systems. Fibre-optic cables are also immune to electrical noise when run parallel to mains cables and, therefore, the present rules of segregation and screening may change in the future. There is no spark risk if the cable is accidentally cut and, therefore, such circuits are intrinsically safe. Intrinsic safety is described later in this chapter under the heading 'Hazardous area installations'.

Optical fibre cables are communication cables made from optical-quality plastic, the same material from which spectacle lenses are manufactured. The energy is transferred down the cable as digital pulses of laser light, rather than current flowing down a copper conductor in electrical installation terms. The light pulses stay within the fibre-optic cable because of a scientific principle known as 'total internal refraction' which means that the laser light bounces down the cable and when it strikes the outer wall it is always deflected inwards and, therefore, does not escape out of the cable, as shown in Figure 3.20.

The biggest advantage of fibre-optic cables is the speed with which they are able to transfer data. In 2014, researchers developed a new fibre-optic technology capable of transferring data at a rate of 255 terabits per second (Tbps) – more data than the total traffic flowing across the internet at peak time. At such speeds, it is possible to transfer a 1 GB movie in 0.03 milliseconds, or a one terabyte file – 1,000 GB of data – in just 0.03 seconds.

Commercial communication providers offer packages ranging from 300 MB/sec to 1 GB/sec so while this speed is not yet available to the public, the potential for huge development is there.

The cables are very small because the optical quality of the conductor is very high and signals can be transmitted over great distances. They are cheap to produce and lightweight because these new cables are made from high-quality plastic and not high-quality copper. Single-sheathed cables are often called 'simplex' cables

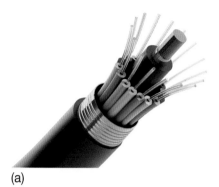

(a)

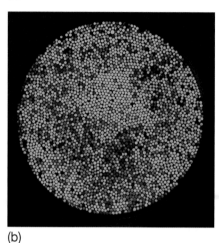

(b)

Figure 3.19 a) Optical fibre cable b) cross-section of optical fibre cable.

Definition

Optical fibre cables are communication cables made from optical-quality plastic, the same material from which spectacle lenses are manufactured. The energy is transferred down the cable as digital pulses of laser light, as against current flowing down a copper conductor in electrical installation terms.

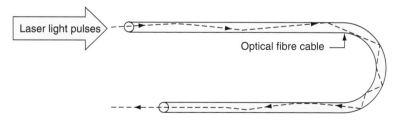

Figure 3.20 Digital pulses of laser light down an optical fibre cable.

Figure 3.21 Category 6 network cables.

and twin-sheathed cables 'duplex', that is, two simplex cables together in one sheath. Multi-core cables are available containing up to 24 single fibres.

Fibre-optic cables look like steel wire armour cables (but of course they are lighter) and should be installed in the same way, and given the same level of protection, as SWA cables. Avoid tight-radius bends if possible and kinks at all costs. Cables are terminated in special joint boxes known as splicing which ensure cable ends are cleanly cut and butted together to ensure the continuity of the light pulses.

Fibre-optic cables are Band I circuits when used for data transmission and must therefore be segregated from other mains cables to satisfy the IET Regulations 528.1.

The testing of fibre-optic cables requires special instruments to measure the light attenuation (i.e. light loss) down the cable. Finally, when working with fibre-optic cables, electricians should avoid direct eye contact with the low-energy laser light transmitted down the conductors.

Data cables

Category 5 cable (cat 5), cat 5e and cat 6 cables are twisted pair cables used for computer networks and carrying telephony and video signals. The cable consists of four pairs of conductors, twisted in pairs to minimize interference. It is usually terminated in punch down, insulation displacement connections to the RJ45 specification. The length of the data circuit is restricted to 100 m and is the fixed wiring usually restricted to 90 m to allow for the length of the patch leads to be added.

Temperature limits of cables

While there are many cable insulation materials, they are often divided into two types, thermoplastic or thermosetting. The current capacity of a cable is determined by the type of insulation as the build-up of heat will have a detrimental effect on the insulation.

- **Thermoplastic** materials are composed of chains of molecules (polyethylene for example). When heat is applied the energy will allow the bonds to separate and the material can flow (melt) and be reformed.
- **Thermosetting** materials are formed when materials such as polyethylene undergo specific heating or chemical processes. During this process the individual chains become cross-linked by smaller molecules making a rigid structure. Thermosetting materials cannot be melted or remoulded.

Due to the structure, thermosetting materials are generally more heat-resistant and have greater strength.

BS 7671 gives current carrying capacities based on a maximum conductor temperature of:

- Thermoplastic 70 °C
- Thermosetting 90 °C

Assessment criteria 3.3

Select an appropriate type of wiring system for a given environment

Assessment criteria 3.4

State the types of wiring systems and associated equipment used in different installations

Choosing an appropriate wiring system

An electrical installation is made up of many different electrical circuits, lighting circuits, power circuits, single-phase domestic circuits and three-phase industrial or commercial circuits.

Figure 3.22 Electrical panels come in all shapes and sizes according to their intended use.

Whatever the type of circuit, the circuit conductors are contained within cables or enclosures.

Part 5 of the IET Regulations tells us that electrical equipment and materials must be chosen so that they are suitable for the installed conditions, taking into account temperature, the presence of water, corrosion, mechanical damage, vibration or exposure to solar radiation. Therefore, PVC insulated and sheathed cables are suitable for domestic installations but, for a cable requiring mechanical protection and suitable for burying underground, a PVC/SWA cable would be preferable. These two types of cable are shown in Figure 3.15 and 3.16.

MI cables are waterproof, heatproof and corrosion-resistant with some mechanical protection. These qualities often make it the only cable choice for hazardous or high-temperature installations such as oil refineries, chemical works, boiler houses and petrol pump installations. An MI cable with terminating gland and seal is shown in Figure 3.17.

Wiring systems and enclosures

The final choice of a wiring system must rest with those designing the installation and those ordering the work, but whatever system is employed, good workmanship by skilled or instructed persons and the use of proper materials is essential for compliance with the IET Regulations (IET Regulation 134.1.1). The necessary skills can be acquired by an electrical trainee who has the correct attitude and dedication to his craft.

PVC insulated and sheathed cable installations and PVC insulated and sheathed wiring systems are used extensively for lighting and socket installations in domestic dwellings. Mechanical damage to the cable caused by impact, abrasion, penetration, compression or tension must be minimized during installation (IET Regulation 522.6.1). The cables are generally fixed using plastic clips incorporating a masonry nail, which means the cables can be fixed to wood, plaster or brick with almost equal ease. Cables should be run horizontally or vertically, not diagonally, down a wall. All kinks should be removed so that the cable is run straight and neatly between clips fixed at equal distances providing adequate support for the cable so that it does not become damaged by its own weight (IET Regulation 522.8.4 and Table D1 of the *On-Site Guide*). Where cables are bent, the radius of the bend should not cause the conductors to be damaged (IET Regulation 522.8.3 and Table D5 of the *On-Site Guide*).

Terminations or joints in the cable may be made in ceiling roses, junction boxes, or behind sockets or switches, provided that they are enclosed in a non-ignitable material, are properly insulated and are mechanically and electrically secure (IET Regulation 526). All joints must be accessible for inspection, testing and maintenance when the installation is completed (IET Regulation 526.3) unless designed to be maintainance free.

Where PVC insulated and sheathed cables are concealed in walls, floors or partitions, they must be provided with a box incorporating an earth terminal at each outlet position. PVC cables do not react chemically with plaster, as do some cables, and consequently PVC cables may be buried under plaster. Further protection by channel or conduit is only necessary if mechanical protection from nails or screws is required or to protect them from the plasterer's trowel.

However, IET Regulation 522.6.202 now tells us that where PVC cables are to be embedded in a wall or partition at a depth of less than 50 mm they should be run along one of the permitted routes shown in Figure 3.24. Figure 3.23 shows a typical PVC installation. To identify the most probable cable routes, IET Regulation 522.6.202 tells us that outside a zone formed by a 150 mm border all around a wall edge, cables can only be run horizontally or vertically to a point or accessory if they are contained in a substantial earthed enclosure, such as a conduit, which can withstand nail penetration, as shown in Figure 3.25.

Where the accessory or cable is fixed to a wall which is less than 100 mm thick, protection must also be extended to the reverse side of the wall if a position can be determined.

Where none of this protection can be complied with, the cable must be given additional protection with a 30 mA residual current device (RCD) (IET Regulation 522.6.202).

The 3rd Amendment to the 17th Edition Regulations introduced a new Regulation 521.11.201 which requires wiring systems in escape routes be supported in a such a manner that they will not prematurely collapse in the event of a fire.

The 18th Edition of the regulations at 521.10.202 now requires the same fire proof support for all systems throughout the installation, not just escape routes.

However, let us be clear about this new Regulation; it is only those wiring systems that might collapse before or during the period when the rescue services would enter a building, that require a fixing which will not collapse prematurely. So, cables installed in metal trunking, or conduit, or cables laid on top of cable tray or ladder rack will require no extra fixing considerations because they are

Table 3.1 Spacing of cable supports

Overall diameter of cable (mm)	Maximum spacings of clips								
	PVC sheathed cables				Armoured cables		Mineral insulated copper sheathed cables	
	Generally		In caravans					
	Horizontal (mm)	Vertical (mm)	Horizontal (mm)	Vertical (mm)	Horizontal (mm)	Vertical (mm)	Horizontal (mm)	Vertical (mm)
1	2	3	4	5	6	7	8	9
Not exceeding 9	250	400	250 (for all sizes)	400 (for all sizes)	–	–	600	800
Exceeding 9 and not exceeding 15	300	400			350	450	900	1200
Exceeding 15 and not exceeding 20	350	450			400	550	1500	2000
Exceeding 20 and not exceeding 40	400	550			450	600	–	–

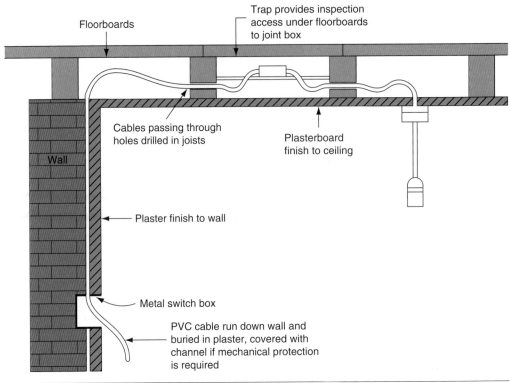

Figure 3.23 A concealed PVC sheathed wiring system.

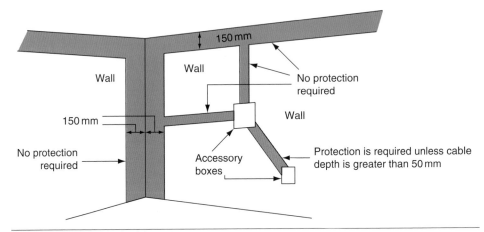

Figure 3.24 Permitted cable routes.

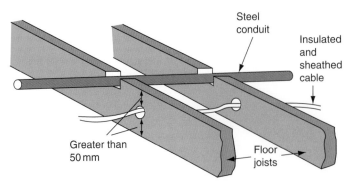

Notes:

1. Maximum diameter of hole should be 0.25 × joist depth.
2. Holes on centre line in a zone between 0.25 and 0.4 × span.
3. Maximum depth of notch should be 0.125 × joist depth.
4. Notches on top in a zone between 0.1 and 0.25 × span.
5. Holes in the same joist should be at least 3 diameters apart.

Figure 3.25 Correct installation of cables in floor joists.

adequately supported. Also, wiring systems fixed within the fabric of the building which is not liable to premature collapse in the event of a fire are safe in these circumstances.

Note 3 to the Regulation tells us that it is the cable systems fixed with plastic clips, or inside plastic trunking which now require further support. These systems can fail when subject to either direct flame or the hot products of combustion, leading to wiring systems hanging down and causing an entanglement risk as a result of the fire.

This makes it impossible for us to use non-metallic cable clips, cable ties or plastic trunking as the only means of support for PVC wiring systems. The regulation tells us that where non-metallic cable systems are used, **a suitable means of fire-resistant support and retention must be used** to prevent cables falling down in the event of a fire. Note 4 of the Regulation advises that suitably spaced steel or copper clips, saddles or ties are examples that will meet the requirements of this Regulation.

Where cables pass through walls, floors and ceilings the hole should be made good with incombustible material such as mortar or plaster to prevent the spread of fire (IET Regulations 527.1.1 and 527.2.1). Cables passing through metal boxes should be bushed with a rubber grommet to prevent abrasion of the cable.

Holes drilled in floor joists through which cables are run should be 50 mm below the top or 50 mm above the bottom of the joist to prevent damage to the cable by nail penetration (IET Regulation 522.6.201 (i)), as shown in Figure 3.25. PVC cables should not be installed when the surrounding temperature is below 0 °C or when the cable temperature has been below 0 °C for the previous 24 hours because the insulation becomes brittle at low temperatures and may be damaged during installation.

Try this

Definitions

In the margin write down a short definition of a 'skilled or instructed person'.

Conduit installations

A conduit is a tube, channel or pipe in which insulated conductors are contained. The conduit, in effect, replaces the PVC outer sheath of a cable, providing mechanical protection for the insulated conductors. A conduit installation can be rewired easily or altered at any time, and this flexibility, coupled with mechanical protection, makes conduit installations popular for commercial and industrial applications. There are three types of conduit used in electrical installation work: steel, PVC and flexible.

Steel conduit

Steel conduits are made to a specification defined by BS 4568 and are either heavy gauge welded or solid drawn. Heavy gauge is made from a sheet of steel welded along the seam to form a tube and is used for most electrical installation work. Solid drawn conduit is a seamless tube which is much more expensive and only used for special gas-tight, explosion-proof or flameproof installations.

Conduit is supplied in lengths up to 3.75 m and typical sizes are 16, 20, 25 and 32 mm. Conduit tubing and fittings are supplied in a black enamel finish for

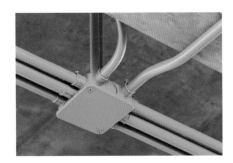

Figure 3.26 Steel conduits.

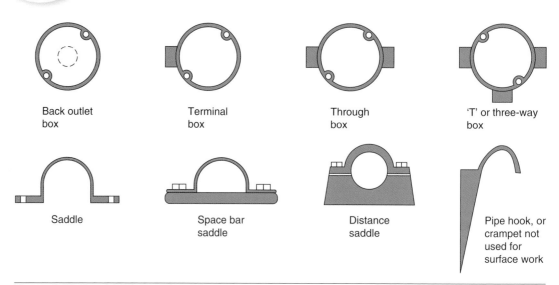

Figure 3.27 Conduit fittings and saddles.

internal use or hot galvanized finish for use on external or damp installations. A wide range of fittings is available and the conduit is fixed using saddles or pipe hooks, as shown in Figure 3.27.

Metal conduits are threaded with stocks and dies and bent using special bending machines. The metal conduit is also utilized as the CPC and, therefore, all connections must be screwed up tightly and all burrs removed so that cables will not be damaged as they are drawn into the conduit. Metal conduits containing a.c. circuits must contain line and neutral conductors in the same conduit to prevent eddy currents from flowing, which would result in the metal conduit becoming hot (IET Regulations 521.5.1, 522.8.1 and 522.8.11).

PVC conduit

PVC conduit used on typical electrical installations is heavy gauge standard impact tube manufactured to BS 4607. The conduit size and range of fittings are the same as those available for metal conduit. PVC conduit is most often joined by placing the end of the conduit into the appropriate fitting and fixing with a PVC solvent adhesive. PVC conduit can be bent by hand using a bending spring of the same diameter as the inside of the conduit. The spring is pushed into the conduit to the point of the intended bend and the conduit then bent over the knee. The spring ensures that the conduit keeps its circular shape. In cold weather, a little warmth applied to the point of the intended bend often helps to achieve a more successful bend.

The advantages of a PVC conduit system are that it may be installed much more quickly than steel conduit and is non-corrosive, but it does not have the mechanical strength of steel conduit. Since PVC conduit is an insulator it cannot be used as the CPC and a separate earth conductor must be run to every outlet.

It is not suitable for installations subjected to temperatures below 25 °C or above 60 °C. Where luminaires are suspended from PVC conduit boxes, precautions must be taken to ensure that the lamp does not raise the box temperature or that the mass of the luminaire supported by each box does not exceed the maximum recommended by the manufacturer (IET Regulations 522.1 and 522.2). PVC conduit also expands much more than metal conduit and so long runs require an expansion coupling to allow for conduit movement and to help prevent distortion during temperature changes.

All conduit installations must be erected first before any wiring is installed (IET Regulation 522.8.2). The radius of all bends in conduits must not cause the cables to suffer damage, and therefore the minimum radius of bends given in Table D5 of the *On-Site Guide* applies (IET Regulation 522.8.3). All conduits should terminate in a box or fitting and meet the boxes or fittings at right angles, as shown in Figure 3.28. Any unused conduit box entries should be blanked off and all boxes covered with a box lid, fitting or accessory to provide complete enclosure of the conduit system. Conduit runs should be separate from other services, unless intentionally bonded, to prevent arcing from occurring from a faulty circuit within the conduit, which might cause the pipe of another service to become punctured.

When drawing cables into conduit they must first be *run off* the cable drum. That is, the drum must be rotated as shown in Figure 3.30 and not allowed to *spiral off*, which will cause the cable to twist.

Cables should be fed into the conduit in a manner which prevents any cable from crossing over and becoming twisted inside the conduit. The cable insulation must not be damaged on the metal edges of the draw-in box. Cables can be pulled in on a draw wire if the run is a long one. The draw wire itself may be drawn in on a fish tape, which is a thin spring steel or plastic tape.

A limit must be placed on the number of bends between boxes in a conduit run and the number of cables which may be drawn into a conduit to prevent the cables from being strained during wiring. Appendix E of the *On-Site Guide* gives a guide to the cable capacities of conduits and trunking.

Flexible conduit

Flexible conduit manufactured to BS 731-1: 1993 is made of interlinked metal spirals often covered with a PVC sleeving. The tubing must not be relied upon to provide a continuous earth path and, consequently, a separate CPC must be run either inside or outside the flexible tube (IET Regulation 543.2.7).

Flexible conduit is used for the final connection to motors so that the vibrations of the motor are not transmitted throughout the electrical installation and to allow for modifications to be made to the final motor position and drive belt adjustments.

Conduit capacities

Single PVC insulated conductors are usually drawn into the installed conduit to complete the installation. Having decided upon the type, size and number of cables required for a final circuit, it is then necessary to select the appropriate size of conduit to accommodate those cables.

The tables in Appendix E of the *On-Site Guide* describe a 'factor system' for determining the size of conduit required to enclose a number of conductors. The method is as follows:

- Identify the cable factor for the particular size of conductor; see Table 3.2.
- Multiply the cable factor by the number of conductors, to give the sum of the cable factors.
- Identify the appropriate part of the conduit factor table given by the length of run and number of bends (see Table 3.3).
- The correct size of conduit to accommodate the cables is that conduit which has a factor equal to or greater than the sum of the cable factors.

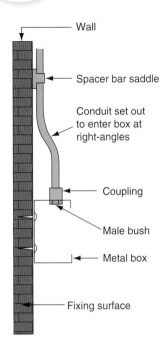

Wall

Spacer bar saddle

Conduit set out to enter box at right-angles

Coupling

Male bush

Metal box

Fixing surface

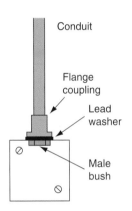

Conduit

Flange coupling

Lead washer

Male bush

Figure 3.28 Terminating conduits.

Definition

Flexible conduit manufactured to BS 731-1: 1993 is made of interlinked metal spirals often covered with a PVC sleeving.

Figure 3.29 Flexible conduit.

Definition

Single PVC insulated conductors are usually drawn into the installed conduit to complete the installation.

Table 3.2 Conduit cable factors

Cable factors for conduit in long straight runs over 3 m, or runs of any length incorporating bends	
Conductor CSA (mm²)	**Cable factor**
1	16
1.5	22
2.5	30
4	43
6	58
10	105
16	145

Table 3.3 Conduit factors

Length of run (m)	Conduit diameter (mm)											
	16	20	25	32	16	20	25	32	16	20	25	32
	Straight				One bend				Two bends			
3.5	179	290	521	911	162	263	475	837	136	222	404	720
4	177	286	514	900	158	256	463	818	130	213	388	692
4.5	174	282	507	889	154	250	452	800	125	204	373	667
5	171	278	500	878	150	244	442	783	120	196	358	643
6	167	270	487	857	143	233	422	750	111	182	333	600
7	162	263	475	837	136	222	404	720	103	169	311	563
8	158	256	463	818	130	213	388	692	97	159	292	529
9	154	250	452	800	125	204	373	667	91	149	275	500
10	150	244	442	783	120	196	358	643	86	141	260	474

Example 1

Six 2.5 mm² PVC insulated cables are to be run in a conduit containing two bends between boxes 10 m apart. Determine the minimum size of conduit to contain these cables.
From Table 3.2:

$$\text{The factor for one 2.5mm}^2 \text{ cable} = 30$$
$$\text{The sum of the cable factors} = 6 \times 30$$
$$= 180$$

From Table 3.3, a 25 mm conduit, 10 m long and containing two bends, has a factor of 260. A 20 mm conduit containing two bends only has a factor of 141 which is less than 180, the sum of the cable factors, and, therefore, 25 mm conduit is the minimum size to contain these cables.

Example 2

Ten 1.0 mm^2 PVC insulated cables are to be drawn into a plastic conduit which is 6 m long between boxes and contains one bend. A 4.0 mm PVC insulated CPC is also included. Determine the minimum size of conduit to contain these conductors.

From Table 3.2:

> The factor for one 1.0 mm cable = 16
> The factor for one 4.0 mm cable = 43
> The sum of the cable factors = (10 × 16) + (1 × 43)
> = 203

From Table 3.3, a 20 mm conduit, 6 m long and containing one bend, has a factor of 233. A 16 mm conduit containing one bend only has a factor of 143 which is less than 203, the sum of the cable factors, and, therefore, 20 mm conduit is the minimum size to contain these cables.

Trunking installations

A trunking is an enclosure provided for the protection of cables which is normally square or rectangular in cross-section, having one removable side. Trunking may be thought of as a more accessible conduit system, and for industrial and commercial installations it is replacing the larger conduit sizes. A trunking system can have great flexibility when used in conjunction with conduit; the trunking forms the background or framework for the installation, with conduits running from the trunking to the point controlling the current-using apparatus. When an alteration or extension is required it is easy to drill a hole in the side of the trunking and run a conduit to the new point. The new wiring can then be drawn through the new conduit and the existing trunking to the supply point.

Trunking is usually supplied in 3 m lengths and various cross-sections measured in millimetres from 50 × 50 up to 300 × 150. Most trunking is available in either steel or plastic.

Metallic trunking

Metallic trunking is formed from mild steel sheet, coated with grey or silver enamel paint for internal use or a hot-dipped galvanized coating where damp conditions might be encountered and made to a specification defined by BS EN 500 85. A wide range of accessories is available, such as 45° bends, 90° bends, tee and four-way junctions, for speedy on-site assembly. Alternatively, bends may be fabricated in lengths of trunking, as shown in Figure 3.32. This may be necessary or more convenient if a bend or set is non-standard, but it does take more time to fabricate bends than merely to bolt on standard accessories.

Insulated non-sheathed cables are permitted in a trunking system which provides at least the degree of protection IPXXD (which means total protection) or IP4X which means protection from a solid object greater than 1.0 mm such as a thin wire or strip. For site fabricated joints such as that shown in Fig 3.32, the installer must confirm that the completed item meets at least IPXXD (IET Regulation 521.10).

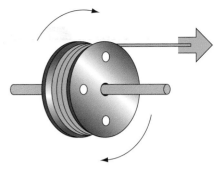

Cables *run off* will not twist; a short length of conduit can be used as an axle for the cable drum

Cables allowed to *spiral off* a drum will become twisted

Figure 3.30 Cables should be run off to avoid twists.

Definition

A *trunking* is an enclosure provided for the protection of cables which is normally square or rectangular in cross-section, having one removable side. Trunking may be thought of as a more accessible conduit system.

Definition

Metallic trunking is formed from mild steel sheet, coated with grey or silver enamel paint for internal use or a hot-dipped galvanized coating where damp conditions might be encountered.

When fabricating bends, the trunking should be supported with wooden blocks for sawing and filing, in order to prevent the sheet-steel from vibrating or becoming deformed. Fish-plates must be made and riveted or bolted to the trunking to form a solid and secure bend. When manufactured bends are used, the continuity of the earth path must be ensured across the joint by making all fixing screw connections very tight, or fitting a separate copper strap between the trunking and the standard bend. If an earth continuity test on the trunking is found to be unsatisfactory, an insulated CPC must be installed inside the trunking. The size of the protective conductor will be determined by the largest cable contained in the trunking, as described in Table 54.7 of the IET Regulations. If the circuit conductors are less than $16\,mm^2$, then a $16\,mm^2$ CPC will be required.

Non-metallic trunking

Trunking and trunking accessories are also available in high-impact PVC. The accessories are usually secured to the lengths of trunking with a PVC solvent adhesive. PVC trunking, like PVC conduit, is easy to install and is non-corrosive.

A separate CPC will need to be installed and non-metallic trunking may require more frequent fixings because it is less rigid than metallic trunking. All trunking fixings should use round-headed screws to prevent damage to cables since the thin sheet construction makes it impossible to countersink screw heads. Non-metal wiring systems must now also meet the new fire proof support Regulation 521.10.202 to prevent cables falling down in the event of a fire. Note 4 of the Regulation advises that suitably spaced steel or copper clips, saddles or ties are examples that will meet this requirement. See 'Wiring Systems and Enclosures' earlier in this chapter.

Mini-trunking

Definition

Mini-trunking is very small PVC trunking, ideal for surface wiring in domestic and commercial installations such as offices.

Mini-trunking is very small PVC trunking, ideal for surface wiring in domestic and commercial installations such as offices. The trunking has a cross-section of $16 \times 16\,mm$, $25 \times 16\,mm$, $38 \times 16\,mm$ or $38 \times 25\,mm$ and is ideal for switch drops or for housing auxiliary circuits such as telephone or audio equipment wiring. The modern square look in switches and sockets is complemented by the mini-trunking which is very easy to install (see Figure 3.31).

Skirting trunking

Definition

Skirting trunking is a trunking manufactured from PVC or steel in the shape of a skirting board and is frequently used in commercial buildings such as hospitals, laboratories and offices.

Skirting trunking is a trunking manufactured from PVC or steel in the shape of a skirting board and is frequently used in commercial buildings such as hospitals, laboratories and offices. The trunking is fitted around the walls of a room at either

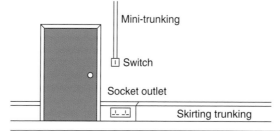

Figure 3.31 Typical installation of skirting trunking and mini-trunking.

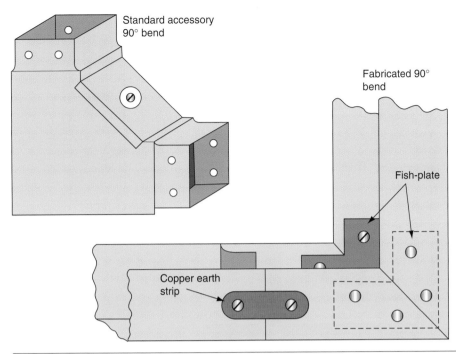

Figure 3.32 *Alternative trunking bends.*

the skirting board level or at the working surface level and contains the wiring for socket outlets which are mounted on the lid, as shown in Figure 3.31.

Where any trunking passes through walls, partitions, ceilings or floors, short lengths of lid should be fitted so that the remainder of the lid may be removed later without difficulty. Any damage to the structure of the buildings must be made good with mortar, plaster or concrete in order to prevent the spread of fire.

Fire barriers must be fitted inside the trunking every 5 m, or at every floor level or room-dividing wall if this is a shorter distance, as shown in Figure 3.33(a).

Where trunking is installed vertically, the installed conductors must be supported so that the maximum unsupported length of non-sheathed cable does not exceed 5 m. Figure 3.33(b) shows cables woven through insulated pin supports, which is one method of supporting vertical cables.

PVC insulated cables are usually drawn into an erected conduit installation or laid into an erected trunking installation. Table E4 of the *On-Site Guide* only gives factors for conduits up to 32 mm in diameter, which would indicate that conduits larger than this are not in frequent or common use. Where a cable enclosure greater than 32 mm is required because of the number or size of the conductors, it is generally more economical and convenient to use trunking.

Trunking capacities

The *ratio* of the space occupied by all the cables in a conduit or trunking to the whole space enclosed by the conduit or trunking is known as the *space factor*.

Where sizes and types of cable and trunking are not covered by the tables in the *On-Site Guide*, a space factor of 45% must not be exceeded. This means that the cables must not fill more than 45% of the space enclosed by the trunking.

The tables take this factor into account.

 Definition

The *ratio* of the space occupied by all the cables in a conduit or trunking to the whole space enclosed by the conduit or trunking is known as the *space factor*.

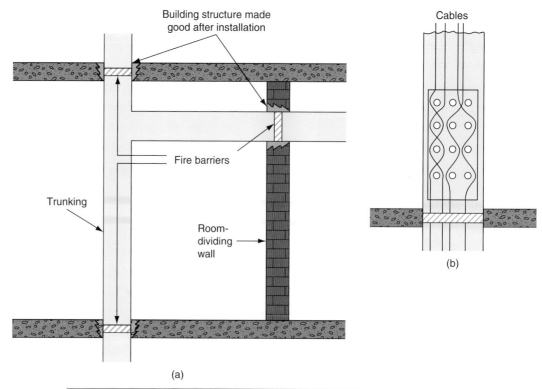

Figure 3.33 Installation of trunking a) fire barriers in trunking and b) cable supports in vertical trunking.

To calculate the size of trunking required to enclose a number of cables:

- Identify the cable factor for the particular size of conductor (see Table 3.5).
- Multiply the cable factor by the number of conductors to give the sum of the cable factors.

Table 3.4 Trunking cable factors

Type of conductor	Conductor CSA (mm²)	PVC cable factor	Thermosetting cable factor
Solid	1.5	8.0	8.6
	2.5	11.9	11.9
Stranded	1.5	8.6	9.6
	2.5	12.6	13.9
	4	16.6	18.1
	6	21.2	22.9
	10	35.3	36.3

Table 3.5 Trunking factors

Dimensions of trunking (mm × mm)	Factor
50 × 38	767
50 × 50	1037
75 × 25	738
75 × 38	1146
75 × 50	1555
75 × 75	2371
100 × 25	993
100 × 38	1542
100 × 50	2091
100 × 75	3189
100 × 100	4252
150 × 38	2999
150 × 50	3091
150 × 75	4743
150 × 100	6394
150 × 150	9697

- Consider the factors for trunking shown in Table 3.5. The correct size of trunking to accommodate the cables is that trunking which has a factor equal to, or greater than, the sum of the cable factors.

Example 3

Calculate the minimum size of trunking required to accommodate the following single-core PVC cables:

20 × 1.5 mm solid conductors

20 × 2.5 mm solid conductors

21 × 4.0 mm stranded conductors

16 × 6.0 mm stranded conductors

(continued)

Example 3 continued

From Table 3.4, the cable factors are:

for 1.5mm solid cable – 8.0

for 2.5mm solid cable – 11.9

for 4.0mm stranded cable – 16.6

for 6.0mm stranded cable – 21.2

The sum of the cable terms is:

$(20 \times 8.0) + (20 \times 11.9) + (21 \times 16.6) + (16 \times 21.2) + 1085.8.$

From Table 3.5, 75 × 38mm trunking has a factor of 1146 and, therefore, the minimum size of trunking to accommodate these cables is 75 × 38mm, although a larger size, say, 75 × 50mm, would be equally acceptable if this was more readily available as a standard stock item.

Segregation of circuits

Where an installation comprises a mixture of low-voltage and very low-voltage circuits such as mains lighting and power, fire alarm and telecommunication circuits, they must be separated or *segregated* to prevent electrical contact (IET Regulation 528.1).

For the purpose of these regulations, various circuits are identified by one of two bands as follows:

- Band I: telephone, radio, bell, call and intruder alarm circuits, emergency circuits for fire alarm and emergency lighting.
- Band II: mains voltage circuits.

When Band I circuits are insulated to the same voltage as Band II circuits, they may be drawn into the same compartment.

When trunking contains rigidly fixed metal barriers along its length, the same trunking may be used to enclose cables of the separate bands without further precautions, provided that each band is separated by a barrier, as shown in Figure 3.34.

Multi-compartment PVC trunking cannot provide band segregation since there is no metal screen between the bands. This can only be provided in PVC trunking if screened cables are drawn into the trunking.

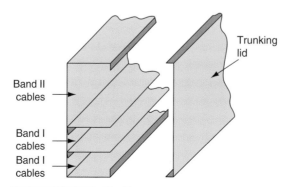

Figure 3.34 Segregation of cables in trunking.

Cable tray installations

Cable tray is a sheet-steel channel with multiple holes. The most common finish is hot-dipped galvanized but PVC-coated tray is also available. It is used extensively on large industrial and commercial installations for supporting MI and SWA cables which are laid on the cable tray and secured with cable ties through the tray holes.

Cable tray should be adequately supported during installation by brackets which are appropriate for the particular installation. The tray should be bolted to the brackets with round-headed bolts and nuts, with the round head inside the tray so that cables drawn along the tray are not damaged.

The tray is supplied in standard widths from 50 to 900 mm, and a wide range of bends, tees and reducers is available. Figure 3.35 shows a factory-made 90° bend at B. The tray can also be bent using a cable tray bending machine to create bends such as that shown at A in Figure 3.35. The installed tray should be securely bolted with round-headed bolts where lengths or accessories are attached, so that there is a continuous earth path which may be bonded to an electrical earth. The whole tray should provide a firm support for the cables, and therefore the tray fixings must be capable of supporting the weight of both the tray and cables.

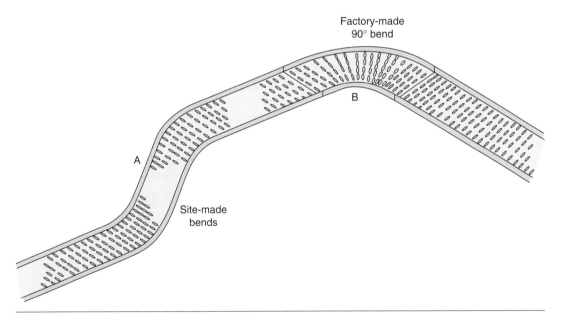

Figure 3.35 Cable tray with bends.

Cable basket installations

Cable basket is a form of cable tray made from a wire mesh material and looks similar to a supermarket trolley. Basket is lighter than tray and is generally used when there are numerous smaller, lighter cables that require a management system.

Cables are generally laid inside the basket rather than pulled in, allowing a quicker, more convenient installation.

Cable basket is much lighter than tray and easier to install; it also makes it easier to carry out alterations to cabling after installation due to the open nature of the basket.

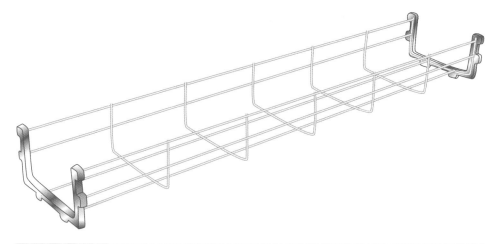

Figure 3.36 Cable basket.

Cable ladder

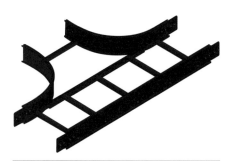

Figure 3.37 Cable ladder.

Cable ladder is at the other end of the scale to cable basket and is a heavy duty version of support, constructed as the name suggests, to look like a ladder.

Cable ladder is used to support heavier cables that do not need a continuous surface to be fixed to, but can span the spaces between the rungs. As these cables are usually high capacity cables, this type of support is usually used in industrial and large commercial installations.

As the cables are capable of carrying high currents, large forces can be exerted on the cables, especially under fault conditions. Due to this, cable clamps are used to attach the cable to the ladder to ensure a sufficiently robust fixing.

Cable ducting

Figure 3.38 Heavy duty corrugated plastic pipes.

Within the electrical industry, underground cable routes often need to be reusable to allow for alterations in the installation. Heavy duty corrugated plastic pipes are installed to facilitate this rather than burying a cable directly into the ground.

Within control panels, a form of trunking with multiple outlets like a comb is also known as ducting. The ducting allows routing of the cabling while also containing it in a neat manner.

Modular wiring systems

Figure 3.39 Cable ducting used in panel wiring.

Modular wiring systems are growing in popularity with designers and installers due to the reduction in installation times that can be up to 70% less than traditional methods. The idea of modular wiring is the provision of a factory manufactured and tested system that is fast to fit. The biggest benefit is that the system can be manufactured off-site in a quality controlled environment and supplied as a pluggable system, reducing on-site connections. These modules join together to form the installation.

Many large projects such as the London Olympics have used modular construction techniques.

Busbar and power track systems

In an electrical system, a busbar is usually a flat, uninsulated copper strip contained inside switchgear or panels for carrying high currents. The flat shape of the conductor gives a large surface area that allows for heat dissipation and for circuits to branch off along the length.

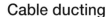

Power track systems are similar to busbar systems in that they have conductors running through the containment and loads can be tapped off at multiple points. Whereas busbar systems are for high current distribution, power track systems are commonly used in offices under floors and in similar installations feeding individual circuits, such as socket outlets in floor boxes, and they provide flexibility for office reorganizations via the final tap off connection, often in flexible conduit.

PVC/SWA cable installations

Steel wire armoured PVC insulated cables are now extensively used on industrial installations and often laid on cable tray. This type of installation has the advantage of flexibility, allowing modifications to be made speedily as the need arises. The cable has a steel wire armouring giving mechanical protection and permitting it to be laid directly in the ground or in ducts, or it may be fixed directly or laid on a cable tray. Figure 3.40 shows a PVC/SWA cable.

It should be remembered that when several cables are grouped together, the current rating will be reduced according to the correction factors given in Appendix 4 (Table 4C1) of the IET Regulations.

The cable is easy to handle during installation, is pliable and may be bent to a radius of 8 times the cable diameter. The PVC insulation would be damaged if installed in ambient temperatures over 70°C or below 0°C, but once installed the cable can operate at low temperatures.

The cable is terminated with a simple gland which compresses a compression ring onto the steel wire armouring to provide the earth continuity between the switchgear and the cable.

Definition

Steel wire armoured PVC insulated cables are now extensively used on industrial installations and often laid on cable tray.

MI cable installations

Mineral insulated cables are available for general wiring as:

- light-duty MI cables for voltages up to 600 V and sizes from 1.0 to 10 mm;

- heavy-duty MI cables for voltages up to 1000 V and sizes from 1.0 to 150 mm.

Figure 3.41 shows an MI cable and termination.

The cables are available with bare sheaths or with a PVC over-sheath. The cable sheath provides sufficient mechanical protection for all but the most severe situations, where it may be necessary to fit a steel sheath or conduit over the cable to give extra protection, particularly near floor level in some industrial situations.

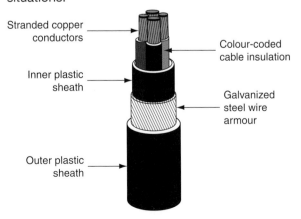

Stranded copper conductors

Colour-coded cable insulation

Inner plastic sheath

Galvanized steel wire armour

Outer plastic sheath

Figure 3.40 A four-core PVC/SWA cable.

The cable may be laid directly in the ground, in ducts, on cable tray or clipped directly to a structure. It is not affected by water, oil or the cutting fluids used in engineering and can withstand very high temperatures or even fire. The cable diameter is small in relation to its current-carrying capacity and it should last indefinitely if correctly installed because it is made from inorganic materials.

These characteristics make the cable ideal for Band I emergency circuits, boiler houses, furnaces, petrol stations and chemical plant installations.

The cable is supplied in coils and should be run off during installation and not spiralled off, as described in Figure 3.30 for conduit. The cable can be work hardened if overhandled or overmanipulated. This makes the copper outer sheath stiff and may result in fracture. The outer sheath of the cable must not be penetrated, otherwise moisture will enter the magnesium oxide insulation and lower its resistance. To reduce the risk of damage to the outer sheath during installation, cables should be straightened and formed by hammering with a hide hammer or a block of wood and a steel hammer. When bending MI cables the radius of the bend should not cause the cable to become damaged and clips should provide adequate support (IET Regulation 522.8.5); see Table 3.1.

The cable must be prepared for termination by removing the outer copper sheath to reveal the copper conductors. This can be achieved by using a rotary stripper tool or, if only a few cables are to be terminated, the outer sheath can be removed with side cutters, peeling off the cable in a similar way to peeling the skin from a piece of fruit with a knife. When enough conductor has been revealed, the outer sheath must be cut off square to facilitate the fitting of the sealing pot, and this can be done with a ringing tool. All excess magnesium oxide powder must be wiped from the conductors with a clean cloth. This is to prevent moisture from penetrating the seal by capillary action.

Cable ends must be terminated with a special seal to prevent the entry of moisture. Figure 3.41 shows a brass screw-on seal and gland assembly, which allows termination of the MI cables to standard switchgear and conduit fittings.

The sealing pot is filled with a sealing compound, which is pressed in from one side only to prevent air pockets from forming, and the pot closed by crimping home the sealing disc with an MI crimping tool such as that shown in Figure 4.4.

Such an assembly is suitable for working temperatures up to 105 °C. Other compounds or powdered glass can increase the working temperature up to 250 °C.

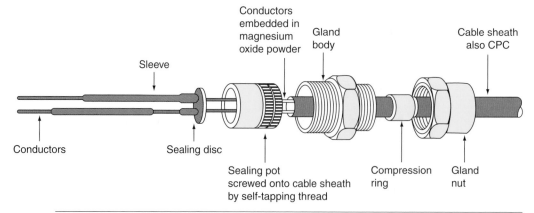

Figure 3.41 MI cable with terminating seal and gland.

The conductors are not identified during the manufacturing process and so it is necessary to identify them after the ends have been sealed. A simple continuity or polarity test, as described later in this chapter, can identify the conductors which are then sleeved or identified with coloured markers.

Connection of MI cables can be made directly to motors, but to absorb the vibrations, a 360° loop should be made in the cable just before the termination. If excessive vibration is expected, the MI cable should be terminated in a conduit through box and the final connection made by flexible conduit.

Copper MI cables may develop a green incrustation or patina on the surface, even when exposed to normal atmospheres. This is not harmful and should not be removed. However, if the cable is exposed to an environment which might encourage corrosion, an MI cable with an overall PVC sheath should be used.

Selecting the appropriate type of wiring system for the environment

Chapter 52 of BS 7671 places requirements on the designer to select the appropriate wiring system for the type of installation giving consideration to:

* cables and conductors;
* their connections, terminations and/or joints;
* their associated supports or suspensions; and
* their enclosures or methods of protection against external influences.

Appendix C of the *On-Site Guide* gives guidance on the application of cables for fixed wiring and flexible cables in Tables C1 and C2.

Assessment criteria 3.5

Recognize the requirements of industrial plugs, sockets and couplers

Industrial plugs, sockets and couplers

In 1968, the UK adopted the IEC 309 standard as BS 4343, and in 1999 replaced it with the European equivalent BS EN 60309.

Electricians often refer to these plugs as 'Commando plugs' (refers to the MK Electric Company Commando range of connectors). The standard covers plugs, socket-outlets and couplers for industrial purposes. IEC 60309-2 specifies a range of mains power connectors with circular housings, and different numbers and arrangements of pins for different applications. The 16A single and three-phase variants are commonly used throughout Europe at campsites, marinas, workshops and farms. The colour of an IEC 60309 plug or socket indicates its voltage rating and the most common colours in use are yellow (110V), blue (230V) and red (400v).

Figure 3.42 Industrial plugs for outdoor use must be more secure than those for domestic use.

Cables to BS 7919

Flexible cable, manufactured to BS 7919, Arctic grade flex, was specifically designed for use at 110V a.c. from centre tapped transformers (55V – 0 – 55V).

The practice of using a 110V centre tapped transformer is a UK practice, hence the lack of European harmonization of the standard.

The key feature of this cable is that it is designed to be suitable for installation and handling down to a temperature of −25°C , e.g. suitable for construction site installations.

As the standard applies to 110V flexes only, strictly speaking, only a yellow flex should be referred to as an 'Artic' cable, however manufacturers supply other colours for other voltages, such as blue for 230V, and often mark these as 'Artic' cables.

Assessment criteria 4.1

Identify the general layout and the equipment at the service position

Electricity supply systems

The British government agreed on 1 January 1995, that the electricity supplies in the United Kingdom would be harmonized with those of the rest of Europe. Thus, the voltages used previously in low-voltage supply systems of 415V and 240V have become 400V for three-phase supplies and 230V for single-phase supplies. The Electricity Supply Regulations 1988 have also been amended to permit a range of variations from the new declared nominal voltage. Previously it was +/−6%, but from January 1995, the permitted tolerance is the nominal voltage +10% or −6%.

This gives a voltage range of 216 to 253V for a nominal voltage of 230V and 376 to 440V for a nominal voltage of 400V (IET Regulation Appendix 2, point number 14).

Figure 3.43 They may be considered unsightly by some, but pylons are currently the most economical way of transferring electricity from one place to another.

It is further proposed that the tolerance levels will be adjusted to 10% of the declared nominal voltage. All European Union countries will adjust their voltages to comply with a nominal voltage of 230V single phase and 400V three phase.

The supply to a domestic, commercial or small industrial consumer's installation is usually protected at the incoming service cable position with a 100A high breaking capacity (HBC) fuse. The maximum, that is, worst case value of external earth fault loop impedance outside of the consumer's domestic installation is:

- 0.8Ω for cable sheath earth supplies (TN-S system);
- 0.35Ω for protective multiple earthing (PME) supplies (TN-C-S system);
- 21.0Ω excluding the consumer's earth electrode for no earth supplies (TT system).

The maximum, that is, worst case value of prospective short-circuit current is 16kA at the supply terminals (see *Electrician's Guide to the Building Regulations* Part P, Chapter 3).

Other items of equipment at this position are the energy meter and the consumer's distribution unit, providing the protection for the final circuits and the earthing arrangements for the installation.

An efficient and effective earthing system is essential to allow protective devices to operate. The limiting values of earth fault loop impedance are given in Tables 41.2 to 41.4 and Chapter 54 of the IET Regulations. Wiring Systems in Part 3 gives details of the earthing arrangements to be incorporated into the supply system to meet the requirements of the regulations. Five systems are described in the definitions but only the TN-S, TN-C-S and TT systems are suitable for public supplies.

A system consists of an electrical installation connected to a supply. Systems are classified by a capital letter designation.

The supply earthing

The supply earthing arrangements are indicated by the first letter, where

T means one or more points of the supply are directly connected to earth and

I means the supply is not earthed or one point is earthed through a fault-limiting impedance.

The installation earthing

The installation earthing arrangements are indicated by the second letter, where

T means the exposed conductive parts are connected directly to earth, and

N means the exposed conductive parts are connected directly to the earthed point of the source of the electrical supply.

The earthed supply conductor

The earthed supply conductor arrangements are indicated by the third letter, where S means a separate neutral and protective conductor and C means that the neutral and protective conductors are combined in a single conductor.

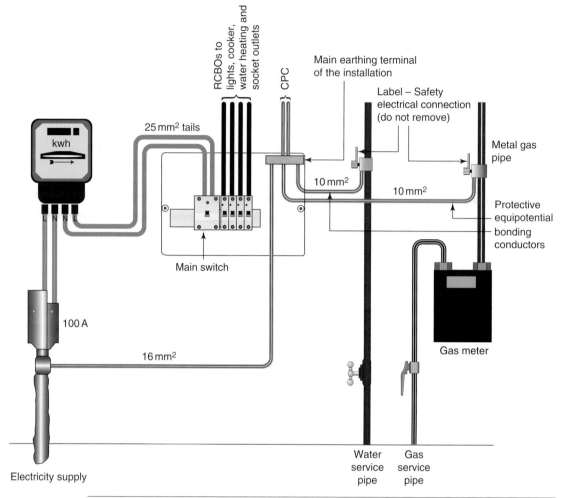

Figure 3.44 Cable sheath earth supplies (TN-S system) showing earthing and bonding arrangements.

Cable sheath earth supply (TN-S system)

This is one of the most common types of supply system found in the United Kingdom where the electricity companies' supply is provided by underground cables. The neutral and protective conductors are separate throughout the system. The protective earth conductor (PE) is the metal sheath and armour of the underground cable, and this is connected to the consumer's main earthing terminal. All extraneous conductive parts of the installation, gas pipes, water pipes and any lightning protective system are connected to the protective conductor via the main earthing terminal of the installation. The arrangement is shown in Figure 3.44, and in Figure 2.1 of the *On-Site Guide*.

Protective multiple earthing supplies (TN-C-S system)

This type of underground supply is becoming increasingly popular to supply new installations in the United Kingdom. It is more commonly referred to as protective multiple earthing (PME). The supply cable uses a combined protective earth and neutral conductor (PEN conductor). At the supply intake point, a consumer's main earthing terminal is formed by connecting the earthing terminal to the neutral conductor. All extraneous conductive parts of the installation, gas pipes, water pipes and any lightning protective system are then connected to the main earthing terminals. Thus, line to earth faults are effectively converted into line to

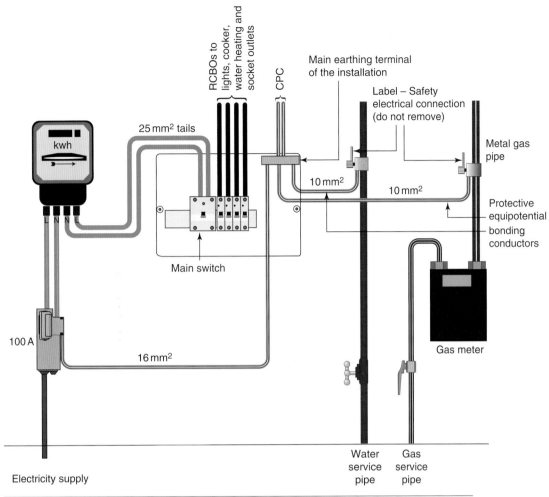

Figure 3.45 Protective multiple earthing supply (TN-C-S system) showing earthing and bonding arrangements.

neutral faults. The arrangement is shown in Figure 3.45, and in Figure 2.1 of the *On-Site Guide*.

No earth provided supplies (TT system)

This is the type of supply more often found when the installation is fed from overhead cables. The supply authorities do not provide an earth terminal and the installation's circuit protective conductors (CPCs) must be connected to earth via an earth electrode provided by the consumer. IET Regulation 542.2.3 lists the type of earth rod, earth plate or earth tapes recognized by BS 7671. An effective earth connection is sometimes difficult to obtain and in most cases a residual current device (RCD) is provided when this type of supply is used. The arrangement is shown in Figure 3.46.

Figures 3.44, 3.45 and 3.46 show the layout of a typical domestic service position for these three supply systems. There are two other systems of supply, the TN-C and IT systems, but they do not comply with the supply regulations and therefore cannot be used for public supplies. Their use is restricted to private generating plants. For this reason, I shall not include them here but they can be seen in Part 3 of the IET Regulations.

Figures 3.44 to 3.46 show circuits protected by RCBOs (a device combining the features of both a circuit breaker and residual current device). The use of RCBOs

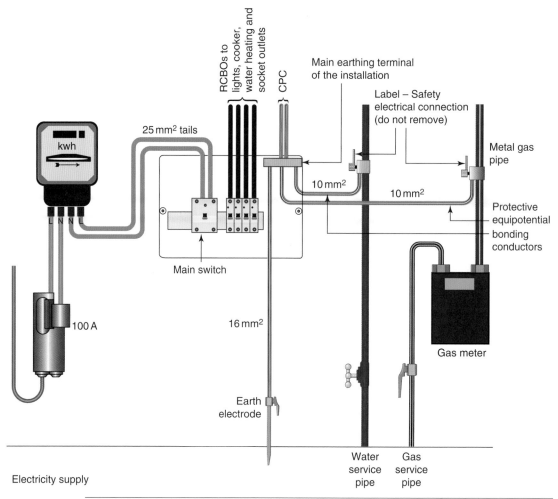

Figure 3.46 No supply earth provided (TT systems) showing earthing and bonding arrangements.

will minimize inconvenience and disruption in the event of a fault occurring, because only the faulty circuit will be disconnected by the RCBO. However, alternative consumer unit arrangements using CBs and RCDs are permissible and are shown in the *On-Site Guide* in Section 3.6.3.

Figures 3.44 to 3.46 show protective equipotential bonding conductors connected to gas and water supplies. This is always necessary if the incoming supplies are metallic. However, the 18th Edition of the IET Regulations at 411.3.1.2 tells us that metal pipes entering a building having **an insulated section at their entry to the building need not** be connected to the equipotential bonding of the installation.

However, the premises' gas, water and electricity supplies must be bonded on the consumers hard metal pipe work, at the point of entry to the building IET Regulation 544.1.2, as shown in Figures 3.44 to 3.46 of this book, and Figures 2.1 (i) to 2.1 (iii) of the *On-Site Guide*.

Function of components

Section 2 of the *On-Site Guide* outlines the components that will be found at the intake of a typical domestic supply intake position.

Distributors cut-out

The incoming supply cable is terminated into the distribution network operator's cut-out. The cut-out fuse will be sealed in position to prevent unintentional withdrawal or connection to the supply before the metering equipment. The fuse in the cut-out provides protection up to the consumer's main switch and limits the current available that could be drawn.

Electricity meter

The electricity meter is the property of the supplier and is a calibrated meter that records consumption in KW hours to allow a bill to be prepared. The meter is sealed to prevent any unauthorized person interfering with it.

By 2020, all homes will have 'smart' meters installed. These meters are designed to work with micro-generation technology and give a real-time display of your consumption and the cost. The meter will also have the ability to send the readings directly to the supply company, removing estimated bills. As the consumer will be able to monitor their usage, it is hoped that the information will help consumers to reduce their consumption, lowering bills and demand nationally.

Meter tails

The meter tails are part of the consumer's installation and should be insulated and sheathed and a minimum size of 25 mm^2. The colour of the insulation should indicate the supply polarity while the sheath is usually grey. Distributors may specify the maximum length of the tails that they will accept. The tails may feed directly into a consumer unit or through an isolator switch as shown in Fig 3.47.

Isolator switch

In recent years, electricity suppliers have started to install an electricity isolator switch between the meter and the consumer unit to allow the installation to be isolated without withdrawing the cut-out fuse.

Consumer's controlgear

A consumer unit to BSEN 60439-3 is used on domestic single-phase installations up to 100 A. The unit is an enclosure consisting of various components to allow for isolation, overcurrent protection and usually earth fault protection.

- Double pole isolator: a main switch to turn off the entire consumer unit. As an isolator, it is not designed to be operated on-load and the individual circuits should be isolated before turning the isolator off.
- Fuses/circuit breakers/RCBOs: these devices provide overcurrent protection and the installation should be designed in such a way that it is split into circuits allowing for maintenance, testing and protection to minimize inconvenience when operated.
- Residual current device: required by the regulations for safety reasons, the RCD provides additional protection against electric shock by monitoring earth leakage.
- Recent fire statistics have shown that a large number of domestic fires involved plastic consumer units as the source of the fire. Consumer units are often located at the entrance or exit door of the home or under the stairs, raising the possibility that a fire starting as a result of faulty wiring could block the emergency exit route. Regulation 421.1.201 now requires that consumer

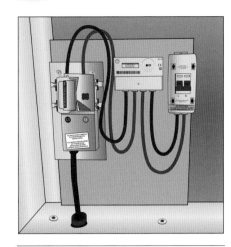

Figure 3.47 Cut-out.

Figure 3.48 A 'smart' meter.

Figure 3.49 One circuit out of several can be turned off using this circuit breaker.

Figure 3.50 Circuit breakers.

units be manufactured from non-combustible material, for example, metal, or be enclosed in a non-combustible enclosure.

- The 18th Edition of the regulations at 421.1.7, now also **recommends** the use of arc fault detection devices (AFDD) in the consumer unit as a means of providing additional protection against fire.
- Wherever RCDs are installed, a note must be fixed near to each RCD stating 'This device must be tested 6-monthly' (IET Regulation 514.12.2). Previously the test period was 3-monthly.

Assessment criteria 5.1

Distinguish the different circuits and wiring layouts used for lighting

---

Assessment criteria 5.2

Identify the different components that can be used in lighting circuits

---

One-way lighting circuit

Lighting circuits may be wired as either one-way, two-way or intermediate switching circuits. The simplest form of lighting circuit is a one-way controlled light. This is where only one switch is used to operate the lights, for example in a bedroom or bathroom, where only one door provides access to the room. In this circuit, the line conductor has a single pole overcurrent protective device (fuse or circuit breaker) fitted in line with a switch. The switch provides functional switching (on and off). A one-way switch simply opens or closes the circuit as shown in Figure 3.52. This circuit may be wired using a joint box as shown in Figure 3.8, or using the 'loop in' method shown in Figure 3.53.

Two-way lighting circuit

A two-way lighting circuit is used where two separate switching positions are required such as a staircase with a switch at either end. The overcurrent protective device is fitted in the line conductor before it feeds the switches, controlling the supply to the light. A two-way switch differs from the one-way

One-way Switch

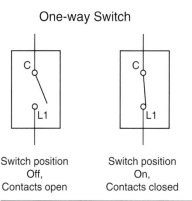

Switch position	Switch position
Off,	On,
Contacts open	Contacts closed

Figure 3.51 A one-way switch.

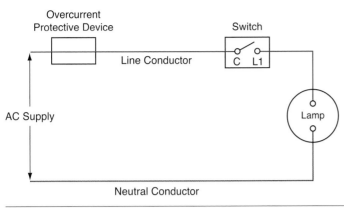

Figure 3.52 A simple one-way lighting circuit.

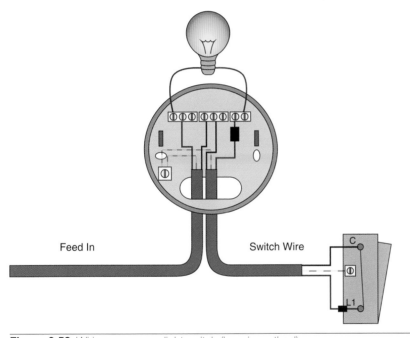

Figure 3.53 Wiring a one-way light switch (loop in method).

switch because the feed in is connected to either one or the other of the two outlet terminals. The switch acts as a changeover, switching the connections between the common and position 1 or the common and position 2, dependent on the switch position as shown in Figure 3.54. The wiring diagram for a one-way converted to a two-way circuit is shown in Figure 3.10.

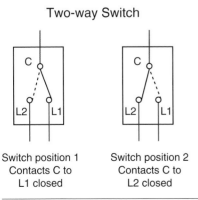

Figure 3.54 Two-way switch connections.

Intermediate switching

On a lighting circuit where the light requires three or more switches, an intermediate switch is used. This may be a staircase with a landing part way or a long corridor, for example. The wiring is carried out in the same way as a two-way installation.

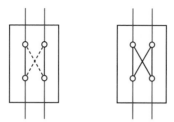

Figure 3.55 An intermediate switch.

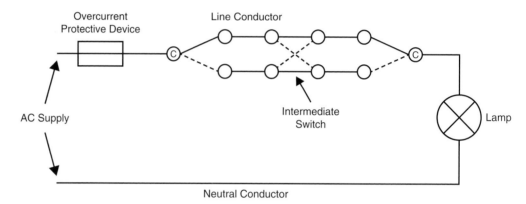

Figure 3.56 A circuit diagram for wiring an intermediate using 3 core plus c.p.c. cable.

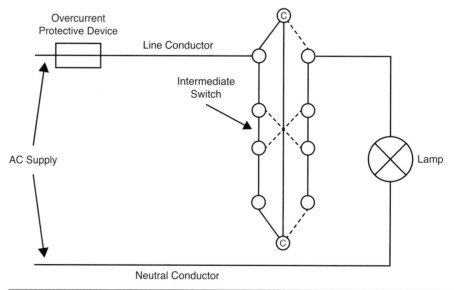

Figure 3.57 Wiring an intermediate switch.

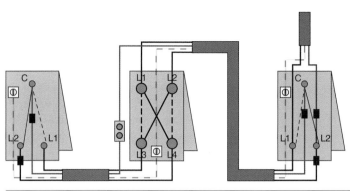

Figure 3.58 Intermediate switch connections in multicore cables.

Junction boxes

When wiring a lighting circuit, the designer will look at the fittings and decide if it is possible or suitable to terminate multiple cables at the accessory. It may be decided that making the joints elsewhere in a junction box is preferable and a single cable run to the switches or fittings for ease of terminations. Junction boxes with fixed terminals or adaptable boxes housing connector blocks, often referred to as RB4s could be used. BS 7671 regulation 526.3 requires all screw terminals to be accessible for maintenance once installed and this must be considered when designing such a circuit.

SELV lighting

Many spotlight type fittings work using SELV (separated extra low voltage). For the system to be classified as SELV, the voltage must not exceed 50V a.c. or 120V d.c. and the circuit protective conductor may not be connected to the load. This is achieved by use of a transformer to BS EN 61558-2-6, where the low voltage (230V) circuit is connected to the primary side and the ELV fitting is connected to the secondary side. Plugs and sockets used for the connections must not be interchangeable so as to avoid the fittings being connected to a higher voltage than they are designed for.

Many SELV systems use LED lighting due to its high efficiency, however not all SELV transformers are dimmable.

Assessment criteria 6.1

Define the requirements of standard ring final socket circuits

Whereas most equipment is wired on a radial circuit, it is common practice in the UK to wire BS 1363 13A socket outlets on a ring final circuit. BS 7671 describes a ring final circuit as 'starting and finishing at the distribution board, where it is connected to a 30A or 32A overcurrent protective device'.

The *On-Site Guide* describes this arrangement as an A1 circuit and while it does not limit the number of sockets that can be connected to this arrangement, it does give a maximum floor area that the circuit may feed as 100m^2.

The minimum cable size for the live conductors of a ring final circuit is given as 2.5 mm² when using thermoplastic cables (1.5 mm for mineral insulated cable as it has a higher current carrying capacity); at first glance this looks like the cable is rated lower than the protective device and will not comply to other regulations. However, due to the ring arrangement, the current will split within the circuit and this design is acceptable. However, care should be taken when designing the circuit to provide reasonable sharing of the load in each leg of the ring.

As stated above, the number of socket outlets on a ring final circuit is unlimited but the load will determine if more than one circuit is required. In this case, the designer should aim to distribute permanently connected equipment across the circuits to avoid overloading any one and to reduce inconvenience in the case of a fault or when maintenance is being carried out.

Number of socket outlets

When installing socket outlets they should be numerous enough and positioned such that all equipment can be connected conveniently without the need for extension leads. The *On-Site Guide* provides Table H7 in Appendix H to give guidance on the minimum numbers required in assorted locations.

Spurs

A socket that is not connected within the ring but fed via a single cable is known as a spur.

The total number of un-fused spurs connected to a ring should not exceed the total number of sockets and stationary equipment connected directly to the ring.

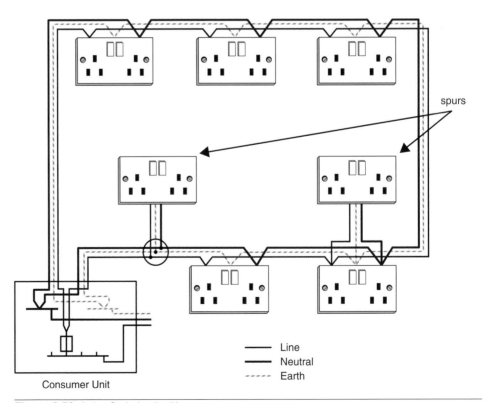

spurs

Line
Neutral
Earth

Consumer Unit

Figure 3.59 A ring final circuit with spurs.

Figure 3.60 Switched fused connection unit.

A non-fused spur should only feed one single or one twin socket outlet or a piece of permanently connected equipment to avoid overloading the cable feeding it. The connection to the ring final circuit should be made at the terminals of an existing socket outlet, the origin of the circuit or in a junction box, and the size of the live conductors should be 2.5 mm² minimum.

When a supply to more than one outlet is required and it is not possible to incorporate the addition into the ring, a fused connection unit (FCU) may be used. The cartridge fuse in the FCU (maximum 13A) will protect the devices connected to it.

The FCU may be connected directly into the ring or as a spur and incorporates a BS EN 1362 fuse. As the largest BS EN 1362 fuse is 13A, this is the maximum load that can be drawn and the number of sockets fed by the FCU is only limited by the load being connected not exceeding 13A. The total number of FCU connected to the ring final circuit is unlimited.

Assessment criteria 6.2

Define the requirements of standard radial final socket circuits

The *On-Site Guide* also recognizes that socket outlets may be wired on radial final circuits. Table H2.1 classifies the circuits as A2 and A3 where an A2 circuit is protected by a 30A or 32A overcurrent protective device, wired in 4 mm² thermoplastic cable (2.5 mm² for MI cable) and a maximum floor area of 75 m².

The A3 circuit uses 2.5 mm² cable (1.5 mm² for MI) and therefore the overcurrent protective device is reduced to 20A. This circuit has a maximum floor area of 50 m².

Table 3.6 General requirements for standard circuits

	Type of circuit Overcurrent protective device		Minimum conductor size (mm²)	Maximum floor area Square Meter
A1	Ring 30 or 32	any type of device	2.5	100
A2	Radial 30 or 32	cartridge fuse or circuit breaker	4	75
A3	Radial 20	any type of device	2.5	50

Assessment criteria 6.3

Describe the standard circuit arrangements for loads and equipment

Common domestic circuits include cookers, showers and immersion heaters. All of these appliances are wired on radial final circuits and the rating of the circuit is determined by an assessment of the current demand of the appliance.

Water-heating circuits

A small, single-point over sink-type water heater may be considered as a permanently connected appliance and so may be connected to a ring circuit through a fused connection unit. A water heater of the immersion type is usually rated at a maximum of 3 kW, and could be considered as a permanently connected appliance, fed from a fused connection unit. However, many immersion heating systems are connected into storage vessels of about 150 litres in domestic installations, and the *On-Site Guide* states that immersion heaters fitted to vessels in excess of 15 litres should be supplied by their own circuit (*On-Site Guide*, Appendix H5).

Therefore, immersion heaters must be wired on a separate radial circuit when they are connected to water vessels which hold more than 15 litres. Figure 3.62 shows the wiring arrangements for an immersion heater. Every switch must be a double-pole (DP) switch and out of reach of anyone using a fixed bath or shower when the immersion heater is fitted to a vessel in a bathroom.

Supplementary equipotential bonding to pipework will only be required as an addition to fault protection (IET Regulation 415.2) if the immersion heater vessel is in a bathroom that does not have:

- all circuits protected by a 30 mA RCD; and
- protective equipotential bonding (IET Regulation 411.3.1.2).

Electric space-heating circuits

Electrical heating systems can be broadly divided into two categories: unrestricted local heating and off-peak heating.

Unrestricted local heating may be provided by portable electric radiators which plug into the socket outlets of the installation. Fixed heaters that are wall mounted or inset must be connected through a fused connection

Figure 3.61 Water heater being worked on.

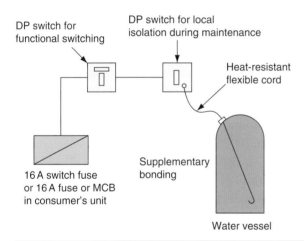

DP switch for
functional switching

DP switch for local
isolation during maintenance

Heat-resistant
flexible cord

16 A switch fuse
or 16 A fuse or MCB
in consumer's unit

Supplementary
bonding

Water vessel

Figure 3.62 Immersion heating wiring.

and incorporate a local switch, either on the heater itself or as a part of the fuse connecting unit. Heating appliances where the heating element can be touched must have a DP switch which disconnects all conductors. This requirement includes radiators which have an element inside a silica-glass sheath.

Off-peak heating systems may provide central heating from storage radiators, ducted warm air or underfloor heating elements. All three systems use the thermal storage principle, whereby a large mass of heat-retaining material is heated during the off-peak period and allowed to emit the stored heat throughout the day. The final circuits of all off-peak heating installations must be fed from a separate supply controlled by an electricity board time clock.

Figure 3.63 Radiator with a digital thermostat.

When calculating the size of cable required to supply a single-storage radiator, it is good practice to assume a current demand equal to 3.4 kW at each point. This will allow the radiator to be changed at a future time with the minimum disturbance to the installation. Each radiator must have a 20 A DP means of isolation adjacent to the heater and the final connection should be via a flex outlet. See Figure 3.64 for wiring arrangements.

Ducted warm air systems have a centrally-sited thermal storage heater with a high storage capacity. The unit is charged during the off-peak period, and a fan drives the stored heat in the form of warm air through large air ducts to outlet grilles in the various rooms. The wiring arrangements for this type of heating are shown in Figure 3.65.

The single-storage heater is heated by an electric element embedded in bricks and rated between 6 and 15 kW depending on its thermal capacity. A radiator of this capacity must be supplied on its own circuit, in cable capable of carrying the maximum current demand and protected by a fuse or circuit breaker (CB) of 30, 45 or 60 A as appropriate. At the heater position, a DP switch must be installed to terminate the fixed heater wiring. The flexible cables used for the final connection to the heaters must be of the heat-resistant type.

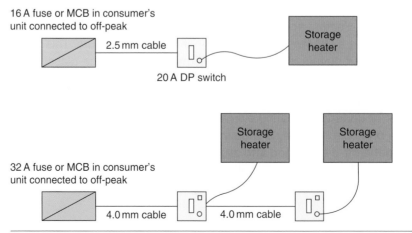

Figure 3.64 Possible wiring arrangements for storage heaters.

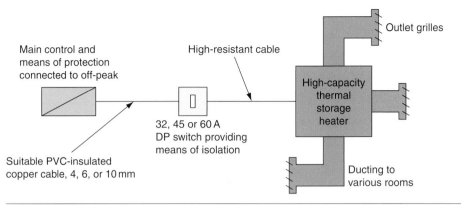

Figure 3.65 Ducted warm air heating system.

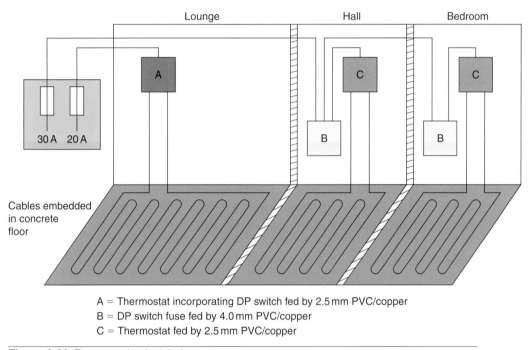

A = Thermostat incorporating DP switch fed by 2.5 mm PVC/copper
B = DP switch fuse fed by 4.0 mm PVC/copper
C = Thermostat fed by 2.5 mm PVC/copper

Figure 3.66 Floor-warming installations.

Floor-warming installations use the thermal storage properties of concrete. Special cables are embedded in the concrete floor screed during construction. When current is passed through the cables they become heated, the concrete absorbs this heat and radiates it into the room. The wiring arrangements are shown in Figure 3.66. Once heated, the concrete will give off heat for a long time after the supply is switched off and is, therefore, suitable for connection to an off-peak supply.

Underfloor heating cables installed in bathrooms or shower rooms must incorporate an earthed metallic sheath or be covered by an earthed metallic grid connected to the protective conductor of the supply circuit (IET Regulation 701.753).

Cooker circuit

A cooker with a rating above 3 kW must be supplied on its own circuit but since it is unlikely that in normal use every heating element will be switched on at the same time, a diversity factor may be applied in calculating the cable size, as detailed in the *On-Site Guide*.

Consider, as an example, a cooker with the following elements fed from a cooker control unit incorporating a 13A socket:

$$4 \times 2 \text{ fast-boiling rings} = 8000 \, W$$
$$1 \times 2 \, kW \text{ grill} = 2000 \, W$$
$$1 \times 2 \, kW \text{ oven} = 2000 \, W$$
$$\text{Total loading} = 12\,000 \, W$$

When connected to 230 V

$$\text{Current rating} = \frac{12\,000}{230} = 52.17 \, A$$

Applying the diversity factor of Table 1A

$$\text{Total current rating} = 52.17 \, A$$
$$\text{First 10 amperes} = 10 \, A$$
$$30\% \text{ of } 42.17 \, A = 12.65 \, A$$
$$\text{Socket outlet} = 5 \, A$$
$$\text{Assessed current demand} = 10 + 12.65 + 5 = 27.65 \, A$$

Therefore, a cable capable of carrying 27.65 A may be used safely rather than a 52.17 A cable.

A cooking appliance must be controlled by a switch separate from the cooker but in a readily accessible position. Where two cooking appliances are installed in one room, such as split-level cookers, one switch may be used to control both appliances provided that neither appliance is more than 2 m from the switch (*On-Site Guide*, Appendix H4)

Figure 3.67 Cooker switches should be in a readily accessible position.

Assessment criteria 7.1

Outline the division of an installation into circuits

Regulation 314 of BS 7671 explains that an installation should be divided into circuits and outlines the reasons as follows:

1 to avoid danger and minimize inconvenience in the event of a fault;

2 facilitate safe inspection, testing and maintenance;

3 take account of hazards that may arise from the failure of a single circuit such as a lighting circuit;

4 reduce the possibility of unwanted tripping of RCDs due to excessive protective conductor currents not due to a fault;

5 mitigate the effects of electromagnetic disturbances;

6 prevent the indirect energizing of a circuit intended to be isolated.

In practice, a domestic dwelling will typically be divided into circuits for power, lighting, water heating and cookers.

In modern properties, ring circuits for the socket outlets may split the house into areas such as upstairs and downstairs or the front and the back of the property and the lighting will usually be divided in a similar way.

Often the kitchen is connected to its own ring final circuit. This is due to the density of high current demand equipment such as a washing machine, a cooker, a kettle and similar equipment. Wired this way, the high demand is removed from other circuits and minimizes the likelihood of an overload.

Connecting highest loads nearest the main switch

When installing overcurrent devices in a consumer unit, it is industry practice to connect the highest-rated device nearest to the main switch and reduce the ratings as you work away from the switch. This is to reduce the stress/loading on the busbar within the consumer unit along its length.

Assessment criteria 7.2

Identify the requirements for polarity on circuits

Polarity requires that all fuses, circuit breakers and switches are connected in the line conductor only, that all socket outlets are correctly wired and that Edison screw-type lamp holders have the centre contact connected to the line conductor (IET Regulation 132.14.1). This is important to ensure that when devices are switched off, the break is in the line conductor to reduce the risk of an electric shock occurring at the load.

Assessment criteria 8.1

Identify the characteristics of earthing systems

As described earlier in this chapter, the definitions describe five systems but only the TN-S, TN-C-S and TT systems are suitable for public supplies.

A system consists of an electrical installation connected to a supply. Systems are classified by a capital letter designation.

The supply earthing

The supply earthing arrangements are indicated by the first letter, where T means one or more points of the supply are directly connected to earth and I means the supply is not earthed or one point is earthed through a fault-limiting impedance.

The installation earthing

The installation earthing arrangements are indicated by the second letter, where T means the exposed conductive parts are connected directly to earth and N means the exposed conductive parts are connected directly to the earthed point of the source of the electrical supply.

The earthed supply conductor

The earthed supply conductor arrangements are indicated by the third letter, where S means a separate neutral and protective conductor and C means that the neutral and protective conductors are combined in a single conductor.

Cable sheath earth supply (TN-S system)

This is one of the most common types of supply system to be found in the United Kingdom where the electricity companies' supply is provided by underground cables. The neutral and protective conductors are separate throughout the system. The protective earth conductor (PE) is the metal sheath and armour of the underground cable, and this is connected to the consumer's main earthing terminal. All exposed conductive parts of the installation, gas pipes, water pipes and any lightning protective system are connected to the protective conductor via the main earthing terminal of the installation. The arrangement is shown in Figure 3.68.

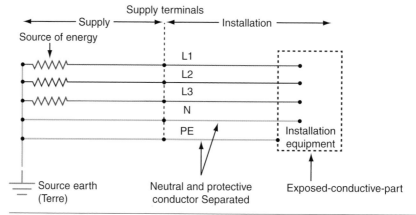

Figure 3.68 TN-S system.

(PME) Protective multiple earthing supply (TN-C-S system)

This type of underground supply is becoming increasingly popular in supplying new installations in the United Kingdom. It is more commonly referred to as protective multiple earthing (PME). The supply cable uses a combined protective earth and neutral (PEN) conductor. At the supply intake point, a consumer's main earthing terminal is formed by connecting the earthing terminal to the neutral conductor.

All exposed conductive parts of the installation, gas pipes, water pipes and any lightning protective system are then connected to the main earthing terminals. Thus, line to earth faults are effectively converted into line to neutral faults. The arrangement is shown in Figure 3.69.

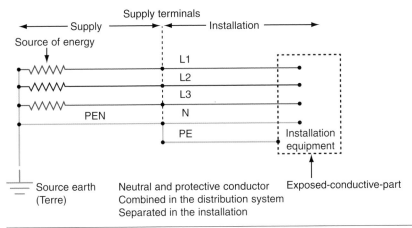

Figure 3.69 TN-C-S system (PME).

No earth provided supply (TT system)

This is the type of supply more often found when the installation is fed from overhead cables. The supply authorities do not provide an earth terminal and the installation's circuit protective conductors must be connected to earth via an earth electrode provided by the consumer. Regulation 542.2.3 lists the type of earth rod, earth plates or earth tapes recognized by BS 7671. An effective earth connection is sometimes difficult to obtain and in most cases a residual current device (RCD) is provided when this type of supply is used. The arrangement is shown in Figure 3.70.

Figures 3.44 to 3.46 show the layout of a typical domestic service position for these three supply systems. They show circuits protected by RCBOs. The use of RCBOs will minimize inconvenience because, in the event of a fault occurring, only the faulty circuit will disconnect. However, alternative consumer unit arrangements using circuit breakers and RCDs are shown in the *On-Site Guide* in Section 3.6.3. The TN-C and IT systems of supply do not comply with the supply regulations and therefore cannot be used for public supplies. Their use is restricted to private generating plants and for this reason I shall not include them here, but they can be seen in Part 2 of the IET Regulations.

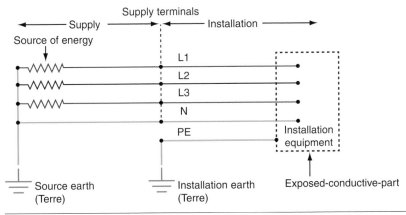

Figure 3.70 TT system.

Assessment criteria 8.2

Outline the purpose of earthing and protective conductors when used for protection

Earthing is required to provide a low impedance path, necessary for the protective device to operate should a fault to earth occur. A low impedance path allows sufficient current to flow to operate the protective device quickly.

low impedance = high current = quick operation of protective device

IET Regulation 411.3.2.2 tells us that for final circuits not exceeding 32 A the maximum disconnection time shall not exceed 0.4 s.

The achievement of these disconnection times is dependent upon the type of protective device used, fuse or circuit breaker, the circuit conductors to the fault and the provision of adequate protective equipotential bonding. The resistance, or we call it the impedance, of the earth fault loop must be less than the values given in Appendix B of the *On-Site Guide* and Tables 41.2 to 41.4 of the IET Regulations.

Protective electrical bonding to earth

The purpose of the bonding regulations is to keep all the exposed metalwork of an installation at the same earth potential as the metalwork of the electrical installation, so that no currents can flow and cause an electric shock. For a current to flow, there must be a difference of potential between two points, but if the points are joined together there can be no potential difference. This bonding or linking together of the exposed metal parts of an installation is known as 'protective equipotential bonding' and gives protection against electric shock.

Assessment criteria 8.3

Recognize the components which provide automatic disconnection of supply

When a fault occurs to earth within a circuit, enough current must flow to operate the protective device. From the point of the fault, where the line conductor has made contact with an earthed metallic part, the current will flow through the circuit protective conductor to the consumer's main earthing terminal ('met'). From here, the current will return to the earthed neutral point of the supply transformer. The route the current takes here depends on the earthing system in use:

* TN-S: the current returns via a separate conductor (often the cable sheath);
* TN-C-S: the current returns via the PEN conductor;
* TT: the current returns via the general mass of earth.

At the transformer's earthed neutral star point, the current flows through the phase winding of the supply transformer and returns to the installation along the line conductor. Within the consumer's installation the current flows through the overcurrent protective device to the point of the fault. This entire circuit is known as the earth fault loop path as shown in Figures 3.71 to 3.73.

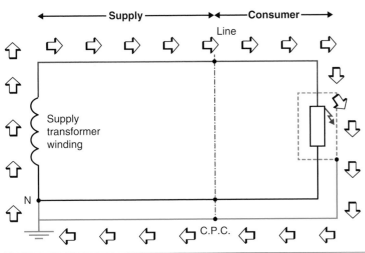

Figure 3.71 TN-S system showing the current path under fault conditions.

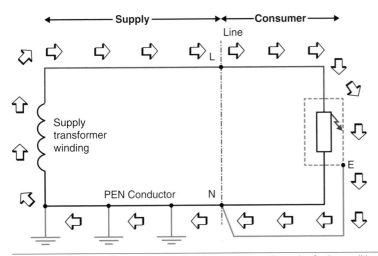

Figure 3.72 TN-C-S system showing the current path under fault conditions.

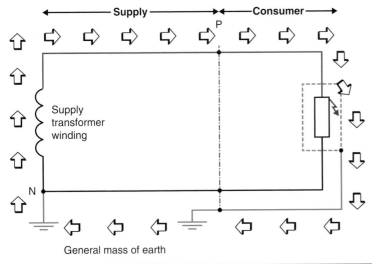

Figure 3.73 TT system showing the current path under fault conditions.

3

Definition

Exposed conductive parts are the metalwork of the electrical installation: the conduit, trunking, metal boxes and equipment that make up the electrical installation.

Assessment criteria 8.4

Recognize an exposed conductive part

BS 7671 defines an exposed conductive part as a conductive part of equipment which can be touched and which is not normally live, but which can become live under fault conditions.

Examples of exposed conductive parts would be metal conduit, metal trunking, metal boxes, metal appliance casings such as cookers and washing machines and metal face plates of switches and socket outlets.

Exposed conductive parts are connected to the main earthing terminal by circuit protective conductors.

Definition

Extraneous conductive parts are the other metal parts which do not form a part of the electrical installation: the structural steelwork of the building, gas, water and central heating pipes and radiators.

Assessment criteria 8.5

Recognize an extraneous conductive part

BS 7671 defines an extraneous conductive part as a conductive part liable to introduce a potential, generally earth potential, and not forming part of the electrical installation.

Examples of extraneous conductive parts are: metallic installation pipes, metallic gas installation pipes, other installation pipe work, for example, heating oil, structural steelwork of the building where rising from the ground and lightning protection systems.

Extraneous conductive parts are connected to the main earthing terminal by bonding conductors.

Assessment criteria 8.6

Distinguish the sections of earth loop impedance path

The path described in assessment criteria 8.6 can be split into three parts:

* Part 1. Ze: all of the impedance (think resistance) path external to the installation, the return from the main earthing terminal to the star point of the transformer, the phase coil and the line conductor to the consumer's installation. This value can be obtained during testing as a live test or by enquiry to the distribution network operator.
* Part 2. (R1 + R2): the resistance of the consumer's line conductor (R1) and the circuit protective conductor (R2) added together. This value is usually obtained when testing the continuity of the circuit protective conductor, the first of the dead tests.
* Part 3. Zs: the entire system, the total impedance of all the component parts added together. This is usually shown as $Zs = Ze + (R1 + R2)$.

The Zs value is compared to the values in Chapter 41 of BS 7671 to ensure the disconnection times of devices will be met.

Assessment criteria 8.7

Identify protective conductors

BS 7671 includes a drawing in Part 2 of earthing and protective conductors to illustrate the correct terminology for each individual conductor. Below is a simplified version.

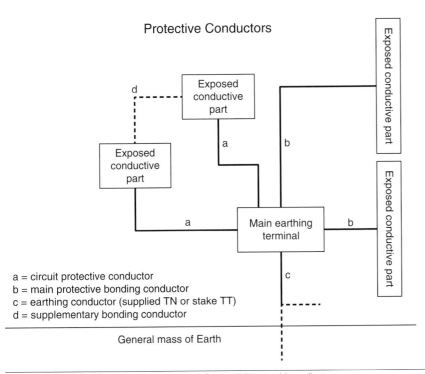

a = circuit protective conductor
b = main protective bonding conductor
c = earthing conductor (supplied TN or stake TT)
d = supplementary bonding conductor

Figure 3.74 Identification of conductors for earthing and bonding.

Assessment criteria 8.8

Outline the general requirements for the installation of main protective bonding

Where earthed electrical equipment may come into contact with the metalwork of other services, they too must be effectively connected to the main protective earthing terminal of the installation (IET Regulations 544).

Other services are described as:

- main water pipes;
- main gas pipes;
- other service pipes and ducting;
- central heating and air-conditioning systems;
- exposed metal parts of the building structure;
- lightning protective conductors.

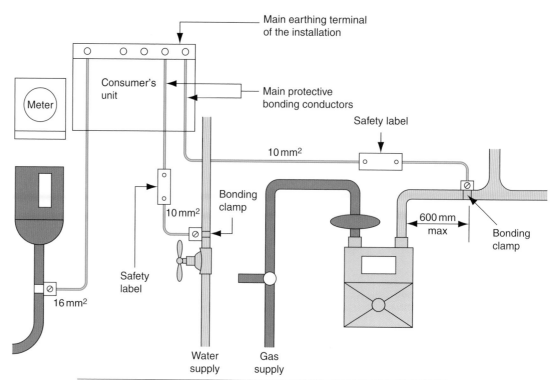

Figure 3.75 Main protective bonding of gas and water supplies.

Protective equipotential bonding should be made to gas and water services at their point of entry into the building, as shown in Figure 3.75, using insulated bonding conductors of not less than half the cross-section of the incoming main earthing conductor. The minimum permitted size is 6 mm² but the cross-section need not exceed 25 mm² (IET Regulations 544). The bonding clamp to BS 951 must be fitted on the consumer's side of the gas meter between the outlet union, before any branch pipework but within 600 mm of the meter (IET Regulation 544.1.2).

A permanent label must also be fixed in a visible position at or near the point of connection of the bonding conductor with the words 'Safety Electrical Connection – Do Not Remove' (IET Regulation 514.13.1) as in Figure 3.76.

The 18th Edition of the IET Regulations has brought in a new regulation at 411.3.1.2 because these days, many main gas and water supplies entering a building are made of an insulating, non-conducting material. The new regulation states that metallic pipes entering the building having an insulated section at their entrance to the building, **need not** be connected to the

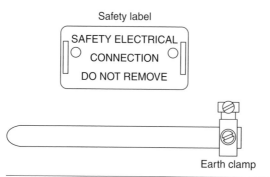

Figure 3.76 Typical earth bonding clamp.

protective equipotential bonding of the installation. This is because an insulated gas or water pipe does not meet the characteristics of an extraneous conductive part, that is a conductive part that is liable to introduce an earth potential into the installation. Insulated pipes reduce this risk and therefore need not be connected to the installations protective equipotential bonding conductors.

However, the premises' gas, water and electricity supplies must be bonded on the consumer's hard metal pipe work, at the point of entry to the building as shown in Figure 3.75 of this book, and figures 2.1 (i) to 2.1 (iii) of the *On-Site Guide* (IET Regulation 544.1.2).

Assessment criteria 9.1

Outline the features of overcurrent and overcurrent protection devices

Protection against overcurrent

Excessive current may flow in a circuit as a result of an overload or a short circuit.

An overload or overcurrent is defined as a current which exceeds the rated value in an otherwise healthy circuit. A short circuit is an overcurrent resulting from a fault of negligible impedance between live conductors having a difference in potential under normal operating conditions. Overload currents usually occur in a circuit because it is abused by the consumer or because it has been badly designed or modified by the installer.

Short circuits usually occur as a result of an accident which could not have been predicted before the event. An overload may result in currents of two or three times the rated current flowing in the circuit, while short-circuit currents may be hundreds of times greater than the rated current. In both cases, the basic requirement for protection is that the circuit should be interrupted before the fault causes a temperature rise which might damage the insulation, terminations, joints or the surroundings of the conductors. If the device used for overload protection is also capable of breaking a prospective short-circuit current safely,

Definition

An *overload current* can be defined as a current which exceeds the rated value in an otherwise healthy circuit.

A *short circuit* is an overcurrent resulting from a fault of negligible impedance connected between conductors.

Figure 3.77 Cartridge fuses.

then one device may be used to give protection from both faults (IET Regulation 432.1). Devices which offer protection from overcurrent are:

* semi-enclosed fuses manufactured to BS 3036;
* cartridge fuses manufactured to BS 88–3: 2010;
* high-breaking capacity fuses (HBC fuses) manufactured to BS 88–2: 2010;
* Circuit breakers (CBs) manufactured to BS EN 60898.

Overcurrent protection

The consumer's mains equipment must provide protection against overcurrent, that is, a current exceeding the rated value (IET Regulation 430.3). Fuses provide overcurrent protection when situated in the line conductors; they must not be connected in the neutral conductor. Circuit breakers may be used in place of fuses, in which case the circuit breaker may also provide the means of isolation, although a further means of isolation is usually provided so that maintenance can be carried out on the circuit breakers themselves.

When selecting a protective device we must give consideration to the following factors:

* the prospective fault current;
* the circuit load characteristics;
* the current-carrying capacity of the cable;
* the disconnection time requirements for the circuit.

The essential requirements for a device designed to protect against overcurrent are:

* it must operate automatically under fault conditions;
* it must have a current rating matched to the circuit design current;
* it must have a disconnection time which is within the design parameters;
* it must have an adequate fault breaking capacity;
* it must be suitably located and identified.

We will look at these requirements below.

An overload may result in currents of two or three times the rated current flowing in the circuit. Short-circuit currents may be hundreds of times greater than the rated current. In both cases, the basic requirements for protection are that the fault currents should be interrupted quickly and the circuit isolated safely before the fault current causes a temperature rise or mechanical effects which might damage the insulation, connections, joints and terminations of the circuit conductors or their surroundings (IET Regulations 131).

The selected protective device should have a current rating which is not less than the full load current of the circuit but which does not exceed the cable current rating. The cable is then fully protected against both overload and short-circuit faults (IET Regulation 435.1). Devices which provide overcurrent protection are:

* High breaking capacity (HBC) fuses to BS 88-2: 2010. These are for industrial applications having a maximum fault capacity of 80 kA.
* Cartridge fuses to BS 88-3: 2010. These are used for a.c. circuits on industrial and domestic installations, having a fault capacity of about 30 kA.
* Cartridge fuses to BS 1362. These are used in 13 A plug tops and have a maximum fault capacity of about kA.

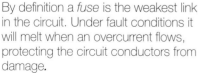

Definition

By definition a *fuse* is the weakest link in the circuit. Under fault conditions it will melt when an overcurrent flows, protecting the circuit conductors from damage.

- Semi-enclosed fuses to BS 3036. These were previously called rewirable fuses and are used mainly on domestic installations having a maximum fault capacity of about 4 kA.
- CBs to BS EN 60898. These are circuit breakers (CBs) which may be used as an alternative to fuses for some installations. The British Standard includes ratings up to 100 A and maximum fault capacities of 9 kA. They are graded according to their instantaneous tripping currents – that is, the current at which they will trip within 100 ms. This is less than the time taken to blink an eye.

Assessment criteria 9.2

Recognize the advantages and disadvantages of different overcurrent protection devices

Semi-enclosed fuses (BS 3036)

The semi-enclosed fuse consists of a fuse wire, called the fuse element, secured between two screw terminals in a fuse carrier. The fuse element is connected in series with the load and the thickness of the element is sufficient to carry the normal rated circuit current. When a fault occurs an overcurrent flows and the fuse element becomes hot and melts or 'blows'.

This type of fuse is illustrated in Figure 3.78. The fuse element should consist of a single strand of plain or tinned copper wire with a diameter appropriate to the current rating of the fuse. This type of fuse was very popular in domestic installations, but is less popular these days because of its disadvantages.

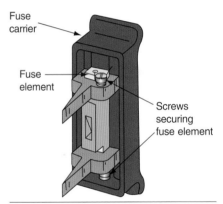

Figure 3.78 A semi-enclosed fuse.

Advantages of semi-enclosed fuses

- They are very cheap compared with other protective devices both to install and to replace.
- There are no mechanical moving parts.
- It is easy to identify a 'blown' fuse.

Disadvantages of semi-enclosed fuses

- The fuse element may be replaced with wire of the wrong size, either deliberately or by accident.
- The fuse element weakens with age due to oxidization, which may result in a failure under normal operating conditions.
- The circuit cannot be restored quickly since the fuse element requires screw fixing.
- They have low breaking capacity since, in the event of a severe fault, the fault current may vaporize the fuse element and continue to flow in the form of an arc across the fuse terminals.
- They are not guaranteed to operate until up to twice the rated current is flowing.
- There is a danger from scattering hot metal if the fuse carrier is inserted into the base when the circuit is faulty.

Cartridge fuses

(BS 88-3: 2012 (previously BS 1361))

The cartridge fuse breaks a faulty circuit in the same way as a semi-enclosed fuse, but its construction eliminates some of the disadvantages experienced with an open-fuse element. The fuse element is encased in a glass or ceramic tube and secured to end-caps which are firmly attached to the body of the fuse so that they do not blow off when the fuse operates. Cartridge fuse construction is illustrated in Figure 3.79.

With larger size cartridge fuses, lugs or tags are sometimes brazed on the end-caps to fix the fuse cartridge mechanically to the carrier. They may also be filled with quartz sand to absorb and extinguish the energy of the arc when the cartridge is brought into operation.

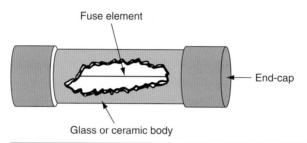

Figure 3.79 Cartridge fuse.

Advantages of cartridge fuses

- They have no mechanical moving parts.
- The declared rating is accurate.
- The element does not weaken with age.
- They have small physical size and no external arcing which permits their use in plug tops and small fuse carriers.
- Their operation is more rapid than semi-enclosed fuses. Operating time is inversely proportional to the fault current, so the bigger the fault current, the quicker the fuse operates.
- They are easy to replace.
- Larger valves have bolt-hole fixings.

Disadvantages of cartridge fuses

- They are more expensive to replace than fuse elements that can be rewired.
- They can be replaced with an incorrect cartridge.
- The cartridge may be shorted out by wire or silver foil in extreme cases of bad practice.
- It is not possible to see if the fuse element is broken.

Circuit breakers (BS EN 60898)

The disadvantage of all fuses is that when they have operated they must be replaced. A circuit breaker (CB) overcomes this problem since it is an automatic switch which opens in the event of an excessive current flowing in the circuit and can be closed when the circuit returns to normal.

A CB of the type shown in Figure 3.80 incorporates a thermal and magnetic tripping device. The load current flows through the thermal and the

Figure 3.80 CBs – B Breaker, fits Wylex standard consumer unit (courtesy of Wylex).

electromagnetic devices in normal operation but under overcurrent conditions they activate and trip the CB.

The circuit can be restored when the fault is removed by pressing the ON toggle. This latches the various mechanisms within the CB and 'makes' the switch contact. The toggle switch can also be used to disconnect the circuit for maintenance or isolation, or to test the CB for satisfactory operation.

Advantages of CBs

- They have factory-set operating characteristics.
- Tripping characteristics and therefore circuit protection is set by the installer.
- The circuit protection is difficult to interfere with.
- The circuit is provided with discrimination.
- A faulty circuit may be quickly identified.
- A faulty circuit may be easily and quickly restored.
- The supply may be safely restored by an unskilled operator.

Disadvantages of CBs

- They are relatively expensive, but look at the advantages to see why they are so popular these days.
- They contain mechanical moving parts and therefore require regular testing to ensure satisfactory operation under fault conditions.

Additional Protection: RCDs

While not an overcurrent protective device, we need to look at the RCD, a device that provides earth leakage protection.

When it is required to provide the very best protection from electric shock and fire risk, earth fault protection devices are incorporated into the installation. The object of the regulations concerning these devices (411.3.2 to 411.3.3) is to remove an earth fault current very quickly, less than 0.4 s for all final circuits not exceeding 32 mA, and limit the voltage which might appear on any exposed metal parts under fault conditions to not more than 50 V. They will continue to provide adequate protection throughout the life of the installation even if the earthing conditions deteriorate. This is in direct contrast to the protection provided by overcurrent devices, which require a low-resistance earth loop impedance path.

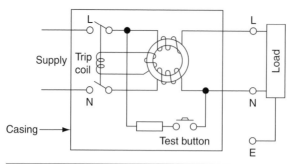

Figure 3.81 Construction of an RCD.

The regulations recognize RCDs as 'additional protection' in the event of failure of the provision for basic protection, fault protection or carelessness by the users of the installation (IET Regulation 415.1.1).

The basic circuit for a single-phase RCD is shown in Figure 3.81. The load current is fed through two equal and opposing coils wound onto a common transformer core. The line and neutral currents in a healthy circuit produce equal and opposing fluxes in the transformer core, which induces no voltage in the tripping coil. However, if more current flows in the line conductor than in the neutral conductor as a result of a fault between live and earth, an out-of-balance flux will result in an e.m.f. being induced in the trip coil which will open the double-pole switch and isolate the load. Modern RCDs have tripping sensitivities between 10 and 30 mA, and therefore a faulty circuit can be isolated before the lower lethal limit to human beings (about 50 mA) is reached.

Consumer units are now supplied which incorporate one or more RCDs, as shown in the *On-Site Guide* Figure 3.6.3. Where RCDs are installed, a note shall be fixed near to the RCD stating 'this device must be tested 6 monthly' (IET Regulation 514.12.2). Previously the test period was 3 months.

RCBO

A residual current-operated circuit breaker with integral overcurrent protection (RCBO) provides protection against overload and/or short circuit. RCBOs give the combined protection of a CB and an RCD in one device.

In a split-board consumer unit, about half of the total number of final circuits are protected by the RCD. A fault on any one final circuit will trip out all of the RCD-protected circuits, which may cause inconvenience.

The RCBO gives the combined protection of a CB plus RCD for each final circuit so protected and, in the event of a fault occurring, only the faulty circuit is interrupted. This arrangement is shown in the *On-Site Guide* in Figure 2.1(i) to 2.1(iii) and in this book at Figures 3.44 to 3.46.

Finally, it should perhaps be said that a fool-proof method of giving protection to people or animals who simultaneously touch both line and neutral has yet to be devised. The ultimate safety of an installation depends upon the skill and experience of the electrical contractor and the good sense of the user.

Assessment criteria 9.3

Recognize the applications of overcurrent protection devices

The selection of protective devices will depend on various factor including:

1. Prospective fault current.
2. Circuit load characteristics.
3. The rated short circuit capacity of a protective device must not be less than the prospective fault current at the point it is installed. The *On-Site Guide* provides Table 7.2.7 (i) where it lists the rated short circuit capacities of overcurrent protective devices.
4. a) BS 3036 semi-enclosed (rewireable) fuses are still permitted, however cartridge fuses are preferred.
   b) Cartridge fuses to BS 1361 (now replaced by BS 88-3) can be found and used in domestic or similar premises.
   c) Cartridge fuses to BS 88 are classified as:
      - gG general application
      - gM motor circuits
      - aM motor circuits.

Characteristics of CBs

CB Type B to BS EN 60898 will trip instantly at between three and five times its rated current and is also suitable for domestic and commercial installations.

CB Type C to BS EN 60898 will trip instantly at between five and ten times its rated current. It is more suitable for highly inductive commercial and industrial loads such as fluorescent lights.

CB Type D to BS EN 60898 will trip instantly at between 10 and 25 times its rated current. It is suitable for motors, welding and X-ray machines where large inrush currents may occur.

Assessment criteria 10.1

State what is meant by diversity factors

The nature of demand

The designer must identify the number of circuits required and the expected load on these circuits.

Table 3.7 Assumed current demand for electrical equipment and circuits

Current using equipment	Assumed current demand
Lighting points	Minimum of 100 watts / lampholder
Discharge lighting such as fluorescent tubes	Lamp watts × 1.8 to take account of the control gear
Electric clock, shaver unit, bell transformer	May be neglected in this assessment
2A socket outlet	0.5 A
Standard household –13 A ring circuit	30 A or 32A – max floor area 100 m^2 wired in 2.5 mm cable
Standard household – 13 A radial circuit	30A or 32A – max floor area 75 m^2 wired in 4.0 mm cable
Standard household – 13 A radial circuit	20A – max floor area 50 m^2 wired in 2.5 mm cable
Cooking appliances	The first 10 A of the rated current plus 30% of the remainder of the rated current plus 5A if control unit incorporates a socket outlet
All other stationary equipment such as shower or immersion heater	British Standard rated current

The total current demand of any final circuit is estimated by adding together the current demands of all points of utilization such as socket and lighting points and equipment outlets.

Figure 3.83 Think of all the electrical devices in your house. Each one will use different amounts of electricity depending on the work being done.

Final circuit current demand

In this chapter we will only look at straightforward household installations but the same principles apply to shops, hotels and guest houses. All of these premises are dealt with in Appendix A of the *On-Site Guide*. Let us begin by looking at the current demand to be assumed for points of utilization given in Table A1 of the *On-Site Guide* and shown here in Table 3.7.

Example 1

Calculate the current demand of a 7.36 KW electric shower connected to the 230 V mains supply.

The power $\quad P = 7.36$ KW or $\quad\quad 7360$ watts

The voltage $\quad\quad V = 230$ volts

Now power = V I $\quad$ so transposing for current $I = \dfrac{Power}{Voltage}$ amps

So the current demand $I = \dfrac{7360}{230} = 32$ amps

Example 2

Calculate the current demand of an electric cooker comprising

A hob with 4 2.5 KW rings = 10 KW

A main oven rated at 2 KW and = 2 KW

A grill/top oven rated at 2 KW = 2 KW Total 14 KW

The cooker control unit incorporates a 13 A socket outlet

As in Example 1, the current $I = \dfrac{Power}{Voltage}$ amps

Therefore $I = \dfrac{14000}{230} = 60.87$ amps

From Table 3.7 above we take the first 10 A of the rated current, plus 30% of the remainder plus 5 A if the control unit incorporates a socket outlet.

Therefore $\quad\quad I = 10A + (\frac{30}{100} \times (60.87 - 10) + 5A$

$\quad\quad\quad\quad\quad I = 10A + (\frac{30}{100} \times 50.87) + 5A$

$\quad\quad\quad\quad\quad I = 10 + 15.26 + 5$

$\quad\quad\quad\quad\quad I = 30.26$ amps

Therefore a 32 amp circuit breaker would adequately supply this circuit.

Example 3

A lighting circuit consists of 10 points. Calculate the current demand.

From Table 3.7 we must assume a minimum of 100 watts per point.

As in the previous examples $I = \dfrac{Power}{Voltage}$ amps

Therefore $\quad\quad I = \dfrac{10 \times 100}{230} = 4.35$ amps

(continued)

Example 3 continued

Therefore a 5 amp fuse or a 6 amp circuit breaker would supply this circuit if ordinary GLS lamps were being used. However, if extra low voltage or discharge lighting is connected to this circuit and protected by a type 'B1 CB we must take account of the inrush current which occurs at switch on and sometimes causes unwanted or nuisance tripping of the CB. To avoid this, the circuit will be adequately protected by a 10 amp CB.

Example 4

The kitchen work surface of a domestic kitchen is to be illuminated by ten 30 watt fluorescent tubes fixed to the underside of the cupboards above the work surface. Calculate the current demand of this circuit.

From Table 3.7 the demand is taken as the lamp watts multiplied by 1.8

As in the previous examples $I = \dfrac{Power}{Voltage}$ amps

Therefore $I = \dfrac{10 \times 30 \times 1.8}{230}$ 2.34 amps

Therefore a 5 amp fuse or a 6 amp CB would supply this circuit.

Example 5

Calculate the current demand of a 3 kW immersion heater installed in a 30 litre water storage vessel. We know from Table 3.7 that we should take the British Standard-rated current for a water heater such as this, or we can calculate the current rating as before.

$I = \dfrac{Power}{Voltage}$ amps   Therefore $I = \dfrac{3000}{230} = 13.04$ amps

A 15 amp fuse or 16 amp CB will supply this circuit. We also know from Appendix H5 of the *On-Site Guide* that water heaters fitted to vessels in excess of 15 litres must be supplied on their own circuit.

Diversity between final circuits

The current demand of a circuit is the current taken by that circuit over a period of time, say 30 minutes. Some loads make a constant demand all the time they are switched on. A 100 watt light bulb will make a constant current demand of (100 watts ÷ 230 volts) 0.43 amperes whenever it is switched on. However, an automatic washing machine is made up of a variable speed motor, a pump and a water heater, all controlled by a programmer. The washer load will not be constant for the whole time it is switched on, but will vary depending upon the wash cycle.

Diversity makes an allowance on the basis that not all of the load or connected items will be in use at the same time.

The design would be wasteful if it did not take advantage of the diversity between different loads. Let us now look at diversity as it applies to a straightforward household installation, but the same principles will apply to shops, offices, business premises and hotels. Appendix A of the *On-Site Guide* gives the diversity for all these different types of premises.

The allowances for diversity shown in Table A2 of the *On-Site Guide* are for very specific situations and can only provide guidance. The diversity allowances for lighting circuits should be applied to 'items of equipment' connected to the consumer unit. However, standard power circuit arrangements for households, as described in Appendix H of the *On-Site Guide*, can be applied to the rated current of the overcurrent protective device for the circuit. It is important to ensure that the consumer's unit is of sufficient rating to take the total load connected without the application of any diversity.

Assessment criteria 10.2

Determine the assumed maximum demand of a circuit after diversity

Example of a calculation applying diversity.

Example 6

Calculate the current demand, including diversity, for a six way consumer unit comprising the following circuits in a domestic dwelling.

Circuit 1. A lighting circuit comprising 10 points as described in example 3, previously having a maximum demand of 4.35 A and protected by a 6 amp CB.

Circuit 2. A lighting circuit comprising ten 30 watt fluorescents as described in example 4, previously having a maximum demand of 2.3 A and protected by a 6 amp CB.

Circuit 3. A thermostatically controlled 3 kW immersion heater as described in example 5, previously having a maximum demand of 13.04 and protected by 16 amp CB.

Circuit 4. A ring circuit of 13 A socket outlets installed in accordance with Section H of the OSG and protected by a 32 amp CB.

Circuit 5. A radial circuit of 13 A socket outlets installed in accordance with section H of the OSG and protected by a 20 amp CB.

Circuit 6. A radial circuit of 13A socket outlets installed in accordance with Section H of the *On-Site Guide* and protected by a 20A circuit breaker.

Table A2 line 9 of the *On-Site Guide* and Table 3.2 of the IET Electrical Installation Design Guide tell us that for standard household circuits as described in Appendix H of the OSG (that is ring and radial circuits), **the allowable diversity is to be calculated as follows, 100% of the current demand of the largest circuit plus 40% of the current demand of every other circuit. The diversity for lighting is 66%.**

(continued)

Example continued

from the question

Circuit 1 has a maximum demand of	4.35 amp
Circuit 2 has a maximum demand of	2.34 amp
Circuit 3 has a maximum demand of	13.04 amp
Circuit 4 has a maximum demand of	32.00 amp
Circuit 5 has a maximum demand of	32.00 amp
Circuit 6 has a maximum demand of	20.00 amp

We can see from the above that the largest circuit in the consumer's unit is 32 amps, so let us nominate circuit 4 as the largest circuit and apply 40% diversity to all other standard circuits.

Circuit 1 = $4.35 \text{ A} \times \dfrac{66}{100} =$ 2.87 A

Circuit 2 = $2.34 \text{ A} \times \dfrac{66}{100} =$ 1.54 A

Circuit 3 = 13.04 A (no diversity)   13.04 A

Circuit 4 = 32.00 A (no diversity)   32 A

Circuit 5 = $32.00 \text{ A} \times \dfrac{40}{100} =$ 12.80 A

Circuit 6 = $20.00 \text{ A} \times \dfrac{40}{100} =$ 8.00 A

Adding the values   25.21 A   +   45.04 A

Total installed demand with diversity = 70.25 amps

Assessment criteria 10.3

Establish design current of a circuit

Assessment criteria 10.4

Identify the rating factors which may affect the conductor size

Conductor size calculations

* The size of a cable to be used for an installation depends upon: the current rating of the cable under defined installation conditions, and
* the maximum permitted drop in voltage as defined by IET Regulation 525.

The factors which influence the current rating are:

1 Design current: cable must carry the full load current.
2 Type of cable: PVC, MICC, copper conductors or aluminium conductors.
3 Installed conditions: clipped to a surface or installed with other cables in a trunking.

4 Surrounding temperature: cable resistance increases as temperature increases and insulation may melt if the temperature is too high.

5 Type of protection: for how long will the cable have to carry a fault current? IET Regulation 525 states that the drop in voltage from the supply terminals to the fixed current-using equipment must not exceed 3% for lighting circuits and 5% for other uses of the mains voltage. That is a maximum of 6.9V for lighting and 11.5V for other uses on a 230V installation. The volt drop for a particular cable may be found from:

$$VD = factor \times design\ current \times length\ of\ run$$

The factor is given in the tables of Appendix 4 of the IET Regulations and Appendix F of the *On-Site Guide*.

The cable rating, denoted It, may be determined as follows:

$$It = \frac{Current\ rating\ of\ protective\ device}{Any\ applicable\ correction\ factors}$$

The cable rating must be chosen to comply with IET Regulation 433.1. The correction factors which may need applying are given below as:

- Ca: the ambient or surrounding temperature correction factor, which is given in Tables 4B1 and 4B2 of Appendix 4 of the IET Regulations.
- Cg: the grouping correction factor given in Tables 4C1 to 4C5 of the IET Regulations and Table 6C of the *On-Site Guide*.
- Cf: the 0.725 correction factor to be applied when semi-enclosed fuses protect the circuit as described in item 5.1.1 of the preface to Appendix 4 of the IET Regulations.
- Ci: the correction factor to be used when cables are enclosed in thermal insulation. IET Regulation Table 523.9 and Table 4A2 gives us three possible correction values:

  1 Where one side of the cable is in contact with thermal insulation, we must read the current rating from the column in the table which relates to reference method A (see Table 3.9).
  2 Where the cable is totally surrounded over a length greater than 0.5m we must apply a factor of 0.5.
  3 Where the cable is totally surrounded over a short length, the appropriate factor given in Table 52.2 of the IET Regulations or Table F2 of the *On-Site Guide* should be applied.

Note: A cable should preferably not be installed in thermal insulations.

Assessment criteria 10.5

Determine the tabulated current-carrying capacity of a cable

Assessment criteria 10.6

Establish voltage drop

Key fact

Volt drop

Maximum permissible volt drop on 230V supplies:

- 3% for lighting = 6.9V
- 5% for other uses = 11.5V

IET Regulation 525 and Table 4A6 of Appendix 4.

Assessment criteria 10.7

State the maximum voltage drop in a consumer's installation

Having calculated the cable rating, the smallest cable should be chosen from the appropriate table which will carry that current. This cable must also meet the voltage drop (IET Regulation 525) and this should be calculated as described earlier. When the calculated value is less than 3% for lighting and 5% for other uses of the mains voltage, the cable may be considered suitable. If the calculated value is greater than this value, the next larger cable size must be tested until a cable is found which meets both the current rating and voltage drop criteria.

Table 3.8 Ambient air temperature correction factors

Type of insulation	Conductor operating temperature	Ambient temperature (°C)			
		25	30	35	40
Thermoplastic (general purpose PVC)	70°C	1.03	1.0	0.94	0.87

Cable size for standard domestic circuits

Appendix 4 of the IET Regulations (BS 7671) and Appendix F of the IET *On-Site Guide* contain tables for determining the current-carrying capacities of conductors which we looked at in the last section. However, for standard domestic circuits, Table 3.11 gives a guide to cable size.

In this table, I am assuming a standard 230V domestic installation, having a sheathed earth or PME supply terminated in a 100A HBC fuse at the mains position. Final circuits are fed from a consumer unit, having Type B CB protection and wired in PVC insulated and sheathed cables with copper conductors having a grey thermoplastic PVC outer sheath or a white thermosetting cable with LSF (low smoke and fume properties). I am also assuming that the surrounding temperature throughout the length of the circuit does not exceed 30°C and the cables are run singly and clipped to a surface.

Table 3.9 Current-carrying capacity of cables

Conductor cross-sectional area	Reference Method A (enclosed in conduit in an insulated wall, etc.)		Reference Method B (enclosed in conduit on a wall or ceiling, or in trunking)		Reference Method C (clipped direct)		Reference Method E (on a perforated cable tray) or in free air	
	Two cables single-phase a.c. or d.c.	Three or four cables three-phase a.c.	One two-core cable, single-phase a.c. or d.c.	One three-core cable or one four-core cable, three-phase a.c.	One two-core cable, single-phase a.c. or d.c.	One three-core cable or one four-core cable, three-phase a.c.	One two-core cable, single-phase a.c. or d.c.	One three-core cable or one four-core cable, three-phase a.c.
1 mm²	2 A	3 A	4 A	5 A	6 A	7 A	8 A	9 A
1	11	10	13	11.5	15	13.5	17	14.5
1.5	14	13	16.5	15	19.5	17.5	22	18.5
2.5	18.5	17.5	23	20	27	24	30	25
4	25	23	30	27	37	32	40	34
6	32	29	38	34	46	41	51	43
10	43	39	52	46	63	57	70	60
16	57	52	69	62	85	76	94	80
25	75	68	90	80	112	96	119	101
35	92	83	111	99	138	119	148	126

Example 1

A house extension has a total load of 6 kW installed some 18 m away from the mains consumer unit for lighting. A PVC insulated and sheathed twin and earth cable will provide a sub-main to this load and be clipped to the side of the ceiling joists over much of its length in a roof space which is anticipated to reach 35°C in the summer and where insulation is installed up to the top of the joists. Calculate the minimum cable size if the circuit is to be protected by a type B CB to BS EN 60898. Assume a TN-S supply, that is, a supply having a separate neutral and protective conductor throughout.

Let us solve this question.

$$\text{Design current, } I_b = \frac{\text{Power}}{\text{Volts}} = \frac{6000\,\text{W}}{230\,\text{V}} = 26.09\,\text{A}$$

Nominal current setting of the protection for this load $I_n = 32$ A.

The cable rating I_t is given by:

$$I_t = \frac{\text{Current rating of protective device } (I_n)}{\text{The product of the correction factors}}$$

The correction factors to be included in this calculation are:

Ca ambient temperature; as shown in Table 3.8 the correction factor for 35°C is 0.94.

Cg grouping factors need not be applied.

Cf, since protection is by CB no factor need be applied.

Ci thermal insulation demands that we assume installed Method A (see Table 3.9).

The design current is 26.09 A and we will therefore choose a 32 A CB for the nominal current setting of the protective device, I_n.

$$\text{Cable rating, } I_t = \frac{32}{0.94} = 34.04\,\text{A}$$

From column 2 in Table 3.9, a 10 mm cable, having a rating of 43 A, is required to carry this current.

Now test for volt drop: from Table 3.10 the volt drop per ampere per metre for a 10 mm cable is 4.4 mV from column 3. So the volt drop for this cable length and load is equal to:

$$4.4 \times 10^{-3}\,\text{V} \times 26.09\,\text{A} \times 18\text{m} = 2.06\,\text{V}$$

Since this is less than the maximum permissible value for a lighting circuit of 6.9 V, a 10 mm cable satisfies the current and drop in voltage requirements when the circuit is protected by an CB. This cable is run in a loft that gets hot in summer and has thermal insulation touching one side of the cable. We must, therefore, use installed reference Method A of Table 3.9. If we were able to route the cable under the floor, clipped direct or in conduit or trunking

(continued)

Example 1 continued

on a wall, we might be able to use a 6mm cable for this load. You can see how the current-carrying capacity of a cable varies with the installed method by looking at Table 3.9. Compare the values in column 2 with those in column 6. When the cable is clipped directly onto a wall or surface the current rating is higher because the cable is cooler. If the alternative route was longer, you would need to test for volt drop before choosing the cable. These are some of the decisions which the electrical contractor must make when designing an installation which meets the requirements of the customer and the IET Regulations.

If you are unsure of the standard fuse and CB rating of protective devices, you can refer to Figure 3A4 of Appendix 3 of the IET Regulations.

Table 3.10 Voltage drop in cables factor

Voltage drop (per ampere per metre)			Conductor operating temperature: 70°C
Conductor cross-sectional area (mm²) 1	Two-core cable, d.c. (mV/A/m) 2	Two-core cable, single-phase a.c. (mV/A/m) 3	Three- or four-core cable, three-phase (mV/A/m) 4
1	44	44	38
1.5	29	29	25
2.5	18	18	15
4	11	11	9.5
6	7.3	7.3	6.4
10	4.4	4.4	3.8
16	2.8	2.8	2.4
25	1.75	1.75	1.50
35	1.25	1.25	1.10

Table 3.11 Cable size for standard domestic circuits

Type of final circuit	Cable size (twin and earth)	MCB rating, Type B (A)	Maximum floor area covered by circuit (m²)	Maximum length of cable run (m)
Fixed lighting	1.0	6	–	40
Fixed lighting	1.5	6	–	60
Immersion heater	2.5	16	–	30
Storage radiator	2.5	16	–	30
Cooker (oven only)	2.5	16	–	30
13 A socket outlets (radial circuit)	2.5	20	50	30
13 A socket outlets (ring circuit)	2.5	32	100	90
13 A socket outlets (radial circuit)	4.0	32	75	35
Cooker (oven and hob)	6.0	32	–	40
Shower (up to 7.5 kw)	6.0	32	–	40
Shower (up to 9.6 kw)	10	40	–	40

Assessment criteria 10.8

Interpret the requirements for the cable capacities of containment

Conduit capacities

Single PVC insulated conductors are usually drawn into the installed conduit to complete the installation. Having decided upon the type, size and number of cables required for a final circuit, it is then necessary to select the appropriate size of conduit to accommodate those cables.

The tables in Appendix E of the *On-Site Guide* describe a 'factor system' for determining the size of conduit required to enclose a number of conductors.

- Identify the cable factor for the particular size of conductor; see Table 3.12.
- Multiply the cable factor by the number of conductors, to give the sum of the cable factors.
- Identify the appropriate part of the conduit factor table given by the length of run and number of bends (see Table 3.13).
- The correct size of conduit to accommodate the cables is that conduit which has a factor equal to or greater than the sum of the cable factors.

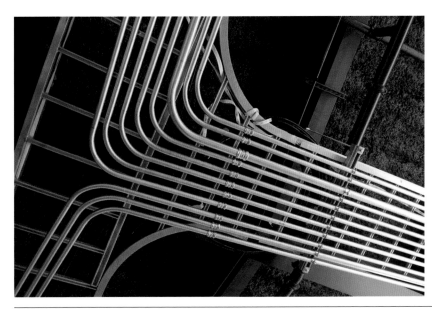

Figure 3.84 An array of newly installed electrical conduits.

Table 3.12 Conduit cable factors

Cable factors for conduit in long straight runs over 3 m, or runs of any length incorporating bends	
Conductor CSA (mm²)	**Cable factor**
1	16
1.5	22
2.5	30
4	43
6	58
10	105
16	145

Trunking capacities

The ratio of the space occupied by all the cables in a conduit or trunking to the whole space enclosed by the conduit or trunking is known as the space factor.

Where sizes and types of cable and trunking are not covered by the tables in the *On-Site Guide*, a space factor of 45% must not be exceeded. This means that the cables must not fill more than 45% of the space enclosed by the trunking. The tables take this factor into account.

To calculate the size of trunking required to enclose a number of cables:

- Identify the cable factor for the particular size of conductor (see Table 3.13).
- Multiply the cable factor by the number of conductors to give the sum of the cable factors.
- Consider the factors for trunking shown in Table 3.14. The correct size of trunking to accommodate the cables is that trunking which has a factor equal to, or greater than, the sum of the cable factors.

Definition

The *ratio* of the space occupied by all the cables in a conduit or trunking to the whole space enclosed by the conduit or trunking is known as the *space factor*.

Table 3.13 Conduit factors

Length of run (m)	Conduit diameter (mm)											
	16	20	25	32	16	20	25	32	16	20	25	32
	Straight				One bend				Two bends			
3.5	179	290	521	911	162	263	475	837	136	222	404	720
4	177	286	514	900	158	256	463	818	130	213	388	692
4.5	174	282	507	889	154	250	452	800	125	204	373	667
5	171	278	500	878	150	244	442	783	120	196	358	643
6	167	270	487	857	143	233	422	750	111	182	333	600
7	162	263	475	837	136	222	404	720	103	169	311	563
8	158	256	463	818	130	213	388	692	97	159	292	529
9	154	250	452	800	125	204	373	667	91	149	275	500
10	150	244	442	783	120	196	358	643	86	141	260	474

Example 1

Six 2.5 mm^2 PVC insulated cables are to be run in a conduit containing two bends between boxes 10 m apart. Determine the minimum size of conduit to contain these cables.

From Table 3.12:

$$\text{The factor for one 2.5mm}^2 \text{ cable} = 30$$
$$\text{The sum of the cable factors} = 6 \times 30$$
$$= 180$$

From Table 3.13, a 25 mm conduit, 10 m long and containing two bends, has a factor of 260. A 20 mm conduit containing two bends only has a factor of 141 which is less than 180, the sum of the cable factors, and, therefore, 25 mm conduit is the minimum size to contain these cables.

Example 2

Ten 1.0 mm² PVC insulated cables are to be drawn into a plastic conduit which is 6 m long between boxes and contains one bend. A 4.0 mm PVC insulated CPC is also included. Determine the minimum size of conduit to contain these conductors.

From Table 3.12:

The factor for one 1.0 mm cable = 16
The factor for one 4.0 mm cable = 43
The sum of the cable factors = (10 × 16) + (1 × 43)
= 203

From Table 3.13, a 20 mm conduit, 6 m long and containing one bend, has a factor of 233. A 16 mm conduit containing one bend only has a factor of 143 which is less than 203, the sum of the cable factors, and, therefore, 20 mm conduit is the minimum size to contain these cables.

Table 3.14 Trunking cable factors

Type of conductor	Conductor CSA (mm²)	PVC cable factor	Thermosetting cable factor
Solid	1.5	8	8.6
	2.5	11.9	11.9
Stranded	1.5	8.6	9.6
	2.5	12.6	13.9
	4	16.6	18.1
	6	21.2	22.9
	10	35.3	36.3

Table 3.15 Trunking factors

Dimensions of trunking (mm × mm)	Factor
50 × 38	767
50 × 50	1037
75 × 25	738
75 × 38	1146
75 × 50	1555
75 × 75	2371
100 × 25	993
100 × 38	1542
100 × 50	2091
100 × 75	3189
100 × 100	4252
150 × 38	2999
150 × 50	3091
150 × 75	4743
150 × 100	6394
150 × 150	9697

Example 3

Calculate the minimum size of trunking required to accommodate the following single-core PVC cables:

   20 × 1.5 mm solid conductors
   20 × 2.5 mm solid conductors
   21 × 4.0 mm stranded conductors
   16 × 6.0 mm stranded conductors

From Table 3.14, the cable factors are:

   for 1.5 mm solid cable – 8.0
   for 2.5 mm solid cable – 11.9
   for 4.0 mm stranded cable – 16.6
   for 6.0 mm stranded cable – 21.2

The sum of the cable terms is:

$(20 \times 8.0) \times (20 \times 11.9) \times (21 \times 16.6) \times (16 \times 21.2) = 1085.8$. From Table 3.15, 75 × 38 mm trunking has a factor of 1146 and, therefore, the minimum size of trunking to accommodate these cables is 75 × 38 mm, although a larger size, say, 75 × 50 mm, would be equally acceptable if this was more readily available as a standard stock item.

When you have completed the questions, check out the answers at the back of the book.

Note: more than one multiple-choice answer may be correct.

Learning outcome 1

1   Failure to refer to the Wiring Regulations could result in which of the following:
    a.   non-compliance with regard to health and safety
    b.   compliance with regard to health and safety
    c.   non-compliance with regard to the building regulations
    d.   compliance with regard to building regulations

2   Which of the following is regarded as non-statutory?
    a.   HASWA 1974
    b.   EWR 1989
    c.   BS 7671
    d.   Working at Height Regulations 2005

3   Which of the following came about through an Act of Parliament?
    a.   Health and Safety Act 1974
    b.   *On-Site Guide*
    c.   Guidance Note 3
    d.   BS 7671

Learning outcome 2

4   A double socket outlet is shown with a scale dimension of 15 mm. Using a scale of 1:100, calculate the actual dimension of the socket.
    a.   1.5 m
    b.   15 m
    c.   1500 cm
    d.   150 mm

5   What symbol is shown here?
    a.   One-way lighting switch
    b.   Two-way lighting switch
    c.   Three-way lighting switch
    d.   Single-way lighting switch

6   Which of the following indicates how a circuit operates?
    a.   Site diagram
    b.   Layout diagram
    c.   Circuit diagram
    d.   Block diagram

7   Which of the following would be beneficial in order to determine how a piece of equipment works?
    a.   Code of practice
    b.   BS 7671
    c.   *On-Site Guide*
    d.   User guide

8   Which of the following would you refer to if you were installing unfamiliar equipment?
    a.   Code of practice
    b.   EAWR
    c.   *On-Site Guide*
    d.   Manufacturer's instructions

9   Familiarizing yourself with which document means that you are following a code of practice?
    a.   Electricity at Work Regulations 1989
    b.   Health and Safety Act 1974
    c.   BS 7671
    d.   Building Regulations

Learning outcome 3

10 An appropriate wiring method for a domestic installation would be:
    a.   a metal conduit installation
    b.   a trunking and tray installation
    c.   PVC singles
    d.   PVC flat profile cables

11 An appropriate wiring method for an underground feed to a remote building would be:
    a.   a metal conduit installation
    b.   a trunking and tray installation
    c.   PVC cables
    d.   PVC/SWA cables

12 Due to its non-ageing property which cable type is often used in listed buildings?
    a.   Mineral insulated
    b.   SWA
    c.   PVC/PVC
    d.   PVC singles

13 A conductor used in the transfer of data is:
a. FP cable
b. fibre-optic
c. mineral insulated
d. PVC/PVC flat profile cable

14 What voltage is associated with this image?
a. 230 V
b. 400 V
c. 50 V
d. 110 V

15 In high temperature environments, a cheaper alternative to mineral insulated cable would be?
a. FP
b. Fibre-optic
c. SWA
d. PVC flex

16 Apart from the colour what identifies this industrial commando plug as being three-phase?
a. The shape of the pins
b. The colour of the pins
c. The size of the pins
d. The number of pins

17 Which of the following is the most appropriate containment system for a low current application such as commercial lighting?
a. Busbar systems
b. Flexible conduit
c. Cable tray
d. Power track

18 Which of the following would be most appropriate regarding the final connection supplying a motor circuit?
a. PVC conduit
b. Plastic conduit
c. Metal conduit
d. Flexible conduit

19 Which type of cable is most appropriate regarding the final connection to an immersion heater?
a. Flexible heat resistant cable
b. Flexible mineral insulated
c. Flexible Arctic cable
d. Flexible SWA

20 Which of the following is the most appropriate containment system for heavy current applications?
a. Busbar systems
b. Flexible conduit
c. Cable tray
d. Power track

21 A PVC conduit installation would be suitable for the following type of installation:
   a. commercial
   b. domestic
   c. horticultural
   d. industrial

22 A steel conduit installation would be suitable for the following type of installation:
   a. commercial
   b. domestic
   c. horticultural
   d. industrial

23 A steel trunking installation would be suitable for the following type of installation:
   a. commercial
   b. domestic
   c. horticultural
   d. industrial

Learning outcome 4

24 For a modern domestic property, the main service fuse should be rated at:
   a. 80A
   b. 100A
   c. 60A
   d. 120A

25 For a modern domestic property, the meter tails should be?
   a. 25mm²
   b. 16mm²
   c. 10mm²
   d. 20mm²

26 For a modern domestic property, which of the following is correct?
   a. Main earthing conductor = 16mm², main protective bonding conductor = 16mm²
   b. Main earthing conductor = 16mm², main protective bonding conductor = 10mm²
   c. Main earthing conductor = 10mm², main protective bonding conductor = 16mm²
   d. Main earthing conductor = 10mm², main protective bonding conductor = 10mm²

Learning outcome 5

27 The term 'strapper cable' is associated with:
   a. power circuits
   b. ring circuits
   c. one-way lighting
   d. two-way lighting

28 What is the normal size and rating of a lighting cable and protective device?
   a. 6 mm^2 and 6 A
   b. 2.5 mm^2 and 6 A
   c. 4 mm^2 and 6 A
   d. 1.5 mm^2 and 6 A

29 If cables are buried in a wall to more than 50 mm in depth, what additional action should be taken?
   a. None
   b. Fit an RCD
   c. Fit an HRC fuse
   d. Fit a thermal protective relay

Learning outcome 6

30 The maximum floor area that a ring circuit can serve is:
   a. 50 m^2
   b. 75 m^2
   c. 100 m^2
   d. 25 m^2

31 Each power socket within a ring circuit can have:
   a. one unfused spur installed with a 4 mm cable
   b. one fused spur installed with a 4 mm cable
   c. one unfused spur installed with a 2.5 mm cable
   d. one fused spur installed with a 6 mm cable

32 In accordance with the Wiring Regulations all power sockets rated at 20 A or less must be given additional protection. This means that:
   a. a 30 mA RCD must be in place
   b. a 100 mA RCD must be in place
   c. a fuse must be fitted
   d. a circuit breaker must be fitted

Learning outcome 7

33 A kitchen is normally fed with its own ring circuit because:
   a. it is normally bigger in size than other rooms
   b. it is normally smaller in size than other rooms
   c. it normally houses a lot of current drawing equipment
   d. it normally houses less current drawing equipment

34 Which of the following applies to polarity?
   a. the inner contact of an Edison lamp holder is connected to the neutral conductor
   b. protective devices can only be connected in the line conductor
   c. circuit switching is carried out by making and braking the line conductor
   d. circuit switching is carried out by making and braking the c.p.c

35 Within a consumer unit, the highest rated circuit breaker should be positioned:
    a. next to the main switch
    b. at the opposite end to the main switch
    c. next to the main RCD
    d. next to the second RCD

36 When wiring up a split-board consumer unit, the downstairs lights should be positioned alongside:
    a. the upstairs lights
    b. the upstairs sockets
    c. the downstairs sockets
    d. the kitchen ring circuit

Learning outcome 8

37 Metal conduit and trunking elements of an electrical installation are:
    a. earth conductors
    b. bonding conductors
    c. exposed conductive parts
    d. extraneous conductive parts

38 The gas, water and central heating pipes of the building not forming a part of the electrical installation are called:
    a. earthing conductors
    b. bonding conductors
    c. exposed conductive parts
    d. extraneous conductive parts

39 The act of connecting the exposed conductive parts of the installation to the main earthing terminal of the installation is called:
    a. earth
    b. earthing
    c. bonding conductor
    d. circuit protective conductor

40 According to the *On-Site Guide* all final circuits not exceeding 32 A in a building supplied with a 230 V TN earthing supply shall have a maximum disconnection time not exceeding:
    a. 0.2 s
    b. 0.4 s
    c. 5.0 s
    d. unlimited

41 Which of the following symbols represents the total earth loop impedance value?
    a. Zx
    b. Zs
    c. Ze
    d. ZT

42 A TNS earthing system makes use of which element to carry a fault current?
   a. Earthing spike
   b. Earthing sheath
   c. Metal sheath
   d. Metal spike

43 The purpose of earthing is to ensure that the:
   a. resistance path is low, fault current is high and disconnection is quick
   b. resistance path is high, fault current is high and disconnection is quick
   c. resistance path is low, fault current is low and disconnection is slow
   d. resistance path is high, fault current is low and disconnection is slow

44 Which of the following is an extraneous conductive part?
   a. Metal casing of a washing machine
   b. Metal switch back plate
   c. Metal casing of a motor
   d. Metal casing of a gas pipe

Learning outcome 9

45 An industrial welder has been installed in a domestic garage but cannot operate because the type B circuit breaker protecting the circuit operates instantly. To correct this, the circuit breaker needs:
   a. to be replaced with an RCD
   b. to be replaced with a Type C circuit breaker
   c. to be replaced with a Type D circuit breaker
   d. to be replaced with a re-wireable fuse

46 The weakest link in the circuit designed to melt when too much overcurrent flows is one definition of:
   a. fault protection
   b. a circuit protective conductor
   c. a fuse
   d. a consumer unit

47 The main service fuse needs to be able to withstand small supply surges and therefore will be which of the following types?
   a. BS 88-2
   b. BS 88-3
   c. BS 3036
   d. BS-EN 60898

48 Which of the following devices incorporates additional protection?
   a. HRC fuse
   b. Re-wireable fuse
   c. RCBO
   d. CB

49 Which earthing system is not recommended for a petrol station?
   a. TNS
   b. TT
   c. TNCS
   d. PME

Learning outcome 10

50 Calculate the total current demand for a domestic lighting circuit if it contains 20 x 100 W filaments and the factor for diversity is 66%?

   a. 7.8 A

   b. 8.6 A

   c. 6 A

   d. 5.7 A

51 What % volt drop is associated with power circuit?

   a. 3%

   b. 4%

   c. 5%

   d. 6%

52 A cooker can be fitted with a smaller cable after applying?

   a. Discrimination

   b. Diversity

   c. Grouping correction

   d. Temperature correction

53 Refer to Table F6 of the *On-Site Guide* and determine the maximum current associated with a 2.5 mm^2 cable which has been clipped direct?

   a. 17.5 A

   b. 27 A

   c. 37 A

   d. 32 A

54 A domestic instantaneous heater is rated at 3 kW. Given that the voltage is 230 V, work out its design current:

   a. 76.6 A

   b. 0.07 A

   c. 13.04 A

   d. 690 A

55 Refer to Table F6 of the *On-Site Guide* and determine the volt drop factor for a 4 mm^2 cable:

   a. 18 mV/A/m

   b. 11 mV/A/m

   c. 7.3 mV/A/m

   d. 4.4 mV/A/m

56 Which of the following is the correction factor regarding ambient temperature?

   a. Cf

   b. Ci

   c. Ca

   d. Cg

57 Which of the following is the symbol regarding the current-carrying capacity of a cable?
   a. Ib
   b. I^2
   c. Iz
   d. Ie

58 Briefly explain why an electrical installation needs protective devices.

59 List the four factors on which the selection of a protective device depends.

60 List the five essential requirements for a device designed to protect against overcurrent.

61 Briefly describe the action of a fuse under fault conditions.

62 State the meaning of 'discrimination' as applied to circuit protective devices.

63 Use a sketch to show how 'effective co-ordination' can be applied to a piece of equipment connected to a final circuit.

64 List typical 'exposed parts' of an installation.

65 List typical 'extraneous parts' of a building.

66 Use a sketch to show the path taken by an earth fault current.

67 Use bullet points and a simple sketch to briefly describe the operation of an RCD.

68 State the need for RCDs in an electrical installation:
   a. supplying socket outlets with a rated current not exceeding 20 A, and
   b. for use by mobile equipment out of doors as required by IET Regulation 411.3.3.

69 Briefly describe an application for RCBOs.

70 State the meaning of fault protection.

71 State the meaning of basic protection.

72 State the meaning of a 'polyphase supply system'.

73 Produce a quick coloured sketch of a PVC insulated and sheathed cable and name the parts.

74 Produce a quick coloured sketch of a PVC/SWA cable and name the parts.

75 Produce a quick sketch of an electric circuit and name the five component parts.

76 Give an example of a device or accessory for each component part. For example, the supply might be from the a.c. mains or a battery.

77 In your own words, state the meaning of circuit overload and short-circuit protection. What will provide this type of protection?

78 State the purpose of earthing and earth protection. What do we do to achieve it and why do we do it?

79   In your own words, state the meaning of exposed and extraneous conductive parts and give examples of each.

80   In your own words, state the meaning of earthing and bonding. What types of cables and equipment would an electrician use to achieve earthing and bonding on an electrical installation?

81   In your own words, state what we mean by 'basic protection' and how it is achieved.

82   In your own words, state what we mean by 'fault protection' and how it is achieved.

83   What methods could you use to find and store some information about:
   * health and safety at work
   * British Standards
   * electrical accessories and equipment?

Unit Elec2/04

Chapter 3 checklist

Learning outcome	Assessment criteria – the learner can	Page number
1. Know the legislation, regulations and guidance that apply to electrical installation work.	1.1 Recognize the legal status of documents used in the electrical industry. 1.2 Define the implications of not complying with regulations, documents and guidance.	198 198
2. Know the technical information used in electrical work.	2.1 Identify the purpose of different sources of technical information used in electrical work. 2.2 Identify the different diagrams and drawings used in electrical work. 2.3 Identify graphical symbols used in diagrams and drawings. 2.4 Convert measurements from a scale drawing.	198 201 201 202
3. Understand the properties, applications and limitations of different wiring systems.	3.1 State the properties, applications, advantages and limitations of different cable types. 3.2 State the features, applications, advantages and limitations of different containment systems. 3.3 Select an appropriate type of wiring system for a given environment. 3.4 State the types of wiring systems and associated equipment used in different installations. 3.5 Recognize the requirements of industrial plugs, sockets and couplers.	209 209 214 215 233
4. Know the general layout of equipment at the service position.	4.1 Identify the general layout and the equipment at the service position.	234
5. Understand standard lighting circuits.	5.1 Distinguish the different circuits and wiring layouts used for lighting. 5.2 Identify the different components that can be used in lighting circuits.	240 240
6. Understand standard ring and radial final circuits.	6.1 Define the requirements of standard ring final socket circuits. 6.2 Define the requirements of standard radial final socket circuits. 6.3 Describe the standard circuit arrangements for loads and equipment.	243 245 246
7. Know the basic requirements for circuits.	7.1 Outline the division of an installation into circuits. 7.2 Identify the requirements for polarity on circuits.	250 251
8. Know the importance of earthing and bonding for protection.	8.1 Identify the characteristics of earthing systems. 8.2 Outline the purpose of earthing and protective conductors when used for protection. 8.3 Recognize the components which provide automatic disconnection of supply. 8.4 Recognize an exposed conductive part. 8.5 Recognize an extraneous conductive part. 8.6 Distinguish the sections of earth loop impedance path. 8.7 Identify protective conductors. 8.8 Outline the general requirements for the installation of main protective bonding.	251 254 254 256 256 256 257 257

CHAPTER **4**

EAL Unit Elec2/05A

Electrical installation methods, procedures and requirements

Learning outcomes

When you have completed this chapter you should:

1. Know the methods and the procedures for installing electrical systems and equipment.
2. Understand termination and connection methods.
3. Know the requirements for installing electrical systems and equipment.
4. Know the methods and procedures for testing de-energized circuits.
5. Know how to communicate and work with others.

EAL Electrical Installation Work – Level 2, 2nd Edition 978 0 367 19562 5
© 2019 Linsley. Published by Taylor & Francis. All rights reserved.
www.routledge.com/9780367195618

DOI: 10.1201/9780429203176-4

Marking out the work on-site

When electricians arrive on-site to begin the work, that is, installing the trunking, conduit or cables and boxes to the final fixing positions in the building, they will need to know what is being installed and exactly where it is being installed.

On a larger job they will probably be given a site plan or layout drawing. This is a scale drawing based upon the architect's site plan of the building and will show the position of the electrical equipment to be installed.

The electrical equipment will be identified by graphical symbols on the plan and the drawing will be to scale; if we say the scale is 1:100, then 10 mm or 1 cm on the site plan will represent 1 m in the building. The electrician can take measurements from the drawing and transfer these to the building, probably using a retractable steel rule. The walls of the building can then be marked with a piece of chalk or a marker to indicate the final positions of the electrical equipment which is to be installed.

If the job is small, the electrician's supervisor or foreman may walk them around the site showing them what is to be installed and where. On a very small job, it might be the customer who describes what is to be done by the electrician.

When the electrician knows what is required, the cable routes can be planned and marked out to show where the trunking and conduits, cables or tray will be installed.

Trunking, conduits, cables and tray must be installed securely, but most of all they must be installed horizontally or vertically for a professional finish. A long spirit level is very useful in determining what is horizontal and vertical, although a plumb line can also be used to determine the vertical. A plumb line is a piece of string with a steel weight attached to the bottom. Holding the string up against a wall, the steel weight at the bottom will initially swing like a pendulum, but quickly come to rest in a true vertical line. If a little chalk is first rubbed onto the line, it becomes a chalk line. If an assistant holds the steel weight in its rest position so that the line presses against the wall, and then plucks the string so that it bounces against the wall, the chalk will transfer to the wall, leaving a true vertical line on the wall. When the measuring and marking out is completed, the electrician can begin the process of installing the trunking, conduits, cables or tray.

Tools used to install wiring systems

Craftsmen earn a living by hiring out their skills or selling products made using their skills and expertise. They shape their environment, mostly for the better, improving living standards for themselves and others.

Figure 4.1 A torque screwdriver.

Tools extend the limited physical responses of the human body and therefore good-quality, sharp tools are important to a craftsman. An electrician is no less a craftsman than a wood carver. Both must work with a high degree of skill and expertise, and both must have sympathy and respect for the materials they use. Modern electrical installations using new materials are lasting longer than 50 years. Therefore they must be properly installed. Good design, taking account of the manufacturer's instructions, good workmanship by skilled or instructed persons and the use of proper materials are essential if the installation is to comply with the relevant regulations (IET Regulation 134.1.1) and reliably and safely meet the requirements of the customer for over half a century.

An electrician must develop a number of basic craft skills particular to their own trade, but they also require some of the skills used in many other trades. An electrician's toolkit will reflect both the specific and general nature of the work.

The basic tools required by an electrician are those used in the stripping and connecting of conductors. These are pliers, side cutters, a knife and an assortment of screwdrivers.

The tools required in addition to these basic implements will depend upon the type of installation work being undertaken. When wiring new houses or rewiring old ones, the additional tools required are those usually associated with a bricklayer and joiner.

When working on industrial installations, installing conduit and trunking, the additional tools required by an electrician would more normally be those associated with a fitter or sheet-metal fabricator.

Where special tools are required, for example, those required to terminate mineral insulated (MI) cables or the bending and cutting tools for conduit and cable trays, they will often be provided by an employer, but most hand tools are provided by the electrician.

In general, good-quality tools last longer and stay sharper than those of inferior quality, but tools are very expensive to buy. A good set of tools can be

Safety first

Hand tools:
- keep them clean;
- keep them sharp;
- keep them safe.

Figure 4.2 The tools of an electrician.

Figure 4.3 Some of the essentials.

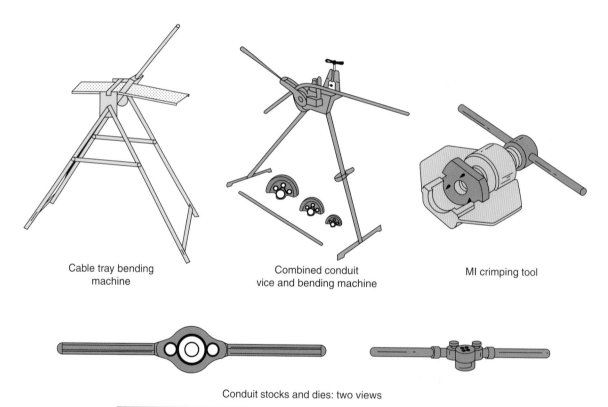

Cable tray bending machine

Combined conduit vice and bending machine

MI crimping tool

Conduit stocks and dies: two views

Figure 4.4 Some special tools required by an electrician engaged in industrial installations.

assembled over the training period if the basic tools are bought first and the extended toolkit acquired one tool at a time.

Another name for an installation electrician is a 'journeyman' electrician and, as the name implies, an electrician must be mobile and prepared to carry his or her tools from one job to another. Therefore, a good toolbox is an essential early investment, so that the right tools for the job can be easily transported.

Tools should be cared for and maintained in good condition if they are to be used efficiently and remain serviceable. Screwdrivers should have a flat, squared-off end and wood chisels should be very sharp. Access to a grindstone will help an electrician to maintain their tools in first-class condition. In addition, wood chisels will require sharpening on an oilstone to give them a very sharp edge.

Electrical power tools (mobile equipment)

Electrical power tools can reduce much of the hard work for any tradesperson and increase productivity. Electrical tools should be maintained in a good condition and be appropriate for the purpose for which they are used.

Many construction sites now insist on low-voltage or battery tools being used, which can further increase safety without any loss of productivity. Some useful electrical tools are shown in Figure 4.5.

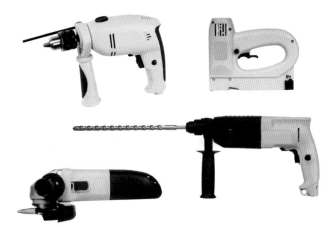

Figure 4.5 Electrical power tools.

Electric drills are probably used most frequently of all electrical tools. They may be used to drill metal or wood. Wire brushes are made which fit into the drill chuck for cleaning the metal. Variable-speed electric drills, which incorporate a vibrator will also drill brick and concrete as easily as wood when fitted with a masonry drill bit.

Hammer drills give between 2,000 to 3,000 impacts per minute and are used for drilling concrete walls and floors.

Cordless electric drills are also available which incorporate a rechargeable battery, usually in the handle. They offer the convenience of electric drilling when an electrical supply is not available or if an extension cable is impractical.

Angle grinders are useful for cutting chases in brick or concrete. The discs are interchangeable. Silicon carbide discs are suitable for cutting slate, marble, tiles, brick and concrete, and aluminium oxide discs for cutting iron and steel such as conduit and trunking.

Jigsaws can be fitted with wood- or metal-cutting blades. With a wood-cutting blade fitted they are useful for cutting across floorboards and skirting boards or any other application where a padsaw would be used. With a metal-cutting blade fitted, they may be used to cut trunking.

When a lot of trunking work is to be undertaken, an electric nibbler is a worthwhile investment. This nibbles out the sheet metal, is easily controllable and is one alternative to the jigsaw.

 Top tip

Drilling brick or stone
When using masonry drills, withdraw several times while drilling to remove dust and reduce the chance of the drill jamming in the hole.

 Top tip

Drilling tiles
When drilling glazed surfaces or tiles, place masking tape on the surface. This allows you to mark the hole centre and prevents the drill bit from skidding.

 Top tip

Drill speed
Remember, the larger the drill bit, the slower the drill speed.

 Safety first

Power tools
Many construction sites now only allow:

- low-voltage tools;
- PAT tested tools;
- battery-powered tools.

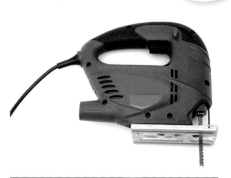

Figure 4.6 Jigsaw.

All tools must be used safely and sensibly. Cutting tools should be sharpened and screwdrivers ground to a sharp square end on a grindstone.

It is particularly important to check that the plug top and cables of handheld electrically-powered mobile equipment and extension leads are in good condition. Damaged plug tops and cables must be repaired before you use them. All electrical power tools of 110 V and 230 V must be inspected and tested with a portable appliance tester (PAT) in accordance with the company's procedures, but probably at least once each year. PAT testing tests the quality of the insulation resistance and the earth continuity. Inspection checks the condition of the plug top, fuse and lead.

Safety first

Site safety

The Health and Safety at Work Act makes *you* responsible for:

- following the site safety rules;
- demonstrating safe working practice;
- wearing or using appropriate PPE.

Try this

Power tools

Look at the power tools at work.

- Do they have a PAT Test label?
- If so, is the PAT Test label 'in date'?

Tools and equipment that are left lying about in the workplace can become damaged or stolen, and may also be the cause of people slipping, tripping or falling. Tidy up regularly and put power tools back in their boxes. **You may have no control over the condition of the workplace in general, but keeping your own work area clean and tidy is the mark of a skilled and conscientious craftsman.**

Figure 4.7 Always tidy up as you go, it not only makes you less likely to lose tools, but it's also safer.

Assessment criteria 1.3

Identify the components of wiring systems

Support and fixing methods for electrical systems

Individual conductors may be installed in trunking or conduit and individual cables may be clipped directly to a surface or laid on a tray using the wiring system which is most appropriate for the particular installation. The installation method chosen will depend upon the contract specification, the fabric of the building and the type of installation – domestic, commercial or industrial.

It is important that the wiring systems and fixing methods are appropriate for the particular type of installation and compatible with the structural materials used in the building construction. The electrical installation must be compatible with the installed conditions, must not damage the fabric of the building, or weaken load-bearing girders or joists.

The installation designer must ask the following questions:

- Does this wiring system meet the contract specification?
- Is the wiring system compatible with this particular installation?
- Do I need to consider any special regulations such as those required by agricultural and horticultural installations, swimming pools or flameproof installations?

- Will this type of electrical installation be aesthetically acceptable and compatible with the other structural materials?

The installation electrician must ask the following questions:

- Am I using materials and equipment which meet the relevant British Standards and the contract specification?
- Am I using an appropriate fixing method for this wiring system or piece of equipment?
- Will the structural material carry the extra load that my conduits and cables will place upon it?
- Will my fixings and fittings weaken the existing fabric of the building?
- Will the electrical installation interfere with other supplies and services?
- Will all terminations and joints be accessible upon completion of the erection period (IET Regulations 513.1 and 526.3)?
- Will the materials being used for the electrical installation be compatible with the intended use of the building?
- Am I working safely and efficiently and in accordance with the IET Regulations (BS 7671)?

A domestic installation usually calls for a PVC insulated and sheathed wiring system. These cables are generally fixed using plastic clips incorporating a masonry nail which means that the cables can be fixed to wood, plaster or brick with almost equal ease. However the 3rd Amendment to the Regulations introduced a new Regulation 521.11.201 which requires wiring systems in escape routes be supported in a such a manner that they will not prematurely collapse in the event of a fire.

The 18th Edition of the regulations at 521.10.202 now requires the same fire proof support for all systems throughout the installation, not just escape routes.

However, it is only those wiring systems that might collapse before or during the period when the rescue services would enter a building, that require a fixing which will not collapse prematurely. So, cables installed in metal trunking, or conduit, or cables laid on top of cable tray or ladder rack will require no extra fixing considerations because they are adequately supported. Also, wiring systems fixed within the fabric of the building which is not liable to premature collapse in the event of a fire is safe in these circumstances.

Note 3 to the regulation tell us it is the cable systems fixed with plastic clips or inside plastic trunking which will now require our consideration. These systems can fail when subject to either direct flame or the hot products of combustion leading to wiring systems hanging down and causing an entanglement risk as a result of the fire.

This makes it impossible for us to use non metallic cable clips, cable ties or plastic trunking as the only means of support for PVC wiring system. The regulation tells us that where non-metallic cable systems are used, **a suitable means of fire-resistant support and retention must be used** to prevent cables falling down in the event of a fire. Note 4 of the Regulation advises that suitably spaced steel or copper clips, saddles or ties are examples that will meet the requirements of this Regulation.

Cables must be run straight and neatly between clips fixed at equal distances which provide adequate support for the cable so that it does not

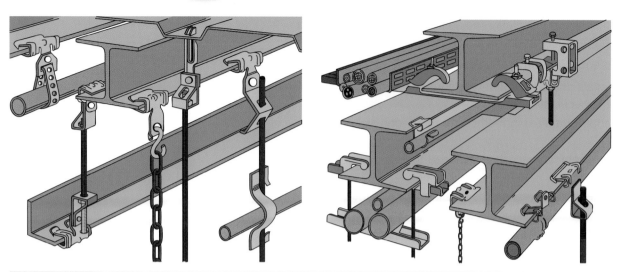

Figure 4.8 Some manufactured girder supports for electrical equipment.

become damaged by its own weight (IET Regulation 522.8.4), as shown in Table 3.1.

A commercial or industrial installation might call for a conduit or trunking wiring system. A conduit is a tube, channel or pipe in which insulated conductors are contained. The conduit, in effect, replaces the PVC outer sheath of a cable, providing mechanical protection for the insulated conductors. A conduit installation can be rewired easily or altered at any time and this flexibility, coupled with mechanical protection, makes conduit installations popular for commercial and industrial applications. Steel conduits and trunking are, however, much heavier than single cables and, therefore, need substantial and firm fixings and supports. A wide range of support brackets is available for fixing conduit, trunking and tray installations to the fabric of a commercial or industrial installation. Some of these are shown in Figure 4.8.

When a heavier or more robust fixing is required to support cabling or equipment, a nut and bolt or screw fixing is called for. Wood screws may be screwed directly into wood but when fixing to stone, brick or concrete it is first necessary to drill a hole in the masonry material which is then plugged with a material (usually plastic) to which a screw can be secured.

For the most robust fixing to masonry materials, an expansion bolt such as that made by Rawlbolt should be used.

For lightweight fixings to hollow partitions or plasterboard, a spring toggle can be used. Plasterboard cannot support a screw fixing directly into itself but the spring toggle spreads the load over a larger area, making the fixing suitable for light loads.

Let us look in a little more detail at individual joining, support and fixing methods.

Joining materials

Plastic can be joined with an appropriate solvent. Metal may be welded, brazed or soldered, but the most popular method of on-site joining of metal on electrical installations is by nuts and bolts or rivets.

A nut and bolt joint may be considered a temporary fastening since the parts can easily be separated if required by unscrewing the nut and removing the bolt. A rivet is a permanent fastening since the parts riveted together cannot be easily separated.

Figure 4.9 Nuts and bolts can easily be separated if required.

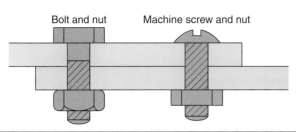

Figure 4.10 Joining of metal.

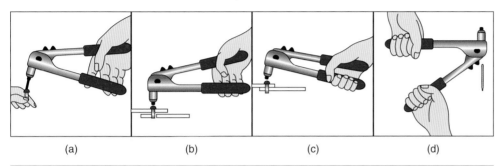

(a)   (b)   (c)   (d)

Figure 4.11 Metal joining with pop rivets.

Two pieces of metal joined by a bolt and nut and by a machine screw and nut are shown in Figure 4.10. The nut is tightened to secure the joint. When joining trunking or cable trays, a round-head machine screw should be used with the head inside to reduce the risk of damage to cables being drawn into the trunking or tray.

Thin sheet material such as trunking is often joined using a pop riveter. Special rivets are used with a hand tool, as shown in Figure 4.11. Where possible, the parts to be riveted should be clamped and drilled together with a clearance hole for the rivet. The stem of the rivet is pushed into the nose bush of the riveter until the alloy sleeve of the rivet is flush with the nose bush (a). The rivet is then placed in the hole and the handles squeezed together (b). The alloy sleeve is compressed and the rivet stem will break off when the rivet is set and the joint complete (c).

To release the broken-off stem piece, the nose bush is turned upward and the handles opened sharply. The stem will fall out and is discarded (d).

Bracket supports

Conduit and trunking may be fixed directly to a surface such as a brick wall or concrete ceiling, but where cable runs are across girders or other steel framework, spring steel clips may be used but support brackets or clips often require manufacturing.

The brackets are usually made from flat iron, which is painted after manufacturing to prevent corrosion. They may be made on-site by the electrician or, if many brackets are required, the electrical contractor may make a working sketch with dimensions and have the items manufactured by a blacksmith or metal fabricator.

The type of bracket required will be determined by the installation, but Figure 4.13 gives some examples of brackets which may be modified to suit particular circumstances.

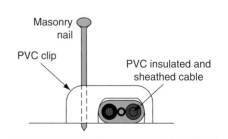

Figure 4.12 PVC insulated and sheathed cable clip.

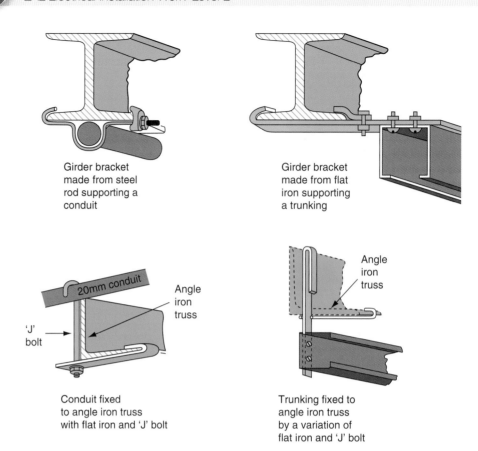

Girder bracket made from steel rod supporting a conduit

Girder bracket made from flat iron supporting a trunking

Conduit fixed to angle iron truss with flat iron and 'J' bolt

Trunking fixed to angle iron truss by a variation of flat iron and 'J' bolt

Figure 4.13 Bracket supports for conduits and trunking.

Assessment criteria 1.4

Recognize fixing and securing methods

Assessment criteria 1.5

Identify an appropriate fixing and securing method

Fixing methods

PVC insulated and sheathed wiring systems are usually fixed with PVC clips in order to comply with IET Regulations 522.8.3 and 4 and are spaced as shown in Table 3.1. The clips are supplied in various sizes to hold the cable firmly, and the fixing nail is a hardened masonry nail. Figure 4.12 shows a cable clip of this type. The use of a masonry nail means that fixings to wood, plaster, brick or stone can be made with equal ease.

When heavier cables, trunking, conduit or luminaires have to be fixed, a screw fixing is often needed. Wood screws may be screwed directly into wood but when fixing to brick, stone, plaster or concrete it is necessary to drill a hole in the masonry material, which is then plugged with a material to which the screw can be secured.

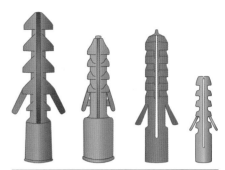 **4**

Plastic plugs

A plastic plug is made of a hollow plastic tube split up to half its length to allow for expansion. Each size of plastic plug is colour coded to match a wood screw size.

A hole is drilled into the masonry, using a masonry drill of the same diameter, to the length of the plastic plug (see Figure 4.14). The plastic plug is inserted into the hole and tapped home until it is level with the surface of the masonry. Finally, the fixing screw is driven into the plastic plug until it becomes tight and the fixture is secure.

Figure 4.14 Screw fixing plastic plugs.

Expansion bolts

The most well-known expansion bolt is made by Rawlbolt and consists of a split iron shell held together by a steel ferrule at one end and a spring wire clip at the other end. Tightening the bolt draws up an expanding bolt inside the split iron shell, forcing the iron to expand and grip the masonry. Rawlbolts are for heavy duty masonry fixings (see Figure 4.15).

A hole is drilled in the masonry to take the iron shell and ferrule. The iron shell is inserted with the spring wire clip end first so that the ferrule is at the outer surface. The bolt is passed through the fixture, located in the expanding nut and tightened until the fixing becomes secure.

Definition

A *plastic plug* is made of a hollow plastic tube split up to half its length to allow for expansion. Each size of plastic plug is colour coded to match a wood screw size.

Definition

The most well-known *expansion bolt* is made by Rawlbolt and consists of a split iron shell held together by a steel ferrule at one end and a spring wire clip at the other end. Tightening the bolt draws up an expanding bolt inside the split iron shell, forcing the iron to expand and grip the masonry. Rawlbolts are for heavy-duty masonry fixings.

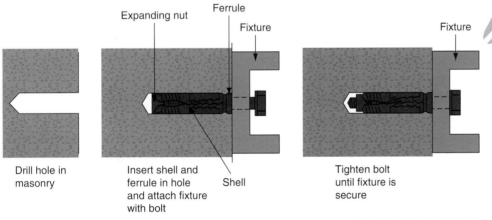

Drill hole in masonry | Insert shell and ferrule in hole and attach fixture with bolt | Tighten bolt until fixture is secure

Figure 4.15 Expansion bolt fixing.

Spring toggle bolts

A spring toggle bolt provides one method of fixing to hollow partition walls which are usually faced with plasterboard and a plaster skimming. Plasterboard and plaster wall or ceiling surfaces are not strong enough to support a load fixed directly into the plasterboard, but the spring toggle spreads the load over a larger area, making the fixing suitable for light loads (see Figure 4.16).

A hole is drilled through the plasterboard and into the cavity. The toggle bolt is passed through the fixture and the toggle wings screwed into the bolt. The toggle wings are compressed and passed through the hole in the plasterboard and into the cavity where they spring apart and rest on the cavity side of the plasterboard. The bolt is tightened until the fixing becomes firm. The bolt of the spring toggle cannot be removed after fixing without the loss of the toggle wings.

If it becomes necessary to remove and refix the fixture, a new toggle bolt will have to be used.

Top tip

Inserting Rawlplugs When placing Rawlplugs into masonry or tile surfaces, ensure the plug is pushed several millimetres past the surface.

Definition

A *spring toggle bolt* provides one method of fixing to hollow partition walls which are usually faced with plasterboard and a plaster skimming.

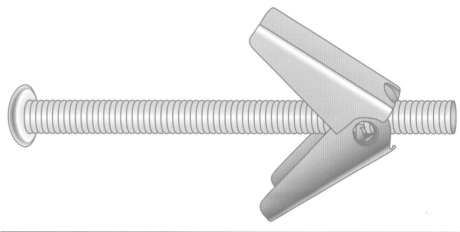

Figure 4.16 Spring toggle bolt fixing.

On-Site Guide information

The IET *On-Site Guide* includes an Appendix D with information on support of cables, conductors and wiring systems.

The tables in the appendix cover:

1 The spacing of supports for thermoplastic, armoured and mineral insulated cables. Smaller cables need clips spaced closer together as they are not as rigid as the cables with larger conductors incorporated in them.
2 The spacing of supports for conduits.
3 The spacing of supports for cable trunking.
4 The minimum internal radii of bends in cables for fixed wiring.

Conduit installations

Definition

A *conduit* is a tube, channel or pipe in which insulated conductors are contained.

A conduit is a tube, channel or pipe in which insulated conductors are contained. The conduit, in effect, replaces the PVC outer sheath of a cable, providing mechanical protection for the insulated conductors. A conduit installation can be rewired easily or altered at any time, and this flexibility, coupled with mechanical protection, makes conduit installations popular for commercial and industrial applications. There are three types of conduit used in electrical installation work: steel, PVC and flexible.

Steel conduit

Steel conduits are made to a specification defined by BS 4568 and are either heavy gauge welded or solid drawn. Heavy gauge is made from a sheet of steel welded along the seam to form a tube and is used for most electrical installation work. Solid drawn conduit is a seamless tube which is much more expensive and only used for special gas-tight, explosion-proof or flameproof installations.

Conduit is supplied in 3.75 m lengths and typical sizes are 16, 20, 25 and 32 mm.

Conduit tubing and fittings are supplied in a black enamel finish for internal use or hot galvanized finish for use on external or damp installations. A wide range of fittings is available and the conduit is fixed using saddles or pipe hooks, as shown in Figure 4.17.

Metal conduits are threaded with stocks and dies and bent using special bending machines. The metal conduit is also utilized as the CPC and, therefore, all

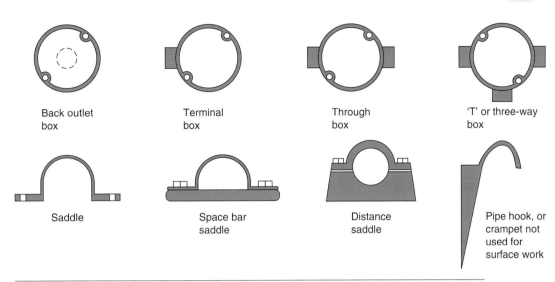

Figure 4.17 Conduit fittings and saddles.

connections must be screwed up tightly and all burrs removed so that cables will not be damaged as they are drawn into the conduit. Metal conduits containing a.c. circuits must contain line and neutral conductors in the same conduit to prevent eddy currents flowing, which would result in the metal conduit becoming hot (IET Regulations 521.5.1, 522.8.1 and 522.8.11).

PVC conduit

PVC conduit used on typical electrical installations is heavy gauge standard impact tube manufactured to BS 4607. The conduit size and range of fittings are the same as those available for metal conduit. PVC conduit is most often joined by placing the end of the conduit into the appropriate fitting and fixing with a PVC solvent adhesive.

PVC conduit can be bent by hand using a bending spring of the same diameter as the inside of the conduit. The spring is pushed into the conduit to the point of the intended bend and the conduit then bent over the knee. The spring ensures that the conduit keeps its circular shape. In cold weather, a little warmth applied to the point of the intended bend often helps to achieve a more successful bend.

Figure 4.18 Flexible metal conduit.

The advantages of a PVC conduit system are that it may be installed much more quickly than steel conduit and is non-corrosive, but it does not have the mechanical strength of steel conduit. Since PVC conduit is an insulator, it cannot be used as the CPC and a separate earth conductor must be run to every outlet. It is not suitable for installations subjected to temperatures below 25 °C or above 60 °C. Where luminaires are suspended from PVC conduit boxes, precautions must be taken to ensure that the lamp does not raise the box temperature or that the mass of the luminaire supported by each box does not exceed the maximum recommended by the manufacturer (IET Regulations 522.1 and 522.2). PVC conduit also expands much more than metal conduit and so long runs require an expansion coupling to allow for conduit movement and to help prevent distortion during temperature changes. Also the 18th Edition of the Regulations at 521.10.202 now requires fire proof support for all wiring systems to prevent cables collapsing in the event of a fire. Suitably spaced steel or copper clips, saddles or ties are examples which will meet this requirement.

All conduit installations must be erected first before any wiring is installed (IET Regulation 522.8.2). The radius of all bends in conduit must not cause the cables

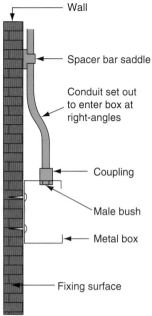

Wall

Spacer bar saddle

Conduit set out to enter box at right-angles

Coupling

Male bush

Metal box

Fixing surface

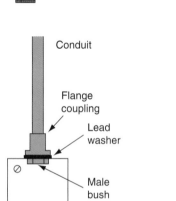

Conduit

Flange coupling

Lead washer

Male bush

Figure 4.19 Terminating conduits.

to suffer damage, and therefore the minimum radius of bends given in Table 4E of the *On-Site Guide* applies (IET Regulation 522.8.3). All conduits should terminate in a box or fitting and meet the boxes or fittings at right angles, as shown in Figure 4.19 . Any unused conduit-box entries should be blanked off and all boxes covered with a box lid, fitting or accessory to provide complete enclosure of the conduit system. Conduit runs should be separate from other services, unless intentionally bonded, to prevent arcing occurring from a faulty circuit within the conduit, which might cause the pipe of another service to become punctured.

Cables should be fed into the conduit in a manner which prevents any cable crossing over and becoming twisted inside the conduit. The cable insulation must not be damaged on the metal edges of the draw-in box. Cables can be pulled in on a draw wire if the run is a long one. The draw wire itself may be drawn in on a fish tape, which is a thin spring steel or plastic tape.

A limit must be placed on the number of bends between boxes in a conduit run and the number of cables which may be drawn into a conduit to prevent the cables being strained during wiring. Appendix E of the *On-Site Guide* gives a guide to the cable capacities of conduits and trunking.

Figure 4.20 PVC conduit.

Flexible conduit

Flexible conduit manufactured to BS 731-1: 1993 is made of interlinked metal spirals, often covered with a PVC sleeving. The tubing must not be relied upon to provide a continuous earth path and, consequently, a separate CPC must be run either inside or outside the flexible tube (IET Regulation 543.2.7).

Flexible conduit is used for the final connection to motors so that the vibrations of the motor are not transmitted throughout the electrical installation and to allow for modifications to be made to the final motor position and drive belt adjustments.

Conduit capacities

Single-PVC insulated conductors are usually drawn into the installed conduit to complete the installation. Having decided upon the type, size and number of cables required for a final circuit, it is then necessary to select the appropriate size of conduit to accommodate those cables.

The tables in Appendix 5 of the *On-Site Guide* describe a 'factor system' for determining the size of conduit required to enclose a number of conductors. The method is as follows:

- Identify the cable factor for the particular size of conductor, see Table 4.1.
- Multiply the cable factor by the number of conductors, to give the sum of the cable factors.
- Identify the appropriate part of the conduit factor table given by the length of run and number of bends, see Table 4.2.
- The correct size of conduit to accommodate the cables is that conduit which has a factor equal to or greater than the sum of the cable factors.

Table 4.1 Conduit factors

Cable factors for conduit in long straight runs over 3 m, or runs of any length incorporating bends	
Conductor CSA (mm²)	**Cable factor**
1	16
1.5	22
2.5	30
4	43
6	58
10	105
16	145

Table 4.2 Conduit factors

Length of run (m)	Conduit diameter (mm)											
	16	**20**	**25**	**32**	**16**	**20**	**25**	**32**	**16**	**20**	**25**	**32**
	Straight				**One bend**				**Two bends**			
3.5	179	290	521	911	162	263	475	837	136	222	404	720
4	177	286	514	900	158	256	463	818	130	213	388	692
4.5	174	282	507	889	154	250	452	800	125	204	373	667
5	171	278	500	878	150	244	442	783	120	196	358	643
6	167	270	487	857	143	233	422	750	111	182	333	600
7	162	263	475	837	136	222	404	720	103	169	311	563
8	158	256	463	818	130	213	388	692	97	159	292	529
9	154	250	452	800	125	204	373	667	91	149	275	500
10	150	244	442	783	120	196	358	643	86	141	260	474

Example 1

Six 2.5 mm² PVC insulated cables are to be run in a conduit containing two bends between boxes 10 m apart. Determine the minimum size of conduit to contain these cables.

From Table 4.1:

$$\text{The factor for one 2.5 mm}^2 \text{ cable} = 30$$
$$\text{The sum of the cable factors} = 6 \times 30$$
$$= 180$$

From Table 4.2, a 25 mm conduit, 10 m long and containing two bends, has a factor of 260. A 20 mm conduit containing two bends only has a factor of 141 which is less than 180, the sum of the cable factors, and, therefore, 25 mm conduit is the minimum size to contain these cables.

Example 2

Ten 1.0 mm² PVC insulated cables are to be drawn into a plastic conduit which is 6 m long between boxes and contains one bend. A 4.0 mm PVC insulated CPC is also included. Determine the minimum size of conduit to contain these conductors.

From Table 4.1:

$$\text{The factor for one 1.0 mm cable} = 16$$
$$\text{The factor for one 4.0 mm cable} = 43$$
$$\text{The sum of the cable factors} = (10 \times 16) + (1 \times 43)$$
$$= 203$$

From Table 4.2, a 20 mm conduit, 6 m long and containing one bend, has a factor of 233. A 16 mm conduit containing one bend only has a factor of 143 which is less than 203, the sum of the cable factors, and, therefore, 20 mm conduit is the minimum size to contain these cables.

Assessment criteria 1.6

Outline the methods and procedures used for the installation of wiring systems

Assessment criteria 1.7

Specify methods and techniques for restoring the building fabric

Trunking installations

A trunking is an enclosure provided for the protection of cables which is normally square or rectangular in cross-section, having one removable side. Trunking may be thought of as a more accessible conduit system and for industrial and commercial installations it is replacing the larger conduit sizes.

A trunking system can have great flexibility when used in conjunction with conduit; the trunking forms the background or framework for the installation, with conduits running from the trunking to the point controlling the current-using apparatus. When an alteration or extension is required it is easy to drill a hole in the side of the trunking and run a conduit to the new point. The new wiring can then be drawn through the new conduit and the existing trunking to the supply point.

Trunking is supplied in 3 m lengths and various cross-sections measured in millimetres from 50 × 50 up to 300 × 150. Most trunking is available in either steel or plastic.

Metallic trunking

Metallic trunking is formed from mild steel sheet, coated with grey or silver enamel paint for internal use or a hot-dipped galvanized coating where damp conditions might be encountered and made to a specification defined by BS EN 500 85. A wide range of accessories is available, such as 45° bends, 90° bends, tee and four-way junctions, for speedy on-site assembly. Alternatively, bends may be fabricated in lengths of trunking, as shown in Figure 4.21. This may be necessary or more convenient if a bend or set is non-standard, but it does take more time to fabricate bends than merely to bolt on standard accessories.

Insulated non-sheathed cables are permitted in a trunking system which provides at least the degree of protection IPXXD (which means total protection) or IP4X which means protection from a solid object greater than 1.0 mm such as a thin wire or strip. For site fabricated joints such as that shown in Figure 4.21,

Definition

A *trunking* is an enclosure provided for the protection of cables which is normally square or rectangular in cross-section, having one removable side. Trunking may be thought of as a more accessible conduit system.

Definition

Metallic trunking is formed from mild steel sheet, coated with grey or silver enamel paint for internal use or a hot-dipped galvanized coating where damp conditions might be encountered.

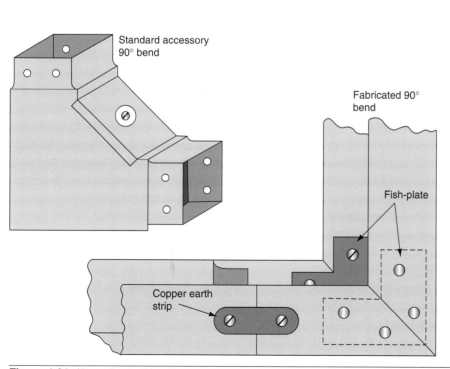

Figure 4.21 Alternative trunking bends.

the installer must confirm that the completed item meets at least IPXXD (IET Regulation 521.10).

When fabricating bends, the trunking should be supported with wooden blocks for sawing and filing, in order to prevent the sheet-steel vibrating or becoming deformed. Fish-plates must be made and riveted or bolted to the trunking to form a solid and secure bend. When manufactured bends are used, the continuity of the earth path must be ensured across the joint by making all fixing screw connections very tight, or fitting a separate copper strap between the trunking and the standard bend. If an earth continuity test on the trunking is found to be unsatisfactory, an insulated CPC must be installed inside the trunking. The size of the protective conductor will be determined by the largest cable contained in the trunking, as described in Table 54.7 of the IET Regulations. If the circuit conductors are less than $16\,\text{mm}^2$, then a $16\,\text{mm}^2$ CPC will be required.

Non-metallic trunking

Trunking and trunking accessories are also available in high-impact PVC. The accessories are usually secured to the lengths of trunking with a PVC solvent adhesive. PVC trunking, like PVC conduit, is easy to install and is non-corrosive. A separate CPC will need to be installed and non-metallic trunking may require more frequent fixings because it is less rigid than metallic trunking.

All trunking fixings should use round-headed screws to prevent damage to cables since the thin sheet construction makes it impossible to countersink screw heads. Also the 18th Edition of the regulations at 521.10.202 makes it impossible to use plastic trunking overhead without suitable metal support. See 'Support and Fixing Methods for Electrical Systems' earlier in this chapter at page 298.

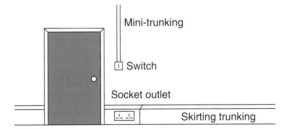

Figure 4.22 Typical installation of skirting trunking and mini-trunking.

Mini-trunking

Mini-trunking is very small PVC trunking, ideal for surface wiring in domestic and commercial installations such as offices. The trunking has a cross-section of $16 \times 16\,\text{mm}$, $25 \times 16\,\text{mm}$, $38 \times 16\,\text{mm}$ or $38 \times 25\,\text{mm}$ and is ideal for switch drops or for housing auxiliary circuits such as telephone or audio equipment wiring. The modern square look in switches and sockets is complemented by the mini-trunking which is very easy to install (see Figure 4.22).

Skirting trunking

Skirting trunking is a trunking manufactured from PVC or steel in the shape of a skirting board and is frequently used in commercial buildings such as hospitals, laboratories and offices. The trunking is fitted around the walls of a room at either the skirting board level or at the working surface level and contains the wiring for socket outlets and telephone points which are mounted on the lid (see Figure 4.22).

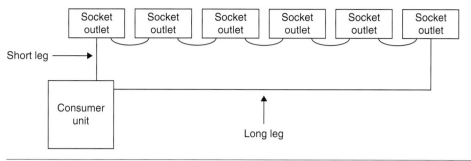

Figure 4.23 The wrong way.

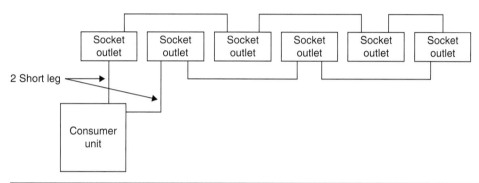

Figure 4.24 Good practice in wiring a ring final circuit.

When wiring a ring final circuit in conduit or trunking, it is good practice to install the wiring so as to feed the sockets alternatively on each leg to avoid a short length to the first socket and a long leg returning from the final socket (see Figures 4.23 and 4.24).

Where any trunking passes through walls, partitions, ceilings or floors, short lengths of lid should be fitted so that the remainder of the lid may be removed later without difficulty, see Figure 4.25. Any damage to the structure of the buildings must be made good with mortar, plaster or concrete in order to prevent the spread of fire.

Fire barriers must be fitted inside the trunking every 5 m, or at every floor level or room dividing wall if this is a shorter distance, as shown in Figure 4.25 (a).

Where trunking is installed vertically, the installed conductors must be supported so that the maximum unsupported length of non-sheathed cable does not exceed 5 m. Figure 4.25 (b) shows cables woven through insulated pin supports, which is one method of supporting vertical cables. PVC insulated cables are usually drawn into an erected conduit installation or laid into an erected trunking installation. Table 5D of the *On-Site Guide* only gives factors for conduits up to 32 mm in diameter, which would indicate that conduits larger than this are not in frequent or common use. Where a cable enclosure greater than 32 mm is required because of the number or size of the conductors, it is generally more economical and convenient to use trunking.

Trunking capacities

The ratio of the space occupied by all the cables in a conduit or trunking to the whole space enclosed by the conduit or trunking is known as the space factor.

Where sizes and types of cable and trunking are not covered by the tables in the *On-Site Guide*, a space factor of 45% must not be exceeded. This means that

Definition

The *ratio* of the space occupied by all the cables in a conduit or trunking to the whole space enclosed by the conduit or trunking is known as the *space factor*.

the cables must not fill more than 45% of the space enclosed by the trunking. The tables take this factor into account.

- To calculate the size of trunking required to enclose a number of cables, identify the cable factor for the particular size of conductor. See Table 4.3.
- Multiply the cable factor by the number of conductors to give the sum of the cable factors.

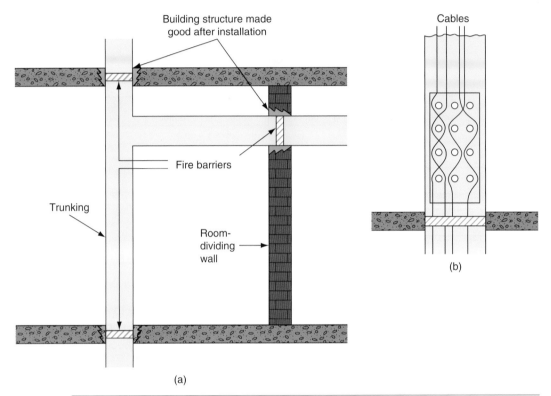

Figure 4.25 Installation of trunking (a) fire barriers in trunking and (b) cable supports in vertical trunking.

Table 4.3 Trunking cable factors

Type of conductor	Conductor CSA (mm²)	PVC cable factor	Thermosetting cable factor
Solid	1.5	8.0	8.6
	2.5	11.9	11.9
Stranded	1.5	8.6	9.6
	2.5	12.6	13.9
	4	16.6	18.1
	6	21.2	22.9
	10	35.3	36.3

- Consider the factors for trunking shown in Table 4.4. The correct size of trunking to accommodate the cables is that trunking which has a factor equal to, or greater than, the sum of the cable factors.

Table 4.4 Trunking factors

Dimensions of trunking (mm × mm)	Factor
50 × 38	767
50 × 50	1037
75 × 25	738
75 × 38	1146
75 × 50	1555
75 × 75	2371
100 × 25	993
100 × 38	1542
100 × 50	2091
100 × 75	3189
100 × 100	4252
150 × 38	2999
150 × 50	3091
150 × 75	4743
150 × 100	6394
150 × 150	9697

Example

Calculate the minimum size of trunking required to accommodate the following single-core PVC cables:

   20 × 1.5 mm solid conductors
   20 × 2.5 mm solid conductors
   21 × 4.0 mm stranded conductors
   16 × 6.0 mm stranded conductors

From Table 4.3, the cable factors are:

   for 1.5 mm solid cable – 8.0
   for 2.5 mm solid cable – 11.9
   for 4.0 mm stranded cable – 16.6
   for 6.0 mm stranded cable – 21.2

(continued)

Example continued

The sum of the cable terms is:

(20 × 8.0) + (20 × 11.9) + (21 × 16.6) + (16 × 21.2) = 1085.8. From Table 4.4, 75 × 38 mm trunking has a factor of 1146 and, therefore, the minimum size of trunking to accommodate these cables is 75 × 38 mm, although a larger size, say, 75 × 50 mm, would be equally acceptable if this was more readily available as a standard stock item.

Segregation of circuits

Definition

The entry of a cable end into an accessory, enclosure or piece of equipment is what we call a *termination*.

Where an installation comprises a mixture of low-voltage and very low-voltage circuits such as mains lighting and power, fire alarm and telecommunication circuits, they must be separated or *segregated* to prevent electrical contact (IET Regulation 528.1).

For the purpose of these regulations various circuits are identified by one of two bands as follows:

- Band I: telephone, radio, bell, call and intruder alarm circuits, emergency circuits for fire alarm and emergency lighting.
- Band II: mains voltage circuits.

When Band I circuits are insulated to the same voltage as Band II circuits, they may be drawn into the same compartment.

When trunking contains rigidly fixed metal barriers along its length, the same trunking may be used to enclose cables of the separate bands without further precautions, provided that each band is separated by a barrier, as shown in Figure 4.26.

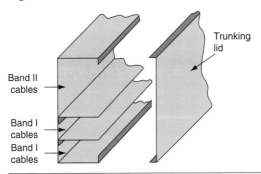

Figure 4.26 Segregation of cables in trunking.

Multi-compartment PVC trunking cannot provide band segregation since there is no metal screen between the bands. This can only be provided in PVC trunking if screened cables are drawn into the trunking.

Assessment criteria 2.1

Describe connection methods

Terminating and connecting conductors

The entry of a cable end into an accessory, enclosure or piece of equipment is called a termination. Section 526 of the IET Regulations tells us that:

1 Every connection between conductors and equipment shall be durable, provide electrical continuity and mechanical strength and protection.
2 Every termination and joint in a live conductor shall be made within a suitable accessory, piece of equipment or enclosure that complies with the appropriate product standard.
3 Every connection shall be accessible for inspection, testing and maintenance.
4 The means of connection shall take account of the number and shape of the wires forming the conductor.
5 The connection shall take account of the cross-section of the conductor and the number of conductors to be connected.
6 The means of connection shall take account of the temperature attained in normal service.
7 There must be no mechanical strain on the conductor connections.

There is a wide range of suitable means of connecting conductors and we shall look at these in a moment. Whatever method is used to connect live conductors, the connection must be contained in an enclosed compartment such as an accessory; for example, a switch or socket box or a junction box. Alternatively, an equipment enclosure may be used; for example, a motor enclosure or an enclosure partly formed by non-combustible building material (IET Regulation 526.5). This is because faulty joints and terminations in live conductors can attain very high temperatures due to the effects of resistive heating. They might also emit arcs, sparks or hot particles with the consequent risk of fire or other harmful thermal effects to adjacent materials.

Types of terminal connection

Junction boxes

Junction boxes are probably the most popular method of making connections in domestic properties. Traditional junction boxes have brass terminals fixed inside a Bakelite container. The two important factors to consider when choosing a junction box are the number of terminals required and the current rating. Socket outlet junction boxes have larger brass terminals than lighting junction boxes. See Figure 4.28

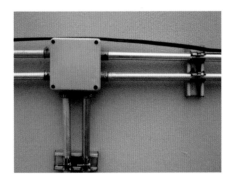

Figure 4.27 Junction box with galvanized conduit pipe connection.

Key fact

Junction boxes are probably the most popular method of making connections in domestic properties.

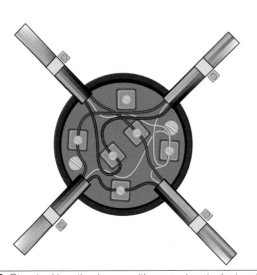

Figure 4.28 Standard junction boxes with screw terminals; junction box must be fixed and cables clamped.

Many manufacturers now produce junction boxes that provide cable clamping where the sheath enters the enclosure and terminals similar to those found in a ceiling rose. These features reduce the strain on the termination if the junction box and its cabling is not fixed to a surface, such as when a downlight is installed into a ceiling. It also enables the electrician to comply with regulation IET 526.2 where the method of connection shall take account of the number and shape of wires forming the conductor.

Strip connectors

Strip connectors or a chocolate block is a very common method of connecting conductors. The connectors are mounted in a moulded plastic block in strips of 10 or 12. The conductors are inserted into the block and secured with the grub-screw. In order that the conductors do not become damaged, the screw connection must be firm but not overtightened. The size used should relate to the current rating of the circuit. Figure 4.29 shows a strip connector.

Pillar terminal

A pillar terminal is a brass pillar with a hole through the side into which the conductor is inserted and secured with a set-screw. If the conductor is small in relation to the hole it should be doubled back. In order that the conductor does not become damaged, the screw connection should be tight but not overtightened. Figure 4.29 shows a pillar terminal.

Screwhead, nut and washer terminals

The conductor being terminated is formed into an eye as shown in Figure 4.29. The eye should be slightly larger than the screw shank but smaller than the outside diameter of the screwhead, nut or washer. The eye should be placed on the screw shank in such a way that the rotation of the screwhead or nut will tend to close the joint in the eye.

Claw washers

In order to avoid inappropriate separation or spreading of individual wires of multiwire, claw washers are used to obtain a good sound connection.

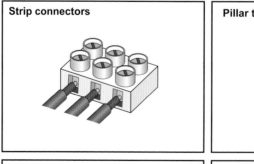
Strip connectors

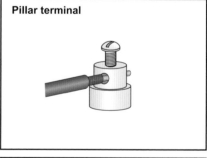
Pillar terminal

Screwhead, nut and washer terminal

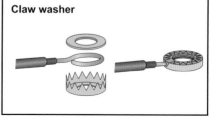
Claw washer

Figure 4.29 Types of terminal.

The looped conductor is laid in the pressing as shown in Figure 4.29, bottom right, a plain washer is placed on top of the conductor and the metal points folded over the washer. When terminating very fine multiwire conductors, see also 526.9 of the IET Regulations.

Crimp terminals

Crimp terminals are made of tinned sheet copper. The chosen crimp terminal is slipped over the end of the conductor and crimped with the special crimping tool. This type of connection is very effective for connecting equipotential bonding conductors to approved earth clamps. Smaller crimp tools are hand operated whereas the larger crimps require hydraulic crimpers to ensure the correct pressure is achieved.

Soldered joints or compression joints

Although the soldering of large underground cables is still common today, joints up to about 100 A are now usually joined with a compression joint. This uses the same principle as for the crimp termination above, it is just a little larger.

If a large SWA cable must be connected and the joint placed in a position which will be inaccessible for future inspection and testing, then a compression joint encased in a resin compound-filled jacket will probably provide a solution. Regulation 526.3 tells us that every connection must be accessible for inspection and testing except for the following:

1 A joint which is designed to be buried underground such as the one described above.
2 A compound-filled or encapsulated joint.
3 A maintenance-free junction box marked with the symbol MF, as shown in Figure 4.30.

The introduction of this maintenance-free junction box was a small but important change made by Amendment No 1: 2011 of the IET Regulations.

There has always been a debate as to when a junction box is accessible for inspection and testing. Is it accessible when installed under floorboards? Is a screwed down floorboard accessible? Does a fitted carpet make a difference, and who will know where it is?

A maintenance-free junction box does not have to be accessible for inspection and testing and this new junction box will provide a solution to those difficult and unavoidable situations where there is doubt as to whether a junction box is inaccessible.

Key fact

Whatever method is used to make the connection in conductors, the connection must be both electrically and mechanically sound if high-resistance joints, corrosion and erosion at the point of termination are to be avoided.

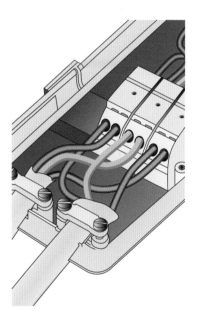

Figure 4.30 A maintenance-free junction box.

Whatever method is used to make the connection in conductors, the connection must be both electrically and mechanically sound if high resistance joints, corrosion and erosion at the point of termination are to be avoided.

Assessment criteria 2.2

Describe the methods for the safe and effective termination and connection of conductors and cables and flexes

Assessment criteria 2.3

Identify the procedures for proving that terminations and connections are electrically and mechanically sound

Assessment criteria 2.4

Define the consequences of terminations not being electrically and mechanically sound

Safe terminations and connections

To ensure that all electrical terminations and connections are safe, the installing electrician should give consideration to the following good practice points:

- all connections must be both electrically and mechanically secure;
- all connections must be long-lasting and not fail quickly;
- the method of connection must take account of:
  - (i) the size of conductor and, therefore, the current carrying capacity of that conductor;
  - (ii) the material of the conductor; copper is a soft metal but aluminium is softer;
  - (iii) the number of conductors being connected;
  - (iv) the temperature to be attained at the point of connection in normal service;
  - (v) the provision of adequate locking arrangements in situations subject to vibration.
- every connection must remain accessible for inspection and testing unless designed to be maintenance-free;
- every connection in a live conductor must be made within:
  - (i) a suitable accessory such as a switch, socket ceiling rose or joint box; or
  - (ii) an equipment enclosure such as a luminaire; or
  - (iii) a non-combustible enclosure designed for this purpose.
- there must be no mechanical strain put on the conductors or connections.

Section 526 of the IET Regulations deals with electrical connections.

Fibre-optic and data cables

Fibre-optic cables used for high speed data transfer can be joined in one of two ways, using connectors for a temporary joint or 'splicing' which creates a permanent joint between them.

Any connection between fibre-optic cables needs to offer minimal losses to ensure the integrity of the data signal being carried and high performance is measured by reducing light loss and back reflection of the light.

A number of factors can cause the loss of light in a connector or splice. The losses are minimized when the two fibre cores are perfectly aligned, the ends of the fibre are properly finished and no dirt is present.

The end finish of the fibre must be properly polished to minimize loss. A rough surface will scatter light and dirt can scatter and absorb light. Since the optical fibre is so small, typical airborne dirt can be a major source of loss. Whenever connectors are not terminated, they should be covered to protect the end of the ferrule from dirt. One should never touch the end of the ferrule, since the oils on one's skin cause the fibre to attract dirt. Before connection and testing, it is advisable to clean connectors with lint-free wipes moistened with cleaning alcohol.

There are many styles of connector for fibre-optic cables but the most popular type, pictured in Figure 4.32, uses the bayonet mount.

When preparing the fibre for termination, the end needs to be flat and square to produce the optimum connection. A tool is used to hold the cable while the end is moved over a fine polishing surface to create this. Once prepared, the end is either crimped or 'glued' into a connector body for the final connection.

Splices are 'permanent' connections between two fibres. There are two types of splices, fusion and mechanical, and the choice is usually based on cost or location, Figure 4.33. Fusion splices are made by 'welding' the two fibres

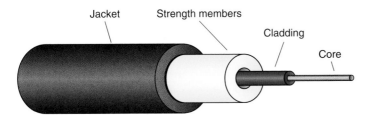

Figure 4.31 Structure of a fibre-optic cable.

Figure 4.32 Bayonet mount connectors.

Fusion splice

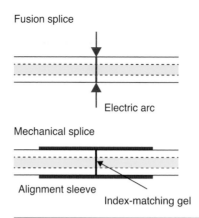

Electric arc

Mechanical splice

Alignment sleeve

Index-matching gel

Figure 4.33 Splices for fibre-optic and data cables.

together, usually by an electric arc. Today's fusion splicers are automated and you would have a hard time making a bad splice. A fusion splicer is expensive, usually in excess of £5,000, however each joint is relatively cheap.

Mechanical splices are alignment devices that hold the ends of two fibres together, with some gel or glue between them. There are a number of types of mechanical splices, like little glass tubes or V-shaped metal clamps. The tools to make mechanical splices are cheap, but the splices themselves are expensive.

Care should always be taken when working with fibre-optic cables, not only to ensure good connections but also to avoid splinters from the core as these are painful and difficult to find and remove.

Twisted pair data cables

A twisted pair cable used for many data and voice installations is a type of cable which is made by putting two separate insulated copper wires together in a twisted pattern and running them parallel to each other. Twisted pair comes with each pair uniquely colour coded when it is packaged in multiple pairs.

Due to its low cost, universal twisted pair (UTP) cabling is used extensively for local-area networks (LANs) and telephone connections.

There are over six different types of UTP categories and, depending on what you want to achieve, you would need the appropriate type of cable. UTP-CAT5E is the most popular UTP cable; it came to replace coaxial cable which was not able to keep up with the continuous need for faster and more reliable networks.

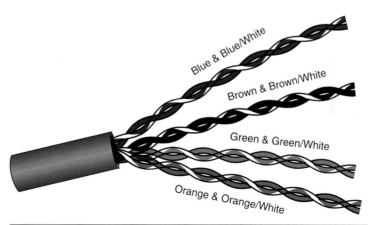

Blue & Blue/White

Brown & Brown/White

Green & Green/White

Orange & Orange/White

Figure 4.34 Typical UTP twisted pair cable.

CAT1 is typically telephone wire. This type of wire is not capable of supporting computer network traffic and is not twisted. It is also used by phone companies who provide ISDN, where the wiring between the customer's site and the phone company's network uses CAT1 cable.

CAT2, CAT3, CAT4, CAT5 and CAT6 are network wire specifications. This type of wire can support computer network and telephone traffic. CAT2 is used mostly for token ring networks, supporting speeds up to 4 Mbps. For higher network speeds (100 Mbps plus) you must use CAT5 wire, but for 10 Mbps CAT3 will suffice. CAT3, CAT4 and CAT5 cable are actually four pairs of twisted copper wires and CAT5 has more twists per inch than CAT3, which means it can run at higher speeds and greater lengths. The 'twist' effect of each pair in the cables will cause any interference presented/picked up on one cable to be cancelled out by the cable's partner, which twists around the initial cable. CAT3 and CAT4 are both used for token ring and have a maximum length of 100 metres.

CAT6 wire was originally designed to support gigabit Ethernet (although there are standards that will allow gigabit transmission over CAT5 wire, this refers to CAT5e). It is similar to CAT5 wire, but contains a physical separator between the four pairs to further reduce electromagnetic interference.

There are two different types of twisted pair cable, unshielded twisted pair (UTP) and shielded twisted pair (STP) and they are used in different kinds of installations.

Shielded twisted pair cable encases the signal-carrying wires in a conducting shield as a means of reducing the potential for electromagnetic interference.

Connecting twisted pair cables

Twisted pair cables are usually connected to RJ45 plugs and sockets. These connections require the use of specialist tools in the form of an insulation displacement push-down tool for the socket plate and a crimper for the plug.

The order of the cables in the plug and socket will be determined by the type of network you are installing and must be followed to allow for data to be successfully transmitted and received by the IT equipment it is connecting.

Steel wire armoured cable (SWA)

Often used for external and underground applications, SWA is designed so that the steel armour surrounding the inner cable offers superior protection against damage when compared with a conventional cable.

Figure 4.35 Phone socket and cable connection.

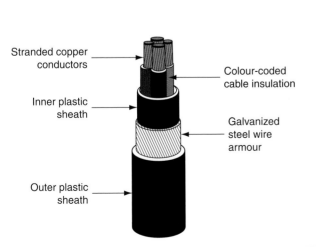

Stranded copper conductors

Colour-coded cable insulation

Inner plastic sheath

Galvanized steel wire armour

Outer plastic sheath

Figure 4.36 Steel wire armoured cable.

Figure 4.37 SWA cable without terminating gland.

An SWA cable is constructed with the insulated conductors encapsulated in a bedding material in a similar construction to a 'flex'. This bedding is then wrapped in a steel wire armour and finally the sheath covers the steel to stop corrosion.

Termination of an SWA cable requires the use of a cable gland, designed to fit the cable. The gland comprises of a main brass body onto which a lock nut is fitted, trapping the steel armour, while the inner cable and conductors pass through to be terminated in the accessory. Once complete, a shroud covers the entire gland providing further protection against damage and the elements.

Smaller cable gland bodies are manufactured with the standard screw tread found on conduit fittings, allowing for easy connection to conduit boxes and enclosures with lock nuts or lock rings.

As these cables are larger than PVC/PVC cables, normal cable clips are not substantial enough to hold the cable in place and cleats are used as they hold the cable in a more substantial manner.

SY cable

SY control cable is a flexible multi-core cable with copper conductors, PVC insulation and a transparent PVC sheath. It also has the added benefit of a galvanized steel wire braid, so it can easily cope with situations of medium to high mechanical stress.

SY cable is used as interconnecting cable for measuring, controlling, signalling and control equipment. This braided control cable is found frequently on assembly and production lines, conveyors, in computer units and machine tool manufacture. The cable's flexible construction makes it a good choice for linking fixed and mobile equipment.

Figure 4.38 Gland used for terminating SY cable.

Like SWA cable, SY cable requires a special gland for terminating it into an accessory. The body encloses a soft rubber 'o' ring which, when compressed by the rear lock nut, closes and holds the cable in the same way as a flexible cable stuffing gland. The braiding is split into two twisted strands and fed through the slots cut into the thread on the body and held in place when the washers and lock nut are fitted.

Mineral-insulated copper-clad cable (MICC)

Mineral insulated copper clad cable (MI or MICC) is constructed with copper conductors inside a seamless copper sheath, insulated by magnesium oxide (a white powder). The cable may or may not be covered by a PVC outer sheath that provides protection against corrosion and can be used for identifying its use, for example red for fire alarms. The copper outer sheath provides the CPC, and the cable is terminated with a pot and sealed with compound and a compression gland.

As the magnesium oxide has a high melting point, 2,852 °C , the cable is capable of carrying a higher current than its thermoplastic PVC/PVC equivalent where the insulation would be damaged by the heat generated.

The construction of the cable means that it is more resistant to fires than PVC/PVC cables and mechanically stronger, being far more resistant to impact. Due to these properties, MI cable is often used for fire alarm installations, oil refineries and chemical works, boiler houses and furnaces, petrol pumps, areas with hazardous environments and older buildings such as churches.

While the cable itself is water resistant, the magnesium oxide is hygroscopic, which means that it readily absorbs moisture from the surrounding air, unless adequately terminated. The termination of an MI cable is a complicated process

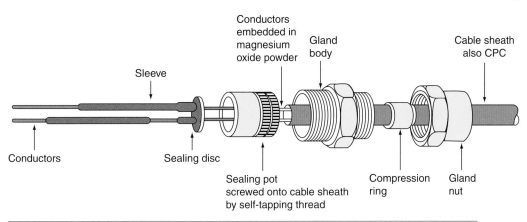

Sleeve

Conductors embedded in magnesium oxide powder

Gland body

Cable sheath also CPC

Conductors

Sealing disc

Sealing pot screwed onto cable sheath by self-tapping thread

Compression ring

Gland nut

Figure 4.39 Mineral insulated cable.

requiring the electrician to demonstrate a high level of practical skill and expertise for the termination to be successful.

FP 200 cable

Figure 4.40 FP 200 cable.

FP 200 cable is similar in appearance to an MI cable in that it is a circular tube, or the shape of a pencil, and is available with a red or white sheath. However, it is much simpler to use and terminate than an MI cable.

The cable is available with either solid or stranded conductors that are insulated with Insudite, a fire-resistant insulation material. The conductors are then screened, by wrapping an aluminium tape around the insulated conductors, that is, between the insulated conductors and the outer sheath. This aluminium tape screen is applied metal side down and in contact with the bare CPC.

The sheath is circular and made of a robust thermoplastic low-smoke, zero halogen material.

FP 200 is available in 2, 3, 4, 7, 12 and 19 cores with a conductor size range from 1.0 to 4.0 mm. The cable is as easy to use as a PVC insulated and sheathed cable. No special terminations are required; the cable may be terminated through a grommet into a knock-out box or terminated through a simple compression gland. The cable is a fire-resistant cable, primarily intended for use in fire alarms and emergency lighting installations, or it may be embedded in plaster.

Assessment criteria 3.1

Outline the requirements for the installation of wiring systems

PVC insulated and sheathed cable installations

PVC insulated and sheathed wiring systems are used extensively for lighting and socket installations in domestic dwellings. Mechanical damage to the cable caused by impact, abrasion, penetration, compression or tension must be minimized during installation (IET Regulation 522.6.1). The cables are generally fixed using plastic clips incorporating a masonry nail, which means the cables can be fixed to wood, plaster or brick with almost equal ease. Cables should be run horizontally or vertically, not diagonally, down a wall. All kinks should be removed so that the cable is run straight and neatly between clips fixed at equal distances providing adequate support for the cable so that it does not become damaged by its own weight (IET Regulation 522.8.4 and Table D1 of the *On-Site Guide*). Where cables are bent, the radius of the bend should not cause the conductors to be damaged (IET Regulation 522.8.3 and Table D5 of the *On-Site Guide*. A table of cable supports is shown in Table 4.5).

Terminations or joints in the cable may be made in ceiling roses, junction boxes, or behind sockets or switches, provided that they are enclosed in a non-ignitable material, are properly insulated and are mechanically and electrically secure (IET Regulation 526). All joints must be accessible for inspection, testing and maintenance when the installation is completed (IET Regulation 526.3).

The 18th Edition of the Regulations at 521.10.202 now requires fire proof support for all wiring systems to prevent cables collapsing in the event of a fire. Suitably spaced steel or copper clips, saddles or ties are examples which will meet this requirement.

Where PVC insulated and sheathed cables are concealed in walls, floors or partitions, they must be provided with a box incorporating an earth terminal at each outlet position. PVC cables do not react chemically with plaster, as do some cables, and consequently PVC cables may be buried under plaster. Further protection by channel or conduit is only necessary if mechanical protection from nails or screws is required or to protect them from the plasterer's trowel.

However, IET Regulation 522.6.201 now tells us that where PVC cables are to be embedded in a wall or partition at a depth of less than 50 mm they should be run along one of the permitted routes shown in Figure 4.42. Figure 4.41 shows a typical PVC installation. To identify the most probable cable routes, IET Regulation 522.6.201 tells us that outside a zone formed by a 150 mm border all around a wall edge, cables can only be run horizontally or vertically to a point or accessory if they are contained in a substantial earthed enclosure, such as a conduit, which can withstand nail penetration, as shown in Figure 4.42.

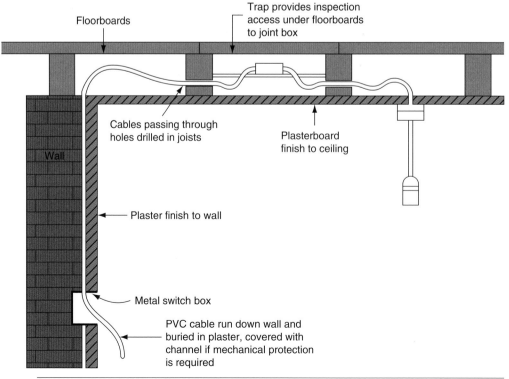

Figure 4.41 A concealed PVC sheathed wiring system.

Table 4.5 Spacing of cable supports

Overall diameter of cable (mm)	Maximum spacings of clips							
	PVC sheathed cables				Armoured cables		Mineral insulated copper sheathed cables	
	Generally		In caravans					
	Horizontal (mm)	Vertical (mm)	Horizontal (mm)	Vertical (mm)	Horizontal (mm)	Vertical (mm)	Horizontal (mm)	Vertical (mm)
1	2	3	4	5	6	7	8	9
Not exceeding 9	250	400	250 (for all sizes)	400 (for all sizes)	–	–	600	800
Exceeding 9 and not exceeding 15	300	400			350	450	900	1200
Exceeding 15 and not exceeding 20	350	450			400	550	1500	2200
Exceeding 20 and not exceeding 40	400	550			450	600	–	–

Where the accessory or cable is fixed to a wall which is less than 100 mm thick, protection must also be extended to the reverse side of the wall if a position can be determined.

Where none of this protection can be complied with and the installation is to be used by ordinary people, the cable must be given additional protection with a 30 mA RCD (IET Regulation 522.6.202).

Where cables pass through walls, floors and ceilings the hole should be made good with incombustible material such as mortar or plaster to prevent the spread of fire (IET Regulations 527.1.2 and 527.2.1). Cables passing through metal boxes should be bushed with a rubber grommet to prevent abrasion of the cable. Holes drilled in floor joists through which cables are run should be 50 mm below the top or 50 mm above the bottom of the joist to prevent damage to the cable by nail penetration (IET Regulation 522.6.100), as shown in Figure 4.44. PVC cables should not be installed when the surrounding temperature is below 0 °C or when the cable temperature has been below 0 °C for the previous 24 hours because the insulation becomes brittle at low temperatures and may be damaged during installation.

Cables in metallic trunking

Regulation 521.5.1 of BS 7671 requires that the line, neutral and CPC conductors of an a.c. circuit in metallic containment, are contained within the same enclosure and if they pass through any holes, that they all pass through the same hole. Due to the alternating current, the magnetic fields created by the conductors will form eddy currents in the metalwork. By passing all of the conductors through

Figure 4.43 Plastering helps prevent the spread of fire.

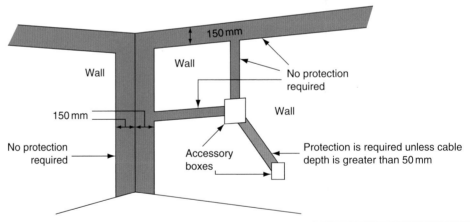

Figure 4.42 Permitted cable routes.

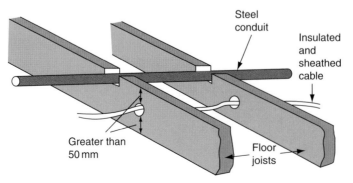

Notes:
1. Maximum diameter of hole should be 0.25 × joist depth.
2. Holes on centre line in a zone between 0.25 and 0.4 × span.
3. Maximum depth of notch should be 0.125 × joist depth.
4. Notches on top in a zone between 0.1 and 0.25 × span.
5. Holes in the same joist should be at least 3 diameters apart.

Figure 4.44 Correct installation of cables in floor joists.

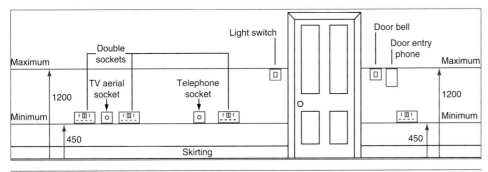

Figure 4.45 Fixing positions of switches and socket outlets.

the same entry, the fields around the line and neutral will cancel each other out and stop the eddy currents being formed. If the cables do enter through separate holes, a slot must be cut between the holes for the same reason.

Heights of switches, socket outlets and controls

Approved document M of the building regulations in England and Wales requires that switches and socket outlets in new dwellings be installed so that they can be easily used by persons with limited reach. In reality, this means that all accessories should be fitted at a minimum height of 450 mm and a maximum of 1200 mm from the finished floor level.

Appendix C of the *On-Site Guide* gives guidance on the protection of cables from materials they may otherwise come into contact with in an installation. Expanded polystyrene, used for insulation, will leach plasticizer from the cables leaving the insulation degraded and brittle. Wood preservative and creosote also contains solvents that will damage PVC cables and they should be covered to protect them until the preservative has dried.

Assessment criteria 3.2

Outline the requirements for circuit segregation

Segregation of circuits

Where an installation comprises a mixture of low-voltage and very low-voltage circuits such as mains lighting and power, fire alarm and telecommunications circuits, they must be separated or segregated to prevent electrical contact (IET Regulation 528.1).

For the purpose of these regulations, various circuits are identified by one of two bands as follows:

- Band I: telephone, radio, bell, call and intruder alarm circuits, emergency circuits for fire alarm and emergency lighting.
- Band II: mains voltage circuits.

When Band I circuits are insulated to the same voltage as Band II circuits, they may be drawn into the same compartment.

When trunking contains rigidly fixed metal barriers along its length, the same trunking may be used to enclose cables of the separate bands without further precautions, provided that each band is separated by a barrier, as shown in Figure 4.46.

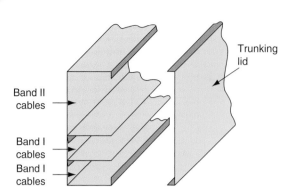

Figure 4.46 Segregation of cables in trunking.

Multi-compartment PVC trunking cannot provide band segregation since there is no metal screen between the bands. This can only be provided in PVC trunking if screened cables are drawn into the trunking.

Assessment criteria 4.1

Select appropriate test instruments for testing de-energized circuits

Approved test instruments

The test instruments and test leads used by the electrician for testing an electrical installation must meet all the requirements of the relevant regulations.

The HSE has published guidance note GS 38 for test equipment used by electricians. The IET Regulations (BS 7671) also specify the test voltage or current required to carry out particular tests satisfactorily. All test equipment must be chosen to comply with the relevant parts of BS EN 61557. All testing must, therefore, be carried out using an 'approved' test instrument if the test results are to be valid. The test instrument must also carry a calibration certificate, otherwise the recorded results may be void. Calibration certificates usually last for a year. Test instruments must, therefore, be tested and recalibrated each year by an approved supplier. This will maintain the accuracy of the instrument to an acceptable level, usually within 2% of the true value.

Modern digital test instruments are reasonably robust, but to maintain them in good working order they must be treated with care. An approved test instrument costs as much as a good-quality camera; it should, therefore, receive the same care and consideration.

Let us now look at the requirements of test meters for testing de-energized circuits.

Continuity tester

To measure accurately the resistance of the conductors in an electrical installation, we must use an instrument which is capable of producing an open circuit voltage of between 4 and 24 V a.c. or d.c., and delivering a short-circuit current of not less than 200 mA (IET Regulation 643.2). The functions of continuity testing and insulation resistance testing are usually combined in one test instrument.

Definition

The *test instruments* and *test leads* used by the electrician for testing an electrical installation must meet all the requirements of the relevant regulations.

Definition

Calibration certificates usually last for a year. Test instruments must, therefore, be tested and recalibrated each year by an approved supplier.

Insulation resistance tester

The test instrument must be capable of detecting insulation leakage between live conductors and between live conductors and earth. To do this and to comply with IET Regulation 612.3, the test instrument must be capable of producing a test voltage of 250, 500 or 1000 V and delivering an output current of not less than 1 mA at its normal voltage.

Let us now consider the tests for de-energized circuits.

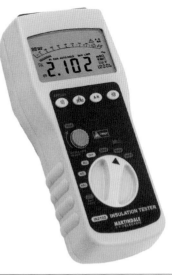

Figure 4.47 Multi-voltage insulation and continuity tester.

Assessment criteria 4.2

Recognize the basic procedures for testing de-energized circuits

Test 1. Testing for continuity of protective conductors, including main and supplementary equipotential bonding (Regulation 643.2)

The object of the test is to ensure that the circuit protective conductor (CPC) is correctly connected, is electrically sound and has a total resistance which is low enough to permit the overcurrent protective device to operate within the disconnection time requirements of IET Regulation 411.4.6, should an earth fault occur. Every protective conductor must be separately tested from the consumer's main protective earthing terminal to verify that it is electrically sound and correctly connected, including the protective equipotential and supplementary bonding conductors, as shown in Figure 4.48. The IET Regulations describe the

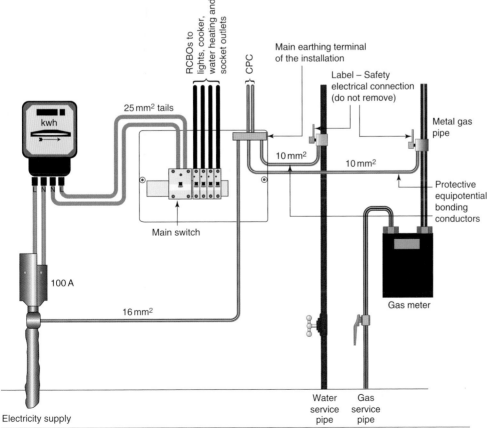

Figure 4.48 Cable sheath earth supplies (TN-S system); showing earthing and main protective equipotential bonding arrangements.

need to consider additional protection by supplementary equipotential bonding in situations where there is a high risk of electric shock, such as kitchens and bathrooms (IET Regulation 415.2).

A d.c. test using an ohmmeter continuity tester is suitable where the protective conductors are of copper or aluminium up to 35 mm

The test is made with the supply disconnected, measuring from the consumer's main protective earthing terminal to the far end of each CPC, as shown in Figure 4.49. The resistance of the long test lead is subtracted from these readings to give the resistance value of the CPC. The result is recorded on an installation schedule such as that given in Appendix 6 of the IET Regulations.

A satisfactory test result for the bonding conductors will be in the order of 0.05 Ω or less (IET Guidance Note 3).

Where steel conduit or trunking forms the protective conductor, the standard test described above may be used, but additionally the enclosure must be visually checked along its length to verify the integrity of all the joints.

If the inspecting engineer has grounds to question the soundness and quality of these joints, then the phase earth loop impedance test should be carried out.

If, after carrying out this further test, the inspecting engineer still questions the quality and soundness of the protective conductor formed by the metallic conduit or trunking, then a further test can be done using an a.c. voltage not greater than 50 V at the frequency of the installation and a current approaching 1.5 times the design current of the circuit, but not greater than 25 A.

This test can be done using a low-voltage transformer and suitably connected ammeters and voltmeters, but a number of commercial instruments are available, such as the Clare tester, which give a direct reading in ohms.

Because fault currents will flow around the earth fault-loop path, the measured resistance values must be low enough to allow the overcurrent protective device to operate quickly. For a satisfactory test result, the resistance of the protective conductor should be consistent with those values calculated for a line conductor of similar length and cross-sectional area. Values of resistance per metre for

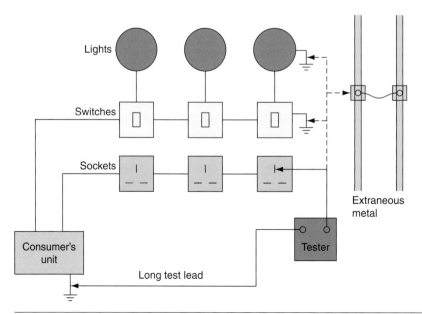

Figure 4.49 Testing continuity of protective conductors.

Example

The CPC for a ring final circuit is formed by a 1.5 mm² copper conductor of 50 m approximate length. Determine a satisfactory continuity test value for the CPC using the value given in Table I1 of the *On-Site Guide*.

Table I1 gives resistance/metre for a 1.5 mm² copper conductor
= 12.10 mΩ/m

Therefore, the resistance of 50 m = $50 \times 12.10 \times 10^{-3}$
= 0.605 Ω

The protective conductor resistance values calculated by this method can only be an approximation since the length of the CPC can only be estimated. Therefore, in this case, a satisfactory test result would be obtained if the resistance of the protective conductor was about 0.6 Ω. A more precise result is indicated by the earth fault-loop impedance test which is carried out later in the sequence of tests.

Table 4.6 Resistance values of some metallic containers

Metallic sheath	Size (mm)	Resistance at 20°C (mΩ/m)
Conduit	20	1.25
	25	1.14
	32	0.85
Trunking	50 × 50	0.949
	75 × 75	0.526
	100 × 100	0.337

copper and aluminium conductors are given in Table I1 of the *On-Site Guide*. The resistances of some other metallic containers are given in Table 4.6.

Test 2. Testing for continuity of ring final circuit conductors (Regulation 643.2)

The object of the test is to ensure that all ring circuit cables are continuous around the ring; that is, that there are no breaks and no interconnections in the ring, and that all connections are electrically and mechanically sound. This test also verifies the polarity of each socket outlet.

The test is made with the supply disconnected, using an ohmmeter as follows:

Disconnect and separate the conductors of both legs of the ring at the main fuse. There are three steps to this test:

Step 1

Measure the resistance of the line conductors (L_1 and L_2), the neutral conductors (N_1 and N_2) and the protective conductors (E_1 and E_2) at the mains position, as shown in Figure 4.50. End-to-end live and neutral conductor readings should be approximately the same (i.e. within 0.05 Ω) if the ring is continuous. The

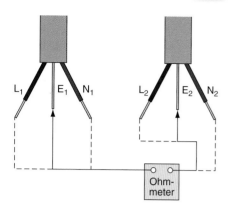

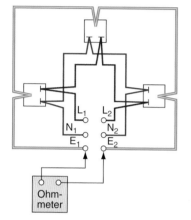

Figure 4.50 Test 2, Step 1 test: measuring the of resistance line, neutral and protective conductors.

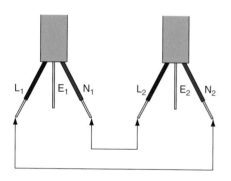

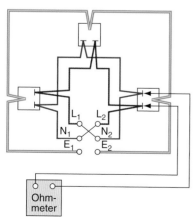

Figure 4.51 Test 2, Step 2 test: connection of mains conductors and test circuit conditions.

protective conductor reading will be 1.67 times as great as these readings if 2.5/1.5 mm cable is used. Record the results on a table such as that shown in Table 4.7.

Step 2

The live and neutral conductors should now be temporarily joined together, as shown in Figure 4.51. An ohmmeter reading should then be taken between live and neutral at every socket outlet on the ring circuit. The readings obtained should be substantially the same, provided that there are no breaks or multiple loops in the ring. Each reading should have a value of approximately half the live and neutral ohmmeter readings measured in Step 1 of this test. Sockets connected as a spur will have a slightly higher value of resistance because they are fed by only one cable, while each socket on the ring is fed by two cables. Record the results on a table such as that shown in Table 4.7.

Step 3

Where the CPC is wired as a ring, for example, where twin and earth cables or plastic conduit is used to wire the ring, temporarily join the live and CPCs together, as shown in Figure 4.52. An ohmmeter reading should then be taken between live and earth at every socket outlet on the ring. The readings obtained should be substantially the same provided that there are no breaks or multiple loops in the ring. This value is equal to R1 + R2 for the circuit. Record the results on an installation schedule such as that given in Appendix 6 of the IET Regulations or a table such as that shown in Table 4.7. The Step 3 value of R1 + R2 should be equal to $(r_1 + r_2)/4$, where r_1 and r_2 are the ohmmeter readings from Step 1 of this test (see Table 4.7).

Test 3. Testing insulation resistance (Regulation 643.3)

The object of the test is to verify that the quality of the insulation is satisfactory and has not deteriorated or short-circuited. The test should be made at the consumer's unit with the mains switch off, all fuses in place and all switches closed. Neon lamps, capacitors and electronic circuits should be disconnected, since they will respectively glow, charge up and be damaged by the test.

There are two tests to be carried out using an insulation resistance tester which must have a test voltage of 500 V d.c. for 230 V and 400 V installations. These

Table 4.7 Table which may be used to record the readings taken when carrying out the continuity of ring final circuit conductors tests according to IET Regulation 643.2

Test	Ohmmeter connected to	Ohmmeter readings	This gives a value for
Step 1	L_1 and L_2		r_1
	N_1 and N_2		
	E_1 and E_2		r_2
Step 2	Live and neutral at each socket		
Step 3	Live and earth at each socket		$R_1 + R_2$
As a check $(R_1 + R_2)$ value should equal $(r_1 + r_2)/4$.			

are line and neutral conductors to earth and between line conductors. The procedures are:

Line and neutral conductors to earth:

1 Remove all lamps.
2 Close all switches and circuit-breakers.
3 Disconnect appliances.
4 Test separately between the line conductor and earth and between the neutral conductor and earth, for every distribution circuit at the consumer's unit, as shown in Figure 4.53a. Record the results on a schedule of test results such as that given in Appendix 6 of the IET Regulations.

Between line conductors:

1 Remove all lamps.
2 Close all switches and circuit-breakers.
3 Disconnect appliances.
4 Test between line and neutral conductors of every distribution circuit at the consumer's unit, as shown in Figure 4.53b, and record the results.

The insulation resistance readings for each test must be not less than 1.0 MΩ for a satisfactory result (IET Regulation 643).

Where the circuit includes electronic equipment which might be damaged by the insulation resistance test, a measurement between all live conductors (i.e. live and neutral conductors connected together) and the earthing arrangements may be made. The insulation resistance of these tests should be not less than 1.0 MΩ (IET Regulation 643).

Although an insulation resistance reading of 1.0 MΩ complies with the regulations, the IET guidance notes tell us that new installations can be expected to give much higher values, and that a reading of less than 2 MΩ might indicate a latent, but not yet visible fault in the installation. In these cases, each circuit should be tested separately to obtain a reading greater than 2 MΩ.

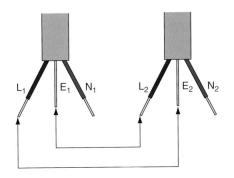

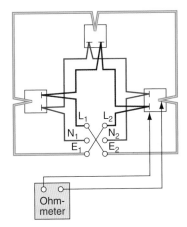

Figure 4.52 Test 2, Step 3 test: connection of mains conductors and test circuit conditions.

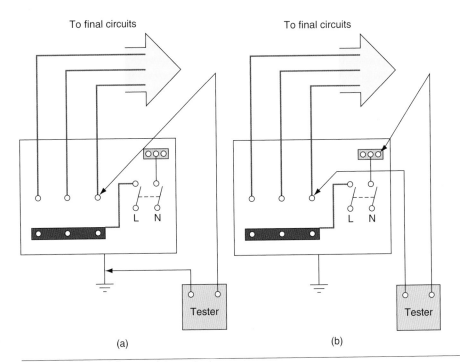

Figure 4.53 Insulations resistance test.

Test 4. Testing polarity (Regulation 643.6)

The object of this test is to verify that all fuses, circuit-breakers and switches are connected in the line or live conductor only, that all socket outlets are correctly wired and that Edison screw-type lamp holders have the centre contact connected to the live conductor. It is important to make a polarity test on the installation since a visual inspection will only indicate conductor identification.

The test is done with the supply disconnected using an ohmmeter or continuity tester as follows:

1 Switch off the supply at the main switch.
2 Remove all lamps and appliances.
3 Fix a temporary link between the line and earth connections on the consumer's side of the main switch.
4 Test between the 'common' terminal and earth at each switch position.
5 Test between the centre pin of any Edison screw lamp holders and any convenient earth connection.
6 Test between the live pin (i.e. the pin to the right of earth) and earth at each socket outlet, as shown in Figure 4.54 and 4.55.

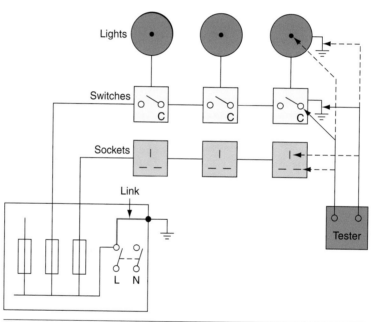

Figure 4.54 Polarity test.

Figure 4.55 Testing at a socket outlet.

Figure 4.56 Secure isolation of a supply.

For a satisfactory test result, the ohmmeter or continuity meter should read very close to zero for each test.

Remove the test link and record the results on a schedule of test results such as that given in Appendix 6 of the IET Regulations.

Test 5. Functional testing (Regulation 643.10)

The object of this test is to ensure that the accessories and equipment do what they are supposed to do within the installation. This is a hands-on test of all switching and control devices forming a part of the installation. For example:

* Do the switch toggles operate as they are supposed to operate?
* Do the electrical isolators, such as those shown in Figure 4.56, switch on and off as they are supposed to do and are you confident that the device is robustly fixed and secure?

Assessment criteria 5.1

Recognize the main roles of the site team

Assessment criteria 5.2

State the importance of company policies and procedures that affect working relationships

Members of the construction team and their role within the industry

An electrician working for an electrical contracting company works as a part of the broader construction industry. This is a multimillion pound industry carrying out all types of building work, from basic housing to hotels, factories, schools, shops, offices and airports. The construction industry is one of the UK's biggest employers, and carries out contracts to the value of about 10% of UK gross national product.

Although a major employer, the construction industry is also very fragmented. Firms vary widely in size, from the local builder employing two or three people to the big national companies employing thousands. Of the total workforce of the construction industry, 92% are employed in small firms of fewer than 25 people.

The yearly turnover of the construction industry is about £35 billion. Of this total sum, about 60% is spent on new building projects and the remaining 40% on maintenance, renovation or restoration of mostly housing. The electrical industries play an important role in all these various construction projects, supplying essential electrical services to meet the needs of those who will use the completed building.

The building team

The construction of a new building is a complex process which requires a team of professionals working together to produce the desired results. We can call this team of professionals the building team, and their interrelationship can be expressed as in Figure 4.57.

The client is the person or group of people with the actual need for the building, such as a new house, office or factory. The client is responsible for financing all the work and, therefore, in effect, employs the entire building team.

The architect is the client's agent and is considered to be the leader of the building team. The architect must interpret the client's requirements and produce working drawings. During the building process the architect will supervise all aspects of the work until the building is handed over to the client.

The quantity surveyor measures the quantities of labour and materials necessary to complete the building work from drawings supplied by the architect.

A site/project manager will have overall responsibility for the successful planning, execution, monitoring and control of a project, usually working for the

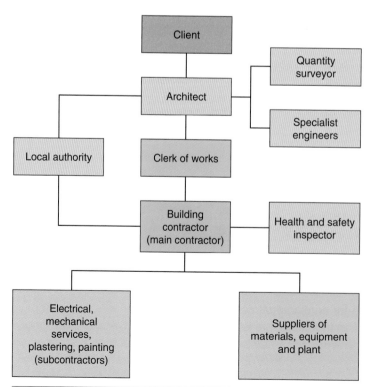

Figure 4.57 The building team.

main construction contractor. They must have a combination of skills including an ability to ask questions of all the team members and resolve conflicts, as well as more general management skills.

Building surveyors provide advice on property and construction, which covers residential, commercial, industrial, leisure and agriculture projects.

They work on the design and development of new buildings as well as the restoration and maintenance of existing ones.

Building surveyors often work to keep buildings in good condition and look for ways to make buildings sustainable. They may be called upon to give evidence in court in cases where building regulations have been breached and as expert witnesses on building defects.

Contracts managers work for the subcontractor and oversee projects, from the start through to completion, ensuring that work is completed on time and within its budget. A contracts manager is sometimes in charge of a single scheme, or may look after several smaller ones.

An estimator in the construction industry is responsible for compiling estimates of how much it will cost to provide a client with products or services. He or she will do this by working out how much a project is likely to cost and quoting accordingly. The job involves assessing material, labour and equipment required and comparing quotes from subcontractors and suppliers.

Construction buyers are responsible for ensuring that the materials required for construction projects are provided to schedule and according to projected budgets. They have a vital part to play in helping ensure the profitability of contracts since they are responsible for ensuring that the most cost-effective and appropriate materials are purchased.

Structural engineers design buildings to withstand stresses and pressures caused by environmental conditions and human use. They ensure buildings and other structures do not deflect, rotate, vibrate excessively or collapse and that they remain stable and secure throughout their use. They work closely with architects and other professional engineers to choose appropriate materials, such as concrete, steel, timber and masonry, to meet design specifications.

Specialist engineers advise the architect during the design stage. They will prepare drawings and calculations on specialist areas of work.

The clerk of works is the architect's 'on-site' representative. He or she will make sure that the contractors carry out the work in accordance with the drawings and other contract documents. They can also agree general matters directly with the building contractor as the architect's representative.

The building control inspector from the local council will ensure that the proposed building conforms to the relevant planning and building legislation.

The health and safety inspectors will ensure that the government's legislation concerning health and safety is fully implemented by the building contractor.

The building contractor will enter into a contract with the client to carry out the construction work in accordance with contract documents. The building contractor is usually the main contractor and may engage subcontractors to carry out specialist services such as electrical installation, mechanical services, plastering and painting.

Subcontractors are individuals or companies employed by the main contractor to perform specific, often specialist tasks as part of the construction process. Specialisms of the subcontractors may include:

- Electrician
- Gas fitter
- Plumber
- H&V Engineer
- RAC Engineer
- Bricklayer
- Carpenter and joiner
- Plasterer
- Tiler
- Decorator
- Groundworker

The electrical team

The electrical contractor is the subcontractor responsible for the installation of electrical equipment within the building.

Electrical installation activities include:

- installing electrical equipment and systems in new sites or locations;
- installing electrical equipment and systems in buildings that are being refurbished because of change of use;
- installing electrical equipment and systems in buildings that are being extended or updated;
- replacement, repairs and maintenance of existing electrical equipment and systems.

An electrical contracting firm is made up of a group of individuals with varying duties and responsibilities. There is often no clear distinction between the duties of the individuals, and the responsibilities carried by an employee will vary from one employer to another. If the firm is to be successful, the individuals must work together to meet the requirements of their customers. Good customer relationships are important for the success of the firm and the continuing employment of the employee.

Definition

The *designer* of any electrical installation is the person who interprets the electrical requirements of the customer within the regulations.

The customer or their representatives will probably see more of the electrician and the electrical trainee than the managing director of the firm and, therefore, the image presented by them is very important. They should always be polite and seen to be capable and in command of the situation. This gives a customer confidence in the firm's ability to meet his or her needs. The electrician and his trainee should be appropriately dressed for the job in hand, which probably means an overall of some kind. Footwear is also important, but is sometimes a difficult consideration for a journeyman electrician. For example, if working in a factory, the safety regulations may insist that protective footwear be worn, but rubber boots with toe protection may be most appropriate for a building site. However, neither of these would be the most suitable footwear for an electrician fixing a new light fitting in the home of the managing director!

The electrical installation in a building is often carried out alongside other trades.

It makes sound sense to help other trades where possible and to develop good working relationships with other employees.

The employer has the responsibility of finding sufficient work for his employees, paying government taxes and meeting the requirements of the Health and Safety at Work Act, described in Chapter 1. The rates of pay and conditions for electricians and trainees are determined by negotiation between the Joint Industry Board and the Trade Unions, which will also represent their members in any disputes. Electricians are usually paid at a rate agreed for their grade; movements through the grades are determined by a combination of academic achievement and practical experience.

The electrical team will consist of a group of professionals and their interrelationship can be expressed as shown in Figure 4.58.

Limits of personal authority

Part 2 of the IET Regulations describes people as:

- **Ordinary persons:** that is, a person who is neither a skilled nor instructed person.
- **Instructed person (electrically):** that is, a person who is adequately advised or supervised by a skilled person to enable that person to perceive risks and to avoid the dangers that electricity may create.
- **Skilled person (electrically):** person who possesses, as appropriate to the nature of the electrical work to be undertaken, adequate education, training and practical skills, and who is able to perceive the risks and avoid the hazards which electricity can create.

When you begin your apprenticeship you will be an ordinary person, one who has no knowledge of the dangers of electricity. In just a few weeks or months you will become an instructed person because you will be doing electrical work under the supervision of your supervisor. While working at college, you will start studying at level 2 and work under the supervision of a more experienced person. As you progress, you will move onto a level 3 course and be expected to complete more complicated tasks and assume more responsibility for your work. When you have completed your apprenticeship you will be a skilled person, able to carry out electrical work safely.

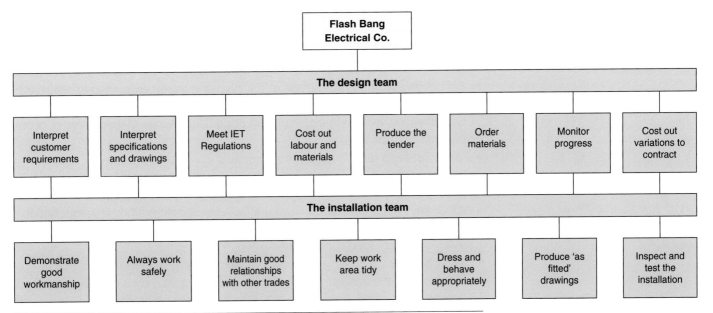

Figure 4.58 The electrical team.

When you start out, you will have little authority and knowledge and so it is reasonable that you are given little responsibility. As you progress through your apprenticeship you will be given more responsibility and authority as your knowledge of working safely and competently increases.

The Health and Safety Regulations make reference to workers having a *duty of care* for the health and safety of themselves and others in the workplace. The Electricity at Work Regulations identify one responsible person on-site as the '*duty holder*'. This recognizes the responsibility of an electrician to take on the control of electrical safety for the whole construction site.

Everyone has a duty of care but not everyone is a duty holder. The duty holder will be the electrician who takes on the responsibility for the electrical health and safety of everyone working on site. The chosen person will certainly be a skilled person, probably someone with great experience and common sense, and whose other responsibilities might be to act as a supervisor, foreman or manager.

Assessment criteria 5.3

Identify the most appropriate communication methods for use in work situations

Communications

When we talk about good communications we are talking about transferring information from one person to another both quickly and accurately. We do this by talking to other people, looking at drawings and plans, and discussing these with colleagues from the same company and with other professionals who have an interest in the same project. The technical information used within our industry comes from many sources. The IET Regulations (BS 7671) are the 'electrician's bible' and form the basis of all our electrical design calculations and installation methods. British Standards, European Harmonised Standards and Codes of Practice provide detailed information for every sector of the electrical industry, influencing all design and build considerations.

Technical information used in the workplace

Technical information is communicated to electrical personnel in lots of different ways. It comes in the form of:

- *Specifications:* these are details of the client's requirements, usually drawn up by an architect. For example, the specification may give information about the type of wiring system to be employed or detail the types of luminaires or other equipment to be used.
- *Manufacturer's data:* if certain equipment is specified, let us say a particular type of luminaire or other piece of equipment, then the manufacturer's data sheet will give specific instructions for its assembly and fixing requirements. It is always good practice to read the data sheet before fitting the equipment. A copy of the data sheet should also be placed in the job file for the client to receive when the job is completed.
- *Reports and schedules:* a report is the written detail of something that has happened or the answer to a particular question asked by another professional person or the client. It might be the details of some problem on-site.

Activity \ Time	1	2	3	4	5	6	7	8	9	10	11	12	13	14
A	■	■												
B	■	■	■	■	■	■	■	■						
C			■	■	■	■								
D					■	■	■	■	■	■	■	■	■	
E	■	■	■	■										
F	■	■	■	■										
G		■	■	■	■	■								
H					■	■	■	■	■					
I		■	■	■	■	■	■							
J	■	■												
K					■	■	■	■						
L									■	■	■			
M												■	■	■

Figure 4.59 A simple bar chart or schedule of work.

If the report is internal to the organization, a handwritten report is acceptable, but if the final report is to go outside the organization, then it must be more formal and typed.

A **schedule** gives information about a programme or timetable of work. It might be a list or a chart giving details of when certain events will take place, for example, when the electricians will start to do the 'first fix' and how many days it will take. A simple **bar chart** is an easy-to-understand schedule of work that shows how different activities interact on a project. Figure 4.59 shows a bar chart or schedule of work where activity A takes two days to complete and activity B starts at the same time as activity A, but carries on for eight days, etc.

User instructions give information about the operation of a piece of equipment. Manufacturers of equipment provide 'user instructions' and a copy should be placed in the job file for the client to receive when the project is handed over.

Those who need or use technical information

Technical information is required by many of the professionals involved in any electrical activity, so who are the key people?

- **The operative:** in our case this will be the skilled electricians actually on-site, doing the job for the electrical company.
- **The supervisor:** he or she may have overall responsibility for a number of electricians on-site and will need the 'big picture'.
- **The contractor:** the main contractor takes on the responsibility for the whole project for the client. The main contractor may take on a subcontractor to carry out some part of the whole project. On a large construction site the electrical contractor is usually the subcontractor.
- **The site agent:** he or she will be responsible for the smooth running of the whole project and for bringing the contract to a conclusion on schedule and within budget. The site agent may be nominated by the architect.
- **The customer or client:** they are also the people ordering the work to be done. They will pay the final bill that pays everyone's wages.

Definition

A *schedule* is a list or programme of planned work.

Definition

Bar chart – the object of any bar chart is to establish the sequence and timing of the various activities involved in the whole job.

Methods of communication

Written instructions or communication are preferable when giving or receiving information. Details can be clearly set out and it is auditable at a later date should any disagreement arise. Nearly all sites will have a telephone available with internet access which means that instant communication via e-mail is possible. A few sites may still use facsimile (fax) machines where the letter is fed into a machine at the sender's end and a duplicate is produced, like a printer, at the receiver's end. If the information is not as time-critical, the information could be posted.

Oral discussions will take place on sites between clients, architects, contractors, subcontractors and colleagues. This may be when an unforeseen problem has arisen and alterations are required, or if the client has a change of mind. It is preferable (some may say essential) to ask for any major changes to be backed-up after the discussion in writing as there is no evidence of the oral discussion. If there are cost implications to any changes, or there is an alteration to the original agreement, it may be necessary at a later date to prove that the alteration was in fact at the request of someone with the authority to ask for the change.

As nearly all people on site now carry mobile telephones, text messages are often used to communicate small messages. Some of the smartphones have larger screens and detailed information can be sent, however most mobile telephones have smaller screens and text messages are used for quick and simple communications.

Figure 4.60 One of the benefits of smartphones is that photographs can be sent on to stakeholders if there are any queries on-site.

Assessment criteria 5.4

Identify the actions to take when conflicts arise

Conflicts between people

Wherever there are people, there will always be conflict and disagreements. Managers and supervisors have to deal with conflict in the workplace every day. Everyone believes they are doing their best work and at times this can lead to disputes with other team members or trades if the interests, needs, goals or values of involved parties interfere with each another.

Often, a conflict is a result of perception, as everyone is trying to achieve the same goal at the end of the project.

Is conflict a bad thing? Not necessarily. While no one wants to have an argument or disagreement, a conflict often presents opportunities for improvement by the requirement to change something and benefit both parties. Ideally, any conflict can be resolved by the parties involved, but on larger sites it may be necessary for the supervisor to referee the disagreement. Therefore, it is important that a manager or supervisor has the skills to resolve problems if the situation should arise.

Managing the situation may take some skilful negotiation by the supervisor; it may be that one of the parties is in the right and the other party needs to accept this. One of the people arguing will be happy at the resolution, however the other person will not, and this could have a negative impact on the relationship between all those involved.

If possible, you should look for a win-win situation where both parties can collaborate and find a mutually beneficial result. Both parties leave the

Figure 4.61 Conflict can present opportunities for improvement, but should be resolved before getting out of hand.

discussion happy with the agreed decision and neither feel they have lost an argument. This method of conflict resolution can help build relationships between workers, as it can help build respect and trust between parties making future collaboration easier.

Sometimes it is not possible for both parties to reach an agreement where they get everything they want. In this situation, compromise is required. Neither party may feel they have won the argument but the agreement may be beneficial to the overall progress of the tasks in hand. The parties involved should see that the manager or supervisor has not favoured either party, but has made a decision to move forward.

Occasionally, it is not possible to accommodate one of the points of view held by one of the parties involved. The manager will need to skilfully negotiate the situation so as not to alienate the individual, while not agreeing with their opinion or point of view.

Assessment criteria 5.5

State the effects that poor communication may have in the workplace

Poor communication

The effects of poor communication to a business or on a project can be major. If good communication does not exist, then tasks or projects may not be completed correctly or may be missed completely. If the communication is poor, then the message received will be hard to understand and can easily lead to confusion, loss of employee cooperation and reduced productivity and profitability. Employees end up working with no clear direction.

In any company, there has to be communication. If the management does not keep in touch with the staff effectively it can lead to demotivation in the business and the business and its employees do not work to their full potential. With the right communication, the company and employees will be able to complete tasks and confusion between workers will be reduced. Good communication helps ensure information is accurate and on time for customers and employees and stops rumours that may demotivate staff.

Within a company, the consequences of poor communication can include:

- Time may be lost as instructions may be misunderstood and jobs may have to be repeated;
- Frustration may develop, as people are not sure of what to do or how to do a task;
- Product may be wasted if a job has to be redone;
- People may feel left out if communication is not open and effective;
- Messages may be misinterpreted or misunderstood, causing bad feelings;
- Increased staff turnover as frustrated employees leave to work elsewhere;
- Poor customer service;
- People's safety may be at risk.

If customers receive poor customer service, then they are unlikely to return and a company will lose repeat business.

Test your knowledge

When you have completed the questions, check out the answers at the back of the book.

Note: more than one multiple-choice answer may be correct.

Learning outcome 1

1. 'Good housekeeping' at work is about:
   a. cleaning up and putting waste in the skip
   b. working safely
   c. making the tea and collecting everyone's lunch
   d. putting tools and equipment away after use

2. To gauge level which of the following would be the most accurate?
   a. Tape measure
   b. Laser line levels
   c. Bubble level
   d. Plumb bob

3. Which hand tools would you use for terminating conductors in a junction box?
   a. A pair of side cutters or knife
   b. A screwdriver
   c. A wood chisel and saw
   d. A tenon saw

4. The type of tool used for removing internal burrs from metal conduit is called a:
   a. wrench
   b. former
   c. reamer
   d. drill

5. Which hand tools would you use to cut across a floorboard before lifting?
   a. A pair of side cutters or knife
   b. A screwdriver
   c. A wood chisel and saw
   d. A tenon saw.

6. Which hand tools would you use to cut and remove a notch in a floor joist?
   a. A pair of side cutters or knife
   b. A screwdriver
   c. A wood chisel and saw
   d. A tenon saw

7. A tracer device is used to:
   a. record voltage through cables
   b. record current through cables
   c. locate existing cables
   d. locate main supply

8. Which of the following is used to support SWA cables?
   a. clips
   b. circlips
   c. cleats
   d. crampets

9. An appropriate fixing for masonry, would be:
   a. gravity toggles
   b. rawl plugs
   c. rawl bolts
   d. crampets

10. An appropriate fixing for plasterboard would be:
    a. spring toggles
    b. raw plugs
    c. Rawlbolts
    d. crampets

11. Which of the following fixing methods would be suitable for securing on long vertical walls?
    a. Cable clip
    b. Rawlbolt
    c. Screw fixing to plastic plug
    d. Gravity toggle

12. The maximum internal bending radius of 2.5 mm² flat profile cable is:
    a. 5 x radius
    b. 3 x radius
    c. 4 x radius
    d. 3 x diameter

13. Regarding bending new and re-used MI cable, which of the following numbers apply?
    a. 6 and 4
    b. 6 and 3
    c. 4 and 6
    d. 3 and 5

14. Restoring the surface after a wall has been chased out is known as:
    a. making full
    b. making out
    c. making good
    d. making firm

Learning outcome 2

15. Loose connections at terminals increase their resistance and result in:
    a. excess current
    b. more heat
    c. protective device cooling
    d. protective device operating

16. What problems are there with using flexible conduit and earthing?
    a. Because of vibration it is hard to wire the cables
    b. Because of vibration it is hard to terminate the cables
    c. Because of vibration it is hard to hang on to the sheath
    d. Because of vibration you cannot use the sheath as a CPC

17. PAT testing is carried out on:
    a. hand tools
    b. domestic appliances only
    c. work electrical tools
    d. electrical equipment (e.g. 110 V transformers)

18. Which of the following results would indicate a good connection?
    a. $0.01\,\Omega$
    b. $240\,\Omega$
    c. $3\,M\Omega$
    d. $100\,M\Omega$

19. If conductors are not clean when making a joint with a crimping tool, the most likely result will be:
    a. a high resistance joint
    b. an electrical fire
    c. a low resistance joint
    d. a loose joint

20. To ensure continuity of CPC when it is supplied through metal trunking itself, all joints should:
    a. be coated in gel
    b. be soldered together
    c. be fitted with continuity strapping
    d. be painted with anti-oxidizing paint

21. A process of tightening a connection to an optimum level using a calibrated wrench is known as:
    a. torque setting
    b. torque made
    c. torque connection
    d. torque key

Learning outcome 3

22. A method of protecting cables as they enter steel trunking is done through using:
    a. blocks
    b. spacers
    c. tape
    d. grommets

23. Conduit ends can be protected against damaging cables by:
    a. reaming
    b. terminating through female bushes
    c. grommet strip
    d. threading

24. When cables are installed in conduit, electromagnetic interference is reduced by ensuring:
    a. line and neutral cables are separated
    b. line and neutral cables are routed together
    c. line and neutral cables are not separated
    d. line and neutral cables are separated by cutting a slot between them

25. Cables of different voltages can be placed together if they are insulated to the:
    a. lowest voltage present
    b. highest voltage present
    c. highest current present
    d. lowest current present

Learning outcome 4

26. What is the first dead test that is carried out on a new lighting circuit?
    a. Insulation resistance
    b. Polarity
    c. Continuity of CPC
    d. Functional test

27. When referring to electrical tools, the initials PAT refer to:
    a. power assisted testing
    b. power appliance testing
    c. portable appliance tools
    d. portable appliance testing

28. A meter is marked 'IR tester'. What function is this?
    a. Sensor testing
    b. Infernal resistance
    c. Internal resistance
    d. Insulation resistance

29. Which of the following instruments is used for carrying out continuity tests?
    a. RCD tester
    b. High resistance ohmmeter
    c. Low resistance ohmmeter
    d. Earth-fault loop impedance tester

30. Test meters used for a periodic inspection must be:
    a. calibrated daily
    b. calibrated weekly
    c. calibrated annually
    d. calibrated monthly

31. Testing a 230 V lighting circuit the minimum acceptable IR value should be:
    a. $0.5\,M\Omega$
    b. $1.0\,M\Omega$
    c. $0.5\,\Omega$
    d. $1.0\,\Omega$

32. The correct method of checking if a circuit is safe or 'dead' is to:
    a. use an approved voltage indicator
    b. use the back of your hand
    c. use a multimeter
    d. use a Martindale

Learning outcome 5

33. Whose responsibility is it to report health and safety problems on-site?
    a. Only supervisors
    b. Only the canteen manager
    c. None of the above
    d. All workers

34. A standard form completed by every employee to inform the employer of the time spent working on a particular site is called a:
    a. job sheet
    b. time sheet
    c. delivery note
    d. daywork sheet

35. A record which confirms that materials ordered have been delivered to site is called a:
    a. job sheet
    b. time sheet
    c. delivery note
    d. daywork sheet

36. A standard form containing information about work to be done usually distributed by a manager to an electrician is called a:
    a. job sheet
    b. time sheet
    c. delivery note
    d. daywork sheet

37 A standard form which records changes or extra work on a large project is called a:
    a. job sheet
    b. time sheet
    c. delivery note
    d. variation order

38. The person who is the company representative responsible for electrical safety in the work environment is called the:

   a. supervisor

   b. duty of care

   c. duty holder

   d. instructed person

39. What kind of chart is shown here?

   a. Gant chart

   b. Time sheet

   c. Memo

   d. Rough drawing

Time / Activity	Day number													
	1	2	3	4	5	6	7	8	9	10	11	12	13	14
A	▓	▓												
B	▓	▓	▓	▓	▓	▓	▓	▓						
C			▓	▓	▓	▓	▓							
D								▓	▓	▓	▓	▓	▓	▓
E	▓	▓	▓	▓	▓	▓								
F	▓	▓												
G		▓	▓	▓	▓	▓	▓	▓	▓	▓				
H						▓	▓	▓	▓	▓	▓			
I		▓	▓	▓	▓	▓	▓	▓						
J	▓	▓	▓	▓	▓	▓	▓							
K							▓	▓	▓	▓	▓	▓	▓	▓
L										▓	▓	▓	▓	▓
M												▓	▓	▓

40. Lack of cooperation with other trades and contractors on-site can lead to:

   a. extremely poor relationships

   b. fewer site meetings

   c. successful job completion

   d. danger to personnel

41. Which of the following would produce technical drawings?

   a. Estimator

   b. Architect

   c. Clerk of works

   d. Project manager

42. Conflicts should be resolved:

   a. straight away

   b. at the end of the day

   c. at the next meeting

   d. at the end of a working week

43. Which of the following is essential for good working relationships?

   a. Positive attitude

   b. Indifferent attitude

   c. Quiet attitude

   d. Negative attitude

44. In the event of any single trade falling behind, corrective action should be to:
    a. arrange a meeting to discuss the issue
    b. ignore the issue
    c. complain about the issue
    d. call for strike action

45. What method would you use to let the office know that the materials you were expecting have not yet arrived?

46. What method would you use to send a long list of materials required for the job you are on to the wholesalers for later delivery to the site? Use bullet points.

47. What are the advantages and disadvantages of having sources of technical information on:
    a. some form of electronic storage system such as a CD, DVD or USB memory stick, or
    b. hard copy such as a catalogue, drawings or *On-Site Guide*. Would it make a difference if you were at the office or on a construction site?

48. State the advantages and disadvantages of:
    a. telephone messages
    b. written messages

49. How does a bar chart help with the organization of a work programme?

50. Why are good relationships important between yourself and the customer and other trades on-site when carrying out work activities?

51. Briefly state why time sheets, fully and accurately completed, are important to:
    a. an employer
    b. an employee

52. State the reasons why you should always present the right image to a client, customer, or his or her representative.

Unit Elec2/05A

Chapter 4 checklist

Learning outcome	Assessment criteria – the learner can	Page number
1. Know the methods and the procedures for installing electrical systems and equipment.	1.1 State the procedures for selecting and safely using equipment for measuring and marking out.	294
	1.2 State the procedures for selecting and safely using tools and equipment.	294
	1.3 Identify the components of wiring systems.	298
	1.4 Recognize fixing and securing methods.	302
	1.5 Identify an appropriate fixing and securing method.	302
	1.6 Outline the methods and procedures used for the installation of wiring systems.	308
	1.7 Specify methods and techniques for restoring the building fabric.	308
2. Understand termination and connection methods.	2.1 Describe connection methods.	314
	2.2 Describe the methods for the safe and effective termination and connection of conductors and cables and flexes.	318
	2.3 Identify the procedures for proving that terminations and connections are electrically and mechanically sound.	318
	2.4 Define the consequences of terminations not being electrically and mechanically sound.	318
3. Know the requirements for installing electrical systems and equipment.	3.1 Outline the requirements for the installation of wiring systems.	323
	3.2 Outline the requirements for circuit segregation.	327
4. Know the methods and procedures for testing de-energized circuits.	4.1 Select appropriate test instruments for testing de-energized circuits.	328
	4.2 Recognize the basic procedures for testing de-energized circuits.	329
5. Know how to communicate and work with others.	5.1 Recognize the main roles of the site team.	335
	5.2 State the importance of company policies and procedures that affect working relationships.	335
	5.3 Identify the most appropriate communication methods for use in work situations.	340
	5.4 Identify the actions to take when conflicts arise.	342
	5.5 State the effects that poor communication may have in the workplace.	343

Answers to 'Test your knowledge' questions

Chapter 1

1. a	2. a	3. d	4. c	5. a, b
6. c, d	7. b, c	8. c, d	9. a, b, d	10. c, d
11. b	12. a	13. a	14. d	15. c
16. a	17. b	18. c	19. b	20. a, b
21. d	22. a	23. b	24. d	25. d
26. d	27. a	28. c	29. a	30. d
31. d	32. b	33. d	34. b	35. b
36. d	37. b	38. b	39. b	40. a, c, d
41. a, b, c	42. a	43. c	44. a	45. a
46. d	47. c	48. c	49. a, b, c, d	50. a
51. c	52. b			
Questions 53 to 71: answers in text of Chapter 1.				

Chapter 2

1. a	2. c	3. b	4. b	5. a
6. d	7. b	8. a, c	9. c	10. b
11. b	12. a	13. d	14. c	15. a, d
16. a	17. a	18. b	19. a, c	20. b, c, d
21. a, c	22. b, d	23. a, b	24. a	25. d
26. a	27. b	28. a, c, d	29. c	30. a
31. c	32. c	33. d	34. a, c	35. d
36. a	37. c	38. a, c	39. d	40. d
41. d	42. a	43. a	44. d	45. b
46. a	47. a, c	48. b, d	49. c	50. a
51. c	52. c	53. c	54. b	55. a
56. a	57. a, b, c	58. a, b, c	59. a	60. c
61. b	62. c	63. a	64. d	65. a, b
66. a, b, c	67. a	68. b	69. a	70. c
71. b	72. b	73. c	74. c	75. a, d

76. c	77. b	78. d	79. b	80. b, c
81. d	82. a	83. a		

Questions 84 to 103: answers in text of Chapter 2.

Chapter 3

1. a, c	2. c	3. a	4. a	5. b
6. c	7. d	8. d	9. c	10. d
11. d	12. a	13. b	14. d	15. a
16. d	17. d	18. d	19. a	20. a
21. a, c	22. a, d	23. a, d	24. b	25. a
26. b	27. d	28. d	29. a	30. c
31. c	32. a	33. c	34. c	35. a
36. b	37. c	38. d	39. b	40. b
41. b	42. c	43. a	44. d	45. c
46. c	47. a	48. c	49. c	50. d
51. c	52. b	53. b	54. c	55. b
56. c	57. c			

Questions 58 to 83: answers in text of Chapter 3.

Chapter 4

1. a, b, d	2. b	3. a, b	4. c	5. d	6. c
7. c	8. c	9. c	10. a	11. d	12. d
13. b	14. c	15. b	16. d	17. c, d	18. a
19. a, b	20. c	21. a	22. d	23. a, b	24. b, c
25. b	26. c	27. d	28. d	29. c	30. c
31. b	32. a	33. d	34. b	35. c	36. a
37. d	38. c	39. a	40. a, d	41. b	42. a
43. a	44. a				

Questions 45 to 52: answers in text of Chapter 4.

APPENDIX
Abbreviations, symbols and codes

Abbreviations used in electronics for multiples and sub-multiples

T	Tera	10^{12}
G	Giga	10^{9}
M	Mega or Meg	10^{6}
k	Kilo	10^{3}
d	Deci	10^{-1}
c	Centi	10^{-2}
m	Milli	10^{-3}
μ	Micro	10^{-6}
n	Nano	10^{-9}
p	Pico	10^{-12}

Terms and symbols used in electronics

Term	Symbol
Approximately equal to	$\cong$
Proportional to	$\propto$
Infinity	∞
Sum of	Σ
Greater than	$>$
Less than	$<$
Much greater than	$\gg$
Much less than	$\ll$
Base of natural logarithms	e
Common logarithms of x	log x
Temperature	θ
Time constant	T
Efficiency	η
Per unit	p.u.

APPENDIX
Formulas for electrical principles

B

- Ohm's law

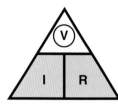 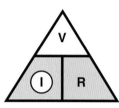 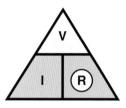

$$\text{V} = I \times R \qquad \text{I} = \frac{V}{R} \qquad \text{R} = \frac{V}{I}$$

- $R = \dfrac{\rho L}{A}$

- $\text{Velocity} = \dfrac{\text{Distance}}{\text{Time}}$

- Force = Mass × Acceleration
- Work done = Force × Distance

- $\text{Mechanical power} = \dfrac{\text{Work done}}{\text{Time taken}}$

- Electrical power

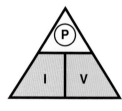 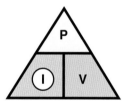 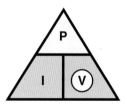

$$\text{P} = I \times V \qquad \text{I} = \frac{P}{V} \qquad \text{V} = \frac{P}{I}$$

- Electrical power = $I^2 \times R$
- Electrical power = $\dfrac{V^2}{R}$
- Effort × Distance to fulcrum = Distance to load × load
- $\text{Efficiency} = \dfrac{\text{Output}}{\text{Input}} \times 100$
- $\text{Tesla (flux density)} = \dfrac{\beta}{\text{Area}}$
- Root mean square, RMS = 0.707 × Peak
- Average value of a sine wave = 0.637 × Peak
- Frequency = 1/Time

- Time = 1/Frequency
- Left hand rule for motors: Force = β (flux density) × I (current) × L (length of conductor)
- Right hand generator rule: e (e.m.f.) = β (flux density) × L (length of conductor) × V (velocity)
- Transformers:

$$\frac{VP}{VS} = \frac{NP}{NS} = \frac{IS}{IP}$$

Electrical principles formulas transposed

1. V = I × R
   - Transpose for $I = \dfrac{V}{R}$
   - Transpose for $R = \dfrac{V}{I}$

2. $R = \dfrac{pL}{A}$
   - Transpose for $p = \dfrac{R \times A}{L}$
   - Transpose for $A = \dfrac{pL}{R}$

3. Velocity $= \dfrac{\text{Distance}}{\text{Time}}$
   - Transpose for Distance = Velocity × Time
   - Transpose for Time $= \dfrac{\text{Distance}}{\text{Velocity}}$

4. Force (newtons) = Mass (kg) × acceleration (9.81)
   - Transpose for Mass $= \dfrac{\text{Force}}{\text{Acceleration}}$
   - Transpose for Acceleration $= \dfrac{\text{Force}}{\text{Mass}}$

5. Work done = Force × Distance
   - Transpose for Force $= \dfrac{\text{Work done}}{\text{Distance}}$
   - Transpose for Distance $= \dfrac{\text{Work done}}{\text{Force}}$

6. Power $= \dfrac{\text{Work done}}{\text{Time taken}}$
   - Transpose for Time taken $= \dfrac{\text{Work done}}{\text{Power}}$
   - Transpose for Work done = Power × Time taken

7. Main formula for Power = Voltage (V) × I (current)
   - Transpose for $I = \dfrac{P}{V}$
   - Transpose for $V = \dfrac{P}{I}$

8. Power= I^2 × R
   - Transpose for $I^2 = \dfrac{\text{Power}}{R}$

- Transpose for R = $\dfrac{\text{Power}}{I^2}$

9. Power = $\dfrac{V^2}{R}$

  - Transpose for $V^2 = P \times R$

  - Transpose for R = $\dfrac{V^2}{P}$

10. Effort × Distance to fulcrum = Distance to load × load

  - Transpose for Effort = $\dfrac{\text{Distance to load} \times \text{load}}{\text{Distance to fulcrum}}$

  - Transpose for Distance to fulcrum = $\dfrac{\text{Distance to load} \times \text{load}}{\text{Effort}}$

  - Transpose for Distance to load = $\dfrac{\text{Effort} \times \text{Distance to fulcrum}}{\text{Load}}$

  - Transpose for Load = $\dfrac{\text{Effort} \times \text{Distance to fulcrum}}{\text{Distance to load}}$

11. Efficiency = $\dfrac{\text{Output} \times 100}{\text{Input}}$

  - Transpose for Output = $\dfrac{\text{Efficiency} \times \text{Input}}{100}$

  - Transpose for Input = $\dfrac{\text{Output} \times 100}{\text{Efficiency}}$

12. Tesla (flux density) = $\dfrac{\beta}{\text{Area}}$

  - Transpose for β = Tesla × Area

  - Transpose for Area = $\dfrac{\beta}{\text{Tesla}}$

13. Left hand rule for motors

    Force = β (flux density) × I (current) × L (length of conductor)

    F = β I L

  - Transpose for I = $\dfrac{F}{\beta L}$

  - Transpose for L = $\dfrac{F}{\beta I}$

  - Transpose for β = $\dfrac{F}{I L}$

14. Root mean square, RMS = 0.707 × Peak

  - Transpose for Peak = $\dfrac{\text{RMS}}{0.707}$

15. Average value of a sine wave = 0.637 × Peak

  - Transpose for Peak = $\dfrac{\text{Average}}{0.707}$

16. Frequency = $\dfrac{1}{T}$ (Periodic time)

  - Transpose for T = $\dfrac{1}{F}$

17. Right hand rule for generators

    e (electromotive force) = β (Flux Density) × L (length of conductor) ×
    V (Velocity)

$e = \beta L V$

- Transpose for $L = \dfrac{e}{\beta V}$

- Transpose for $V = \dfrac{e}{\beta L}$

- Transpose for $\beta = \dfrac{e}{L V}$

18. Transformers

$$\dfrac{VP}{VS} = \dfrac{NP}{NS} = \dfrac{IS}{IP}$$

- Transpose for $VS = \dfrac{VP \times NS}{NP}$

- Transpose for $VP = \dfrac{NP \times VS}{NS}$

- Transpose for $NS = \dfrac{NP \times VS}{VP}$

- Transpose for $NP = \dfrac{VP \times NS}{VS}$

- Transpose for $IS = \dfrac{NP \times IP}{NS}$

- Transpose for $IP = \dfrac{IS \times NS}{NP}$

Glossary of terms

Acceleration Acceleration is the rate of change in velocity with time.

$$\text{Acceleration} = \frac{\text{Velocity}}{\text{Time}} = (\text{m/s}^2)$$

Accident An accident may be defined as an uncontrolled event causing injury or damage to an individual or property.

Alarm call points Manually operated alarm call points should be provided in all parts of a building where people may be present, and should be located so that no one need walk for more than 30 m from any position within the premises in order to give an alarm.

Alloy An alloy is a mixture of two or more metals.

Appointed person An appointed person is someone who is nominated to take charge when someone is injured or becomes ill, including calling an ambulance if required. The appointed person will also look after the first aid equipment, including re-stocking the first aid box.

Approved test instruments The test instruments and test leads used by the electrician for testing an electrical installation must meet all the requirements of the relevant regulations. All testing must, therefore, be carried out using an 'approved' test instrument if the test results are to be valid. The test instrument must also carry a calibration certificate; otherwise the recorded results may be void.

Basic protection Basic protection is provided by the insulation of live parts in accordance with Section 416 of the IET Regulations.

Bonding The linking together of the exposed or extraneous metal parts of an electrical installation.

Bonding conductor A conductor providing protective bonding.

Cable tray Cable tray is a sheet-steel channel with multiple holes. The most common finish is hot-dipped galvanized but PVC-coated tray is also available. It is used extensively on large industrial and commercial installations for supporting MI and SWA cables which are laid on cable tray and secured with cable ties through the tray holes.

Capacitive reactance Capacitive reactance (XC) is the opposition to an a.c. current in a capacitive circuit. It causes the current in the circuit to lead ahead of the voltage.

Centrifugal force Centrifugal force is the force acting away from the centre, the opposite to centripetal force.

Centripetal force Centripetal force is the force acting towards the centre when a mass attached to a string is rotated in a circular path.

Circuit protective conductor (CPC) A protective conductor connecting exposed conductive parts of equipment to the main earthing terminal.

Cohesive or adhesive force Cohesive or adhesive force is the force required to hold things together.

Compact fluorescent lamps (CFLs) CFLs are miniature fluorescent lamps designed to replace ordinary GLS lamps.

Compressive force Compressive force is the force pushing things together.

Conductor A conductor is a material, usually a metal, in which the electrons are loosely bound to the central nucleus. These electrons can easily become 'free electrons' which allows heat and electricity to pass easily through the material.

Conduit	A conduit is a tube, channel or pipe in which insulated conductors are contained.
Corrosion	Corrosion is the destruction of a metal by chemical action.
Delivery notes	A delivery note is used to confirm that goods have been delivered by the supplier, who will then send out an invoice requesting payment.
Duty holder	This phrase recognizes the level of responsibility which electricians are expected to take on as a part of their job in order to control electrical safety in the work environment. Everyone has a duty of care, but not everyone is a duty holder. The person who exercises 'control over the whole systems, equipment and conductors' and is the electrical company's representative on-site is a duty holder.
Earth	The conductive mass of the earth the electrical potential of which is taken as zero.
Earthing	The act of connecting the exposed conductive parts of an installation to the main protective earthing terminal of the installation.
Efficiency of any machine	The ratio of the output power to the input power is known as the efficiency of the machine. The symbol for efficiency is the Greek letter 'eta' (η). In general, $$\eta = \frac{\text{Power output}}{\text{Power input}}$$
Electric current	The drift of electrons within a conductor is known as an electric current, measured in amperes and given the symbol I.
Electric shock	Electric shock occurs when a person becomes part of the electrical circuit.
Electrical force	Electrical force is the force created by an electrical field.
Electrical industry	The electrical industry is made up of a variety of individual companies, all providing a service within their own specialism to a customer, client or user.
Emergency lighting	Emergency lighting is not required in private homes because the occupants are familiar with their surroundings, but in public buildings people are in unfamiliar surroundings. In an emergency people do not always act rationally, but well-illuminated and easily identified exit routes can help to reduce panic.
Emergency switching	Emergency switching involves the rapid disconnection of the electrical supply by a single action to remove or prevent danger.
Escape/standby lighting	Emergency lighting is provided for two reasons: to illuminate escape routes, called 'escape' lighting; and to enable a process or activity to continue after a normal lights failure, called 'standby' lighting.
Expansion bolts	The most well-known expansion bolt is made by Rawlbolt and consists of a split iron shell held together at one end by a steel ferrule and a spring wire clip at the other end. Tightening the bolt draws up an expanding bolt inside the split iron shell, forcing the iron to expand and grip the masonry. Rawlbolts are for heavy-duty masonry fixings.
Exposed conductive parts	The metalwork of an electrical appliance or the trunking and conduit of an electrical system which can be touched because they are not normally live, but which may become live under fault conditions.
Extraneous conductive parts	The structural steelwork of a building and other service pipes such as gas, water, radiators and sinks.
Faraday's law	Faraday's law states that when a conductor cuts or is cut by a magnetic field, an e.m.f. is induced in that conductor.
Fault protection	Fault protection is provided by protective bonding and automatic disconnection of the supply (by a fuse or circuit breaker, CB) in accordance with IET Regulations 411.3 to 6.
Ferrous	A word used to describe all metals in which the main constituent is iron.

Fire	Fire is a chemical reaction which will continue if fuel, oxygen and heat are present.
Fire alarm circuits	Fire alarm circuits are wired as either normally open or normally closed. In a normally open circuit, the alarm call points are connected in parallel with each other so that when any alarm point is initiated the circuit is completed and the sounder gives a warning of fire. In a *normally closed circuit*, the alarm call points are connected in series to normally closed contacts. When the alarm is initiated, or if a break occurs in the wiring, the alarm is activated.
First aid	First aid is the initial assistance or treatment given to a casualty for any injury or sudden illness before the arrival of an ambulance, doctor or other medically qualified person.
First aider	A first aider is someone who has undergone a training course to administer first aid at work and holds a current first aid certificate.
Flashpoint	The lowest temperature at which sufficient vapour is given off from a flammable substance to form an explosive gas–air mixture is called the flashpoint.
Flexible conduit	Flexible conduit manufactured to BS 731-1: 1993 is made of interlinked metal spirals often covered with a PVC sleeving.
Fluorescent lamp	A fluorescent lamp is a linear arc tube, internally coated with a fluorescent powder, containing a low-pressure mercury vapour discharge.
Force	The presence of a force can only be detected by its effect on a body. A force may cause a stationary object to move or bring a moving body to rest.
Friction force	Friction force is the force which resists or prevents the movement of two surfaces in contact.
Functional switching	Functional switching involves the switching on or off, or varying the supply, of electrically operated equipment in normal service.
Fuse	A fuse is the weakest link in the circuit. Under fault conditions it will melt when an overcurrent flows, protecting the circuit conductors from damage.
Gravitational force	Gravitational force is the force acting towards the centre of the earth due to the effect of gravity.
Hazard	A hazard is something with the 'potential' to cause harm, for example, chemicals, electricity or working above ground.
Hazard risk assessment	Employers of more than five people must document the risks at work and the process is known as hazard risk assessment.
Hazardous area	An area in which an explosive gas–air mixture is present is called a hazardous area, and any electrical apparatus or equipment within a hazardous area must be classified as flameproof to protect the safety of workers.
Heating, magnetic or chemical	The three effects of an electric current: when an electric current flows in a circuit it can have one or more of the following three effects: heating, magnetic or chemical.
Impedance	The total opposition to current flow in an a.c. circuit is called impedance and given the symbol Z.
Inductive reactance	Inductive reactance (X_L) is the opposition to an a.c. current in an inductive circuit. It causes the current in the circuit to lag behind the applied voltage.
Inertial force	Inertial force is the force required to get things moving, to change direction or stop.
Inspection and testing techniques	The testing of an installation implies the use of instruments to obtain readings. However, a test is unlikely to identify a cracked socket outlet, a chipped or loose switch plate, a missing conduit-box lid or saddle, so it is also necessary to make a visual inspection of the installation. All existing installations should be periodically

inspected and tested to ensure that they are safe and meet the IET Regulations (IET Regulations 610 to 634).

Instructed person (electrically)	An instructed person (electrically) is a person adequately advised or supervised by electrically skilled persons to be able to perceive risks and avoid the hazards which electricity can create.
Insulator	An insulator is a material, usually a non-metal, in which the electrons are very firmly bound to the nucleus and, therefore, will not allow heat or electricity to pass through it. Good insulating materials are PVC, rubber, glass and wood.
Intrinsically safe circuit	An intrinsically safe circuit is one in which no spark or thermal effect is capable of causing ignition of a given explosive atmosphere.
Intruder alarm systems	An intruder alarm system serves as a deterrent to a potential thief and often reduces home insurance premiums.
Isolation	Isolation is defined as cutting off the electrical supply to a circuit or item of equipment in order to ensure the safety of those working on the equipment by making dead those parts which are live in normal service.
Job sheets	A job sheet or job card carries information about a job which needs to be done, usually a small job.
Lamp	A lamp is a device for converting electrical energy into light energy.
Lever	A lever is any rigid body which pivots or rotates around a fixed axis or fulcrum. Load force × Distance from fulcrum = Effort force × Distance from fulcrum.
Levers and turning force	A lever allows a heavy load to be lifted or moved by a small effort.
Line conductor	A conductor in an AC system for the transmission of electricity other than a neutral or protective conductor.
Luminaire	A luminaire is equipment which supports an electric lamp and distributes or filters the light created by the lamp.
Magnesium oxide	The conductors of mineral insulated metal sheathed (MICC) cables are insulated with compressed magnesium oxide.
Magnetic field	The region of space through which the influence of a magnet can be detected is called the magnetic field of that magnet.
Magnetic force	Magnetic force is the force created by a magnetic field.
Magnetic hysteresis	Magnetic hysteresis loops describe the way in which different materials respond to being magnetized.
Magnetic poles	The places on a magnetic material where the lines of flux are concentrated are called magnetic poles.
Maintained emergency lighting	In a maintained system the emergency lamps are continuously lit using the normal supply when this is available, and change over to an alternative supply when the mains supply fails.
Manual handling	Manual handling is lifting, transporting or supporting loads by hand or by bodily force.
Mass	Mass is a measure of the amount of material in a substance, such as metal, plastic, wood, brick or tissue, which is collectively known as a body. The mass of a body remains constant and can easily be found by comparing it on a set of balance scales with a set of standard masses. The SI unit of mass is the kilogram (kg).
Mechanics	Mechanics is the scientific study of 'machines', where a machine is defined as a device which transmits motion or force from one place to another.
Metallic trunking	Metallic trunking is formed from mild steel sheet, coated with grey or silver enamel paint for internal use or a hot-dipped galvanized coating where damp conditions might be encountered.

Mini-trunking	Mini-trunking is very small PVC trunking, ideal for surface wiring in domestic and commercial installations such as offices.
Mobile equipment	Electrical equipment which is moved while in operation or can be moved while connected to the supply. Previously called portable.
Movement or heat detector	A movement or heat detector placed in a room will detect the presence of anyone entering or leaving that room.
Mutual inductance	A mutual inductance of 1 henry exists between two coils when a uniformly varying current of 1 ampere per second in one coil produces an e.m.f. of 1 volt in the other coil.
Non-ferrous	Metals which do not contain iron are called non-ferrous. They are non-magnetic and resist rusting. Copper, aluminium, tin, lead, zinc and brass are examples of non-ferrous metals.
Non-maintained emergency lighting	In a non-maintained system the emergency lamps are only illuminated if the normal mains supply fails.
Non-statutory regulations and codes of practice	Non-statutory regulations and codes of practice interpret the statutory regulations telling us how we can comply with the law.
Ohm's law	Ohm's law says that the current passing through a conductor under constant temperature conditions is proportional to the potential difference across the conductor.
Optical fibre cables	Optical fibre cables are communication cables made from optical-quality plastic, the same material from which spectacle lenses are manufactured. The energy is transferred down the cable as digital pulses of laser light, as against current flowing down a copper conductor in electrical installation terms.
Ordinary person	An ordinary person is a person who is neither a skilled person nor an instructed person.
Overload current	An overload current can be defined as a current which exceeds the rated value in an otherwise healthy circuit.
Passive infra-red (PIR) detectors	PIR detector units allow a householder to switch on lighting units automatically whenever the area covered is approached by a moving body whose thermal radiation differs from the background.
People	People may be described as an ordinary person, a skilled person, an instructed person or a skilled or instructed person.
Perimeter protection system	A perimeter protection system places alarm sensors on all external doors and windows so that an intruder can be detected as he or she attempts to gain access to the protected property.
Person	A person can be described as ordinary, instructed or skilled depending upon that person's skill or ability.
Personal protective equipment (PPE)	PPE is defined as all equipment designed to be worn, or held, to protect against a risk to health and safety.
Phasor	A phasor is a straight line, having definite length and direction, which represents to scale the magnitude and direction of a quantity such as a current, voltage or impedance.
Plastic plugs	A plastic plug is made of a hollow plastic tube split up to half its length to allow for expansion. Each size of plastic plug is colour-coded to match a wood screw size.
Polyvinylchloride (PVC)	PVC used for cable insulation is a thermoplastic polymer.
Potential difference	The potential difference (p.d.) is the change in energy levels measured across the load terminals. This is also called the volt drop or terminal voltage, since e.m.f. and p.d. are both measured in volts.

Power	Power is the rate of doing work.
	$$\text{Power} = \frac{\text{Work done}}{\text{Time taken}} \text{ (W)}$$
Power factor	Power factor (p.f.) is defined as the cosine of the phase angle between the current and voltage.
Pressure or stress	Pressure or stress is a measure of the force per unit area.
	$$\text{Pressure or stress} = \frac{\text{Force}}{\text{Area}} \text{ (N/m}^2\text{)}$$
Primary cell	A primary cell cannot be recharged. Once the active chemicals are exhausted, the cell must be discarded.
Protective bonding	This is protective bonding for the purpose of safety.
PVC/SWA cable installations	Steel wire armoured PVC insulated cables are now extensively used on industrial installations and often laid on cable tray.
Reasonably practicable or absolute	If the requirement of the regulation is absolute, then that regulation must be met regardless of cost or any other consideration. If the regulation is to be met 'so far as is reasonably practicable', then risks, cost, time, trouble and difficulty can be considered.
Relay	A relay is an electromagnetic switch operated by a solenoid.
Resistance	In any circuit, resistance is defined as opposition to current flow.
Resistivity	The resistivity (symbol ρ – the Greek letter 'rho') of a material is defined as the resistance of a sample of unit length and unit cross-section.
Risk	A risk is the 'likelihood' of harm actually being done.
Risk assessments	Risk assessments need to be suitable and sufficient, not perfect.
Rubber	Rubber is a tough elastic substance made from the sap of tropical plants.
Safety first – isolation	We must ensure the disconnection and separation of electrical equipment from every source of supply and that this disconnection and separation is secure.
Secondary cells	A secondary cell has the advantage of being rechargeable. If the cell is connected to a suitable electrical supply, electrical energy is stored on the plates of the cell as chemical energy.
Secure supplies	A UPS (uninterruptible power supply) is essentially a battery supply electronically modified to provide a clean and secure a.c. supply. The UPS is plugged into the mains supply and the computer systems are plugged into the UPS.
Security lighting	Security lighting is the first line of defence in the fight against crime.
Shearing force	Shearing force is the force which moves one face of a material over another.
Shock protection	Protection from electric shock is provided by basic protection and fault protection.
Short-circuit	A short-circuit is an overcurrent resulting from a fault of negligible impedance connected between conductors.
SI units	SI units are based upon a small number of fundamental units from which all other units may be derived.
Silicon rubber	Introducing organic compounds into synthetic rubber produces a good insulating material such as FP200 cables.
Simple machines	A machine is an assembly of parts, some fixed, others movable, by which motion and force are transmitted. With the aid of a machine we are able to magnify the effort exerted at the input and lift or move large loads at the output.

Single PVC insulated conductors	Single PVC insulated conductors are usually drawn into the installed conduit to complete the installation.
Skilled person (electrically)	A skilled person (electrically) is a person with relevant education's training and practical skills, and sufficient experience to be able to perceive risks and to avoid the hazards which electricity can create.
Skirting trunking	Skirting trunking is a trunking manufactured from PVC or steel in the shape of a skirting board which is frequently used in commercial buildings such as hospitals, laboratories and offices.
Socket outlets	Socket outlets provide an easy and convenient method of connecting portable electrical appliances to a source of supply.
Sounders	The positions and numbers of sounders should be such that the alarm can be distinctly heard above the background noise in every part of the premises.
Space factor	The ratio of the space occupied by all the cables in a conduit or trunking to the whole space enclosed by the conduit or trunking is known as the space factor.
Speed	Speed is concerned with distance travelled and time taken.
Spring toggle bolts	A spring toggle bolt provides one method of fixing to hollow partition walls which are usually faced with plasterboard and a plaster skimming.
Static electricity	Static electricity is a voltage charge which builds up to many thousands of volts between two surfaces when they rub together.
Statutory regulations	Statutory regulations have been passed by Parliament and have, therefore, become laws.
Step-down transformers	Step-down transformers are used to reduce the output voltage, often for safety reasons.
Step-up transformers	Step-up transformers are used to increase the output voltage. The electricity generated in a power-station is stepped up for distribution on the National Grid network.
Switching for mechanical maintenance requirements	The switching for mechanical maintenance requirements are similar to those for isolation except that the control switch must be capable of switching the full load current of the circuit or piece of equipment.
Synthetic rubber	Synthetic rubber is manufactured, as opposed to being produced naturally.
Tensile force	Tensile force is the force pulling things apart.
Thermoplastic polymers	These may be repeatedly warmed and cooled without appreciable changes occurring in the properties of the material.
Thermosetting polymers	Once heated and formed, products made from thermosetting polymers are fixed rigidly. Plug tops, socket outlets and switch plates are made from this material.
Time sheets	A time sheet is a standard form completed by each employee to inform the employer of the actual time spent working on a particular contract or site.
Transformer	A transformer is an electrical machine which is used to change the value of an alternating voltage.
Trap protection	Trap protection places alarm sensors on internal doors and pressure pad switches under carpets on through routes between, for example, the main living area and the master bedroom.
Trunking	A trunking is an enclosure provided for the protection of cables which is normally square or rectangular in cross-section, having one removable side. Trunking may be thought of as a more accessible conduit system.

Velocity

In everyday conversation we often use the word velocity to mean the same as speed, and indeed the units are the same. However, for scientific purposes this is not acceptable since velocity is also concerned with direction.

Visual inspection

The installation must be visually inspected before testing begins. The aim of the visual inspection is to confirm that all equipment and accessories are undamaged and comply with the relevant British and European Standards, and also that the installation has been securely and correctly erected.

Weight

Weight is a measure of the force a body exerts on anything which supports it. Normally it exerts this force because it is being attracted towards the earth by the force of gravity.

Work done

Work done is dependent upon the force applied times the distance moved in the direction of the force. Work done = Force $\times$ Distance moved in the direction of the force (J). The SI unit of work done is the newton metre or joule (symbol J).

Index

For more Routledge books in line with British Standard BS 7671 (the "Wiring Regulations")

Go to www.routledge.com